CONSTRUCTION SAFETY ENGINEERING PRINCIPLES

DESIGNING AND MANAGING SAFER JOB SITES

Books in the McGraw-Hill Construction Series

Defect-Free Buildings: A Construction Manual for Quality Control and Conflict Resolution by Robert S. Mann

McGraw-Hill Construction Locator: Building Codes, Construction Standards, Project Specifications, and Government Regulations by Joseph A. MacDonald

Building Anatomy: An Illustrated Guide to How Structures Work by Iver Wahl

Construction Safety Engineering Principles: Designing and Managing Safer Job Sites by David V. MacCollum

CONSTRUCTION SAFETY ENGINEERING PRINCIPLES

DESIGNING AND MANAGING SAFER JOB SITES

David V. MacCollum

McGraw Hill

New York Chicago San Francisco Lisbon London Madrid Mexico City
Milan New Delhi San Juan Seoul Singapore Sydney Toronto

The McGraw·Hill Companies

Construction Safety Engineering Principles: Designing and Managing Safer Job Sites

1 2 3 4 5 6 7 8 9 0 FGR/FGR 0 1 9 8 7 6

ISBN-13: 978-0-07-148244-8
ISBN-10: 0-07-148244-X

The sponsoring editor for this book was Cary Sullivan, the editorial supervisor was Patty Mon, the production supervisor was George Anderson, the acquisitions coordinator was Laura Hahn, the project manager was Rajni Pisharody, the copy editor was Anthony Green, the proofreader was Bev Weiler, and the indexer was Kevin Broccolli. It was composed and illustrated by International Typesetting and Composition. The art director for the cover was Anthony Landi and the cover design was by Elizabeth Pisacreta.

Printed and bound by Fairfield Graphics

ABOUT THE AUTHOR

David V. MacCollum, PE, CSP, is an award-winning engineer with more than 35 years of experience as a consultant specializing in safety research and technical assistance for high-risk hazards. He is also the author of *Crane Hazards and Their Prevention.* His articles regularly appear in professional and trade journals such as *Professional Safety, Western Construction, National Safety News, Professional Engineer, Journal of Industrial Hygiene,* and *Hazard Prevention.* Mr. MacCollum is the principal founder of the Hazard Information Foundation, Inc., a nonprofit foundation that maintains a resource library of safety and hazard information. He is listed in *Who's Who in Engineering.*

Contents

Acknowledgments

The Center to Protect Workers' Rights played a key role in the development of this book by funding the Hazard Information Foundation, Inc. (HIFI) to write the research report "Inherently Safer Design Principles for Construction," which was completed in August 2005.

During the last century construction projects have become increasingly complex, with serious hazards to overcome. Credit is well-deserved for the countless projects that have been completed without catastrophic consequences due to the dedication of design engineers, construction managers, and safety professionals who held safety as a paramount consideration. The personal efforts of some of these individuals have gained both national and international acclaim for their innovative application of available technology to control hazards. These safety efforts started with the control of malaria during the construction of the Panama Canal and continued with the prevention of falls by means of safety nets during the creation of the Golden Gate Bridge. The list of such lifesaving innovations goes on and on.

Neil Chitwood, chief of the safety branch of the Portland District U.S. Army Corps of Engineers, was a forward thinker who included safety as part of the design of the Dulles Dam on the Columbia River and other Northwest flood control dams in the 1950s. I have been blessed many times over as a result of his mentorship in my early career.

Thanks go to recent forward-thinking educators, including John Gambatese, PhD and Michael Behm, PhD at Oregon State University. These men began research into the identification of design features that would eliminate hazards "upstream" at the time of design.

The doctrine of the concept of safety by design started early with the development of military aircraft during World War II and involved Amos Moses of Boeing, a leader in developing system safety concepts that were published in military specifications in the 1960s. Safety at the time of design began with a convergence of engineers who happened to be involved in both aerospace and construction. Thanks go to the many outspoken engineers who recognized the need for a system to identify construction hazards upstream before workers arrived at a worksite. Some of the most notable engineer advocates for design safety in construction are DOD consultant Richard Hislop, PE; Richard Hughes, PE, the owner of a structural design firm; Rear Admiral Ben Lehman, a leader of marine engineering; Chuck Miller, the first executive manager of the National Transportation Safety Board; and Jeff Vincoli, safety-program evaluator for Bechtel Worldwide. These design-safety leaders recognized that all engineers needed some basic principles to analyze their designs for safety.

Sincere appreciation goes to Corey Heniser at MICO engineering for the use of the photos in Section 15, Eric Gewiss and Stephen Gass of SawStop for the photos featured in Section 18, and James Crowley of Abby Restoration in Quincy, MA for the hair-raising photos in Section 20. Thanks also to Nigel Ellis for the grim reaper depiction in Chapter 1.

Special thanks to Rowena Davis, whose hard work and language skills transformed these concepts into a readable book. This text would not be what it is without her dedication, her attention to detail, and her constant quest for the information behind the concepts. Her untiring efforts made this book a desk reference for developing training for engineers.

David V. MacCollum, PE, CSP

Introduction

Construction has been plagued with serious injuries and deaths for years. Unfortunate incidents have contributed to excessive loss of lives and damage to property, casting a pall over the construction industry. Past efforts in construction safety have usually focused on identifying hazards after workers arrive at the site and tailoring worker behavior in an attempt to avoid injury. Recently new concepts have heralded a change of direction in the industry. Construction planners are increasingly looking "upstream" to remove or control hazards at their source. Identifying and combating the source of hazards through the concept of "inherently safe design" increases the safety of a project before the workers arrive at the job site by preventing hazards before they cause injury. The idea of inherently safe design is quickly gaining momentum to change the very nature of the construction industry. This book offers a practical guide to the basics of early hazard identification to create a safer, more efficient construction project from planning to completion.

Starting in 1985 the International Labour Office in Geneva and the European Foundation for the Improvement of Living and Working Conditions concluded that approximately 60% of all fatal accidents in construction arise from faulty design or insufficient planning. While the exact percentage remains the subject of debate and discussion, a number of this magnitude challenges those who manage the process of design to save lives and money simply by the application of improved engineering practices. Such perspective should be a revelation to designers and engineers, who are accustomed to viewing the majority of accidents in the construction industry as attributable to the routine chaos of the construction site. It also challenges the construction manager to move outside the traditional role as an enforcer of safe work practices at the construction site and play an integral part in project design and engineering. To do so would put more focus on prevention than ever before.

The objective of this book is to develop and expand engineering principles of inherently safer design for construction. To accomplish this task, this book will draw from four pioneering approaches to safer design. These approaches, which formed the cornerstone of modern system safety engineering to prevent hazards by design, are to

- eliminate the hazard by improving the design of the facility to be constructed and the equipment used in the task,
- select safer methodology if possible,
- provide guarding to prevent contact with the hazard,
- provide safety factors to minimize the hazard,
- provide redundancy with several safeguards to confine the hazard.

During World War II Boeing began to examine military aircraft failures and immediately improved the design on the production line, making aircraft safety a function of design. Design-safety concepts were then refined by the military/aerospace industry in the post-World War II period as part of their *system safety* approach to injury and loss prevention. Missile systems and space vehicles cannot tolerate failure, so engineering specialists designed ways to improve the design of each component and develop an analysis of the entire system. In the 1950s the U.S. Army Corps of Engineers in the Portland District of the North Pacific Division embarked upon "Design and Build" projects for hydroelectric flood control and navigation dams on the Columbia River and its tributaries, and instituted a sophisticated program to ensure that safety was included at the time of design. The author of this book, who was personally involved, addressed project design-safety features as well as planning which examined the methods of erection and the type of construction equipment to ensure they were safe for use on a project.

Eliminating the hazards from the worksite before the workers arrive makes the project easier to manage. It is interesting to note that when integrating staging as part of slip forms for concrete casting, labor costs were reduced far beyond the cost of the staging. Safety at the time of design invariably shows a cost benefit. By 1985 the office of the Chief of Engineers incorporated system safety as a corps-wide design requirement to be included on all their projects worldwide.

The next significant design-safety doctrine was developed by the chemical industry with inherently safe design for oil refineries and chemical plants. This focused primarily on energy control. The National Safety Council applied these same concepts and named it "safety by design" for heavy industry for machinery such as punch presses and die-casting machines. Recently Dr. Gambatese of Oregon State University, and a few others in academia, conducted research on "designing for construction safety" in order to extend these ideas into design processes and construction planning. All of these approaches involve the application of engineering principles to achieve fail-safe construction processes as part of design and building features. All of these applications necessarily involve highly focused engineers. Although these concepts have been developed, there is still no published methodology to simplify them into a simple process to make them accessible so that engineers outside of safety focus groups can apply them.

The Hazard Information Foundation, Inc. (HIFI) conducted a research study on Inherently Safe Design Principles for Construction and developed a methodology that reaches beyond these focus groups to make the principles and applications of system safety accessible on all levels of industry. This study was funded by the Center to Protect Workers' Rights (CPWR). The purpose of this book is to expand the study in order to provide an accessible, practical reference for all engineers on how to achieve inherently safer design.

Part I of this book develops methodology for identifying hazards and matching the hazard with design features and safety appliances for the prevention of the hazard. This methodology reduces the complex processes of system safety to five basic principles that provide a practical guideline for engineers to identify and overcome hazards in the design and planning stage. Part I highlights the role of the design engineer to ensure safety. When the engineer places emphasis on hazard prevention by "designing out" the hazards inherent in the facility being built, then the construction processes, methods, machinery, and the circumstances that produce injuries can be drastically reduced while increasing productivity and value.

Part I also provides the engineer with a system for identifying hazards, with an insight into the nature of hazards, and guidance that categorizes the hazards into seven manageable groups. Identification of the different types of hazards in the design stage streamlines the hazard-elimination process by providing guidelines to determine general control measures. This book will provide design engineers, construction managers, and construction safety specialists with simple principles of system safety that have been distilled specifically for construction. Then it will provide a method to quantify the ability of the engineering of design improvements to reliably prevent damage, injury, and death. Part I also provides an overview of how reliability can be used as a tool for developing viable projections of injury-free performance of the construction project.

For instance, before a construction project even begins, the construction manager often faces the potential hazards of faulty design by the architect, defective design of the equipment that must be used, and hazards inherent within the construction site or property. To successfully control these hazards during the project, they must be identified and addressed during the design and planning stages. This book provides a design matrix that can be used as a check sheet to ensure that each potential hazard is identified and controlled before the design has left the drafting room or even progressed past the planning stage.

Parts II and III examine 50 examples of common construction hazards that have been or can be eliminated by design. By eliminating or controlling the hazard, each inherently dangerous machine, process, and facility can be transformed into a safe one. Many of the examples used have been taken from injury litigation to show why the hazard existed and how it could be overcome by engineering innovations, improved methods, and the use of safety appliances. Moreover, each example addresses in detail how the concepts provided in Part I can be applied within the practical context of the construction site. These litigation examples illustrate the limitations of human performance. Removing or controlling hazards reduces the potential for worker injury through human error. To assume that the construction worker will always recognize the hazards and to believe the injury was the result of unsafe practices does nothing to provide a safe workplace. Removing the hazard eliminates the opportunity for worker error. This philosophy clearly supports a paraphrase of a well-known saying that reads, "To err is human, to forgive design." This adage, introduced to safety philosophy by human factors expert Alphonse Chiaponis, emphasizes the necessity of safe design in saving lives. The solution to overcoming unsafe worker practices or acts is to develop an understanding that an unsafe act cannot occur unless a hazard exists. The engineer, in his or her role as a designer or planner who eliminates hazards, becomes the most valuable member of the management team because of their ability to save lives while reducing costs.

In a world of exponential increases in technology, engineers have new and exciting options available to them. Engineers who are designers or construction managers who can "think outside the box" will find many opportunities to reengineer products using new materials and informational devices such as sensors in a way that produces an inherently safe product or process. As an example of using technology to innovatively reengineer for safety, the section on machine guarding in Part II discusses an electronic solution to protect users of unguarded table saws more efficiently than the use of the removable mechanical guards that are currently available.

The hazards listed in Part II were chosen to be as diverse as possible to provide the reader with a wide array of examples and tools to broaden their awareness of what has and can be done, so that they can apply their own imaginations and creativity in order to heighten innovation of engineering design processes for safer construction activities.

Part III illustrates and provides a detailed discussion on how the incorporation of the five principles from Part I can be utilized on both an individual and group level during the design-development and planning process. The 50 examples in this book specifically address these principles as they apply to design and safety planning. Starting literally at the drawing board, Part III outlines the role of the architect in designing out hazards and addresses the duties of the construction manager in ensuring additional hazard prevention by means of safe work equipment and processes. This part also provides useful guidance for the engineer who designs construction equipment for making machines user-friendly for error-free operation. By using these cases as examples, construction management can easily adapt the different components of architectural and construction safety planning to apply to the dynamics of their project. This segment works in concert with specific variables to create a picture of the construction safety planning process from start to finish.

Full adoption of the principles of safe design in the construction industry requires the application of previous construction safety triumphs. The case example in Part II that addresses the hazard of two-blocking shows how litigation prompted widespread manufacturer acceptance of a safety appliance before adoption of standards requiring the appliance. The following body of research illustrates that system-safety concepts for inherently safe design principles provide the best fit for identifying and controlling hazards on the construction site at the time of design and project planning.

Part I Development of the Five Principles for Safety Design in Construction

INTRODUCTION

Inherently safer construction design is not a new concept—its roots are evident from the earliest recorded history. A long historical pathway shows how early designers evolved into today's engineer and developed the ability to remove hazards at the time of design or during construction planning. This pathway led to many inconsistencies and trial-and-error processes, but has always returned to the fundamental conclusion that the removal of hazards by design or planning is the most effective means to prevent construction injuries. Unfortunately, there have always been barriers to designing something right the first time. Most often, the method for ensuring safety is a management assumption that the primary way to achieve safety lies in changing the behavior of the worker. This concept is misguided, because it places the burden of accident prevention on an external source that is not always reliable or controllable. Sound engineering concepts applied to hazard prevention principles are an obvious and effective way to control construction mishaps. It is time to embrace new innovations in hazard elimination using established concepts in design safety.

BACKGROUND AND HISTORY

The history of construction safety starts nearly 4000 years ago with harsh requirements and penalties. In approximately 2000 BC, Hammurabi promulgated a civil code to protect the public from unsafe construction. Among this King of Babylon's many admonitions, the one on construction clearly addresses design safety: "If a builder builds a house for a man and does not make the construction firm, and the house which he built collapses and causes the death of the owner or his family, the builder shall be put to death." This theme is echoed in the Bible, particularly in the Book of Deuteronomy,

22:8, which states: "When you build a new house, be sure to put a railing around the edge of the roof, then you will not be responsible if someone falls off and is killed." It is only natural to believe that builders would seek relief from a penalty of death by pleading the case before the king and seeking alternate recourse. Influential builders would probably have received lesser sentences, giving rise to the current problem of avoidance of liability for unsafe construction.

As the art and science of construction advanced in the Greek and Roman Empires with the formation of guilds under the supervision of a master builder, design became a closely guarded oral tradition. These craft guilds organized lodges to develop apprenticeships so the worker could advance within the craft and become a master in order to oversee the particular trade under the direction of the architect. The building of the great cathedrals was the handiwork of the architect, who oversaw the erection on a daily basis; and was usually a father-to-son activity that may have taken a hundred years or more to complete. With the Renaissance, building design became a uniform style for each classical type of architectural structure. With the coming of the Industrial Revolution, construction methods became a separate function no longer under the control of the architect. The design drawings of whatever was to be built illustrated only the completed structure; the builder, not the architect, became the authority on how the structure would be erected. As structures became more complex, the builder faced many hazards that often were not apparent, certainly not open and obvious. The builder in essence became a contractor who was overwhelmed with complicated construction. This role required making and implementing many more decisions, too often in the face of considerable uncertainty. As a result the builder (contractor) had great difficulty changing the design to afford worker safety. With the separation of design and construction into two different professions, safety issues usually arose only after the worker was in danger, and often after a serious injury or death occurred, instead of at the time of design, when the potential incident might have been prevented.

Early Design Safety Concepts

With the advent of the railroads, the U.S. Railroad Appliance Act was legislated in the 1890s, which required the adoption of self-couplers and Westinghouse air brakes. These requirements saved the lives of hundreds of workers and passengers. This act placed the emphasis on engineering controls to achieve safety. In the early 1900s, a group of engineers who worked for insurance companies formed the American Society of Safety Engineers. These engineers, with training and experience in the disciplines of civil, mechanical, and electrical realms, relied upon engineering controls to prevent injury. A typical example of designing for safety was the innovation of sprinkler systems to prevent the spread of fires in a building. Construction in the early 1900s was exceedingly dangerous, as construction management did not employ engineers to ensure a safe construction site, and rarely designated anyone for this function. In 1904, the state of Wyoming was the first to enact a Professional Engineering license, which established standards of education and experience. These conditions of practice were to protect the public from unsafe and faulty design or supervision. In rapid succession, most states passed similar laws to give engineers the authority to sign and seal their own drawings. This authority made the engineer accountable for the design of the intended structure. The code of ethics of most engineering disciplines requires that the safety of the public is the leading consideration of all design and operational oversight. With these state laws, the engineer is given the primary authority for inherently safer design. When the engineer's role is fulfilled, he or she earns a well-respected social responsibility.

Causes of Liability Laws

Despite this legislation, by 1912 the courts were clogged with a massive number of worker injury lawsuits, which delayed proper treatment of the injured and escalated costs to the contractor. This led to the adoption of state workers' compensation laws. These laws provided immediate medical treatment and minimum compensation, while setting a very high standard of deliberate misconduct to which the employer could be held liable when a serious injury or death resulted from the contractor's unsafe work-site. The standard defining deliberate misconduct should have been much lower and defined as gross negligence. As the workers' compensation law was written, the loop-holes in it created a no-fault insurance policy. With these laws, some states engaged in providing no-fault workers' compensation coverage, or created a new market for insurance companies. Under this coverage, the employer and employee could not be held liable for unsafe conditions on a work site. Negligence became a pass-through business expense, with few penalties for unsafe design or dangerous construction methods.

Privity was a legal concept developed in England to ensure that only the manufacturer of a product would be liable if it failed and caused injury. For instance, the maker of a wheel could be sued if it was found to be defective. If the same product is a component of a machine, the assembler of the machine is exempt from liability. This means that an automaker who uses a defective wheel in an automobile cannot be held liable if the wheel fails and causes injury or death. If a victim of an automobile accident due to a defective tire tried to gain recompense, the auto manufacturer could not be held liable because he did not make the tire. However, since no contract exists between the tire maker and the victim of the accident, recompense does not come from the tire maker either, and the victim is left without an avenue of restitution. This insulating law was overcome in 1916 in the landmark case of McPherson v. Buick Motor Co., when a defective wheel made by a subcontractor created an obligation for Buick to inspect the wheels for safety before installing them on automobiles. The overthrow of privity created an obligation for assemblers to install a product that is safe for its intended use. However, third-party personal-injury liability resulting from the overthrow of privity has not become an entirely successful incentive for construction safety.

Who is Responsible?

In the late 1920s a tugboat towing barges carrying a very valuable cargo left New York harbor and proceeded south past New Jersey, unaware of an incoming storm. In order to save their lives in the increasingly rough seas, the crew cut the barges loose and headed for the safety of a nearby port. The owners of the lost cargo sued the tugboat company because of a failure to have radio receivers that would have warned of the storm. The tugboat company claimed that radios were unreliable, and none of the fleet in the tugboat industry used radios. Justice Learned Hand of the 12th District U.S. Court ruled on this famous T.J. Hooper case[1] , answering the claim of unreliability by replying, "Knowing

[1] T.J. Hooper, Circuit Court of Appeals of the United States, Second Circuit 1932 60F.2d 73, by Justice Learned Hand.

about bad weather part of the time is better than knowing about bad weather none of the time." He further ruled that a whole industry's lag in adopting lifesaving technology is not a valid excuse for its absence. By ruling in favor of the plaintiffs who lost their cargo, he reasoned that a safety device did not have to be completely reliable or perfect in order to be useful. (If a safety appliance is able to prevent injury from a specific hazard in seven out of ten occurrences of a similar circumstance, an improvement of the overall safety of the situation has been made. Unfortunately, safety appliances are often discarded because they are unable to prevent ten injuries out of ten injury-producing circumstances. For the inability of the safety appliance to prevent three injuries, seven injuries occur. This reality can be construed as an antisafety philosophy.) Both the McPhearson and T.J. Hooper cases began to open the door to third-party-liability cases for construction. They affected support of inherently safer construction as an ethical and moral issue, and promoted the responsibility of component manufacturers.

In the 1930s A.W. Heinrich took a great step backward by declaring that 85% of total on-the-job injuries were the fault of unsafe acts by the worker[2]. His focus on worker behavior ignored the removal of the hazards with engineering controls as the first step in injury prevention. In today's world, the complexity of machines makes them almost impossible to be used safely in the total absence of engineering controls. Worker behavior modification cannot be solely responsible for the safe use of an inherently dangerous machine! A machine responsible for recurring injuries must be examined for design flaws and corrected before all injuries will abate. "Downstream" or mid-process corrections will not eliminate the design flaw responsible for injuries. It is not enough to remove the oil slick on the floor, if a mobile machine is still designed to leak oil.

Birth of System Safety

During World War II, when military aircraft were being manufactured by the thousands, Amos Wood of Boeing Aircraft began investigating aircraft crashes to identify hazards and failure modes and inquire into an alternate design to prevent future failures[3]. The findings were adapted in the production line to create safer aircraft. Ross McFarland, a psychologist, also began to examine aircraft crashes to see whether pilot error was a cause. With his data, he developed a design for control panels that would not cause errors. In the years before our entry into World War II, he found that many Navy airplane crashes occurred when the pilot transferred from the land-based trainer aircraft to the carrier-based tactical aircraft. The setting for full throttle on the trainer was in a position all the way back, toward the pilot. In the tactical aircraft, full throttle was achieved by pushing the throttle all the way forward, away from the pilot. The result of this discrepancy was that the pilot was conditioned upon takeoff to push the throttle in the wrong direction and crash into the sea.

---

[2]H.W. Heinrich, *Industrial Accident Prevention,* McGraw-Hill, 1931.

[3]*The Organization and Utilization of an Aircraft Manufacturer's Air Safety Program;* Paper presented at the 14th Annual Meeting of the Aeronautical Sciences, New York, NY, 1946.

At the advent of complex missile systems, there was no way of determining at the time of design whether the missile would work reliably. To develop some realistic probability of either success or failure, a mathematical reliability analysis was developed. By the war's end, design safety had evolved into a system analysis that addressed both human factors and system reliability. By 1963, sufficient experience was gained by the U.S. Air Force to codify this information in to a Mil Spec for all Air Force and missile contractors. In 1969, the Department of Defense (DOD) adopted the Air Force's Mil Spec into Mil Spec 882 for all aircraft and missile systems. This standard is still in effect, with updates and revisions every few years to make it adaptable to new engineering technology.

Construction Safety by Design

After WWII, the U.S. Army Corps of Engineers civil districts embarked upon building flood control, navigation, and power generation dams. This federal government agency recognized that safety needed to be incorporated into the facility design and construction planning. The goal was to reduce both construction and operational injuries. Design safety on all contracts included compliance with their manual, the EM385-1-1, *Safety and Health Requirements Manual.* This manual was first published in 1958, and was most recently updated in November, 2003. For years the Corps' record for construction injuries has been one-fifth of the casualty rate for similar work not under their supervision. Their program required design review of all drawing and specifications by an engineer who specialized in safety engineering. Construction safety planning conferences were required on all contracts as a review of the written plan before the notice to proceed was signed[4]. In the 1980s the U.S. Army Corps of Engineers adopted system safety requirements similar to the DoD System Specification 882 for all of their civil works construction contracts. Now the design of dams and other civil work projects incorporate system safety in the design and planning process.

Concurrently, chemical plant and refinery explosions brought about "Inherently Safer Design" concepts, which were similar to system safety in the elimination of hazards, but were primarily directed toward the controlled release of unwanted energy. In the mid-1990s the National Safety Council funded the Institute of Safety by Design, directed to improve safety in production facilities, auto assembly lines, and machinery manufacturing.

Awakening of Academia

Beginning at the turn of this century, some academics began examining the construction safety process to look "upstream" for the source of hazards in design. In 2000 the Hazard Information Foundation, Inc. (HIFI) was formed to identify and analyze ways hazards could be subdued by engineering controls. General industry has wrongly focused on personal behavior modification, a school of thought that places the root cause of injuries on unsafe acts. This philosophy ignores the fact that a hazard must be present for an unsafe act to occur. The focus of this book is to boost the core value

[4]Testimony of Donald M. Pittinger on H.R. 4652–1990, on The Construction Safety, Health, Education Improvement Act of 1990.

of eliminating hazards with engineering controls to create inherently safer construction with fewer injuries. Both management and engineers recognize that there will always be some contributory factors causing injuries on construction projects. However, by applying the available engineering technology to remove or minimize hazardous conditions at the time of design, and at review of the methods during construction planning, the goal of zero injuries and fatalities can become a reality for most construction projects. The skills of the engineer make the crucial difference in the art of hazard prevention to save lives while reducing construction costs. The next five chapters define and discuss the five principles of safer design and construction planning.

1 Principle One: Definition of a Hazard

To begin to address inherently safer design principles in construction, one must first understand the actual nature of hazards. A specific definition of hazards provides the engineer with a basis to develop a methodology for planning and evaluating the construction process for safety, and ensure for the design of inherently safe construction equipment and other support systems. The undertaking of such construction principles leads to inherently safe operation of a completed facility.

THE NATURE OF HAZARDS

To accomplish this, let's start by defining a hazard in practical terms.

Definition: A hazard is an unsafe physical condition that is always in one of three modes: *Dormant/Latent* (unable to cause harm), *Armed* (can cause harm), *Active* (causing injury, death, and/or damage by releasing unwanted energy, substances, or biological agents, or as a result of defective computations from computer software).

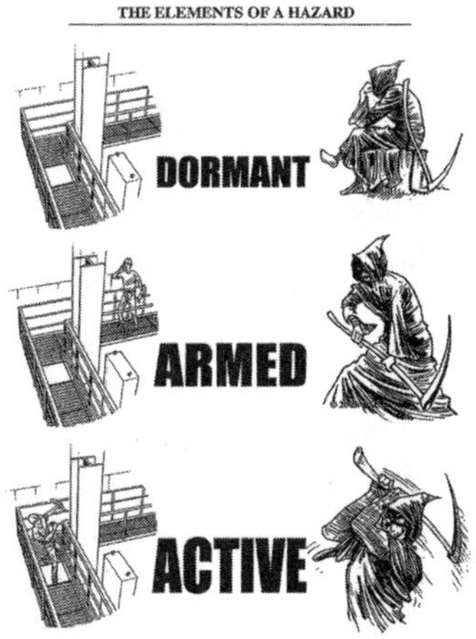

Failure Modes

In greater detail, a dormant/latent hazard is a design defect that is susceptible to a failure mode. Foreseeable misuse should also be considered. For example, a kitchen chair may be used to stand on to reach upper cabinets and needs to be sturdy enough to prevent collapse. The armed hazard is created by a change of circumstances and is ready to cause harm (i.e., the chair may have a big knot on one leg). The active hazard is an armed hazard triggered into action (i.e., when the chair is stepped on, the knot cannot support the additional load, and the chair leg collapses, causing a fall.)

The three modes of a hazard can be further explained by this simple analogy: Icebergs in the North Atlantic present a dormant hazard. The hazard becomes armed as the *Titanic* steams full speed at night into an area where icebergs collect. The hazard becomes active when the *Titanic* strikes an iceberg, resulting in massive loss of life. There are many misconceptions concerning this disaster, as most people will judge the actions of the captain rather than the need for safer design features for the ship. The initial perception that the conduct of the captain was outrageous and reprehensible in regard to the life and safety of the passengers and crew is justified. News reports of the maiden voyage proclaimed the *Titanic* an unsinkable ship. However, even assuming that the captain believed he had a state-of-the-art, unsinkable ship, the decision to steam through a sea known to contain hazardous icebergs was foolish. His actions were due to his presumed, yet erroneous perception that the eleven watertight bulkheads just below the waterline made the ship unsinkable, and the icebergs were not a hazard. In 1910 when the Titanic was being built, there were safer design features readily available in state-of-the-art technology. Design of the battleships of that time included double-compartmented hulls for damage prevention from torpedoes. Had this design feature been utilized in addition to the watertight bulkheads above the waterline, the flooding would have been confined to outer compartments and the *Titanic* would not have sunk. This analogy shows how the public often focuses primarily on the failures of humans, who cause a hazard to become armed and active. Safety engineering looks upstream to the design of the ship, providing it with proper bulkheads, a double hull, and a sufficient number of lifeboats.

FINDING THE CAUSE

In each of the 50 examples presented in Part II (Chapters 8 through 13), the reader will see a short review of real-life occurrences that in most cases were litigated as a third-party personal-injury lawsuit. These examples will disprove the behavior-based-safety idea of classifying the cause of an injury as solely an unsafe act of the victim.

H.W. Heinrich presented a theory of why injuries occur with a five-domino sequence of cause and effect. He used dominoes in an attempt to show the causes of injuries as cumulative, without considering hazards.The following are the principles of cause listed by Heinrich:

1. Ancestry and social environment
2. Fault of person
3. Unsafe act and/or mechanical or physical hazard

4. Accident
5. Injury

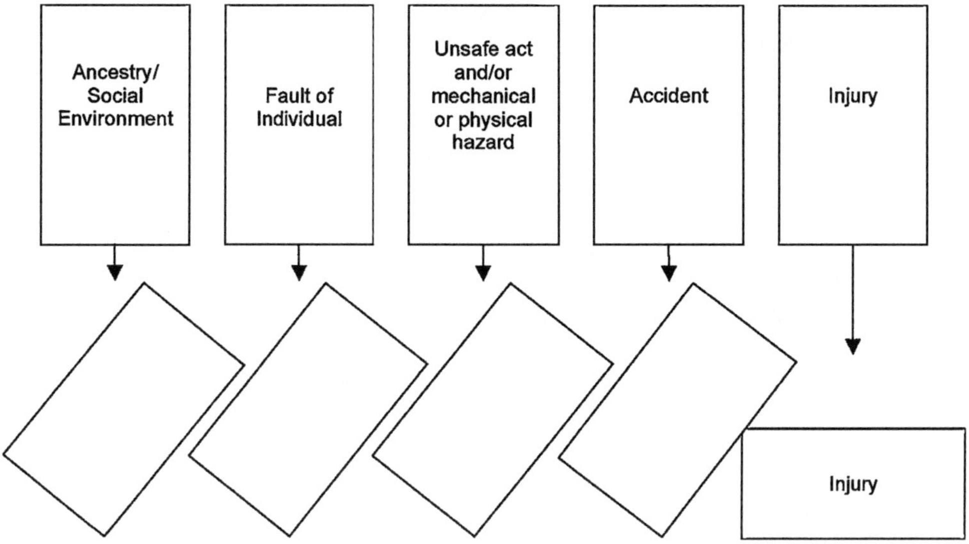

These concepts are erroneous[1] because the emphasis is placed on changing human behavior, not hazard elimination. A more accurate domino effect could be depicted as follows:

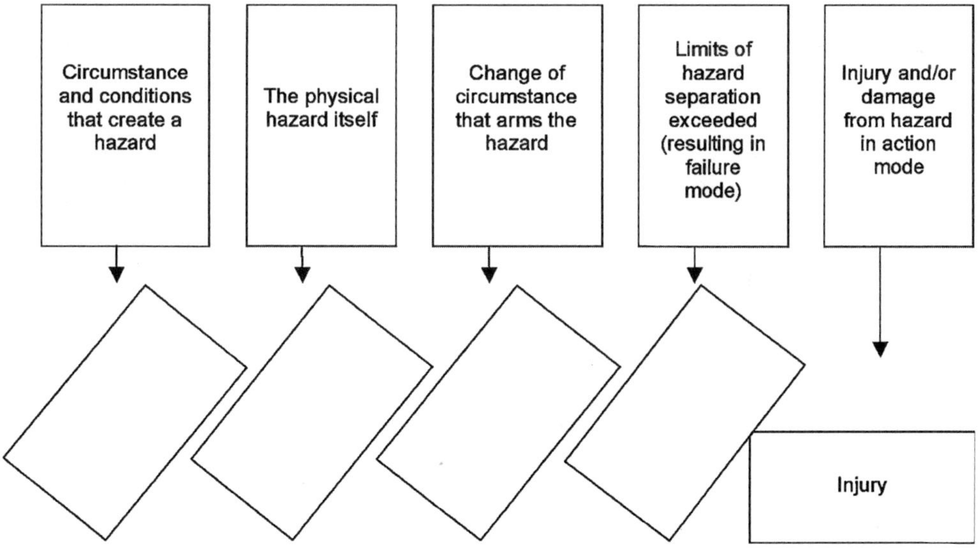

[1] "Warning! Behavior-Based Safety can be Hazardous to Your Health," Jim Howe, United Auto Workers (UAW) Health and Safety Department Publication #14, 1988.

1. Conditions that create a hazard
2. The hazard
3. Change of circumstance that arms the hazard within the limit of the action mode (This line between an armed hazard and an active hazard is called the hazard separation limit. From this point, one misstep can exceed the hazard separation limit and activate the hazard.)
4. The hazard separation limit is exceeded, resulting in a failure mode
5. Injury and/or damage from the hazard in the action mode

This analysis provides the catalyst to look upstream in order to identify the conditions that create a hazard. Once the source of the hazard is apparent, the use of engineering controls to remove the dangerous conditions means personnel are no longer threatened by that particular hazard.

Another approach to charting the three modes of a hazard is to place them into a logic chart that shows both the method of control and the failure mode necessary to set the hazard into the active mode. The illustration below details such a logic chart, which shows the changes of circumstances that can arise and how they affect people, activities, and conditions.

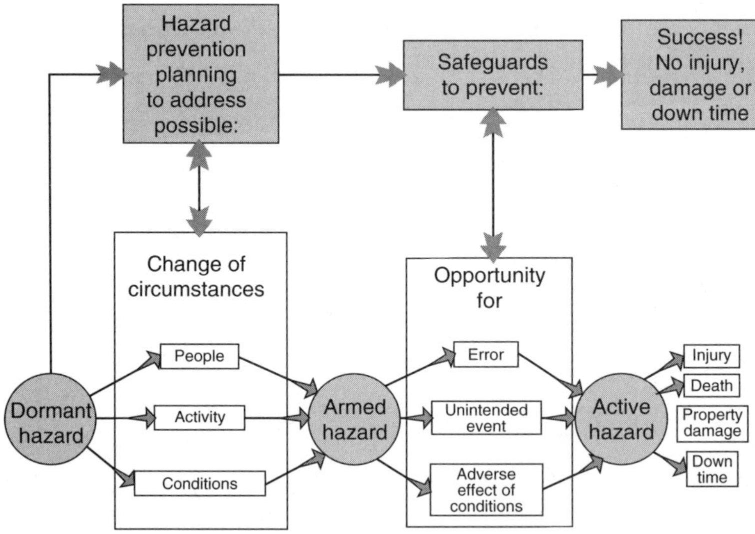

The definition of the hazard reveals that the nature of the hazard, always occurring in three modes, provides insight into most circumstances and conditions that cause injury. The development of inherently safer construction must rely upon a standard of performance that requires the elimination, or the minimization of the harm a hazard can cause. Such a standard also creates the opportunity to examine the worst hazard first, in order to develop priorities that will lead to a reliably disaster-free construction project. This first principle provides a starting point to identify hazards, enabling the engineer to successfully design them out of a construction project in the subsequent steps presented in this book.

2 Principle Two: The Standard of Care

In order to be effective, safety must be converted into a powerful design priority and overriding planning concern. To avoid the hazard, it must rely primarily on *the physical elimination of each hazard, rather than upon human performance*, which is variable and cannot be programmed. Through the evaluation and close scrutiny of each activity, task, or phase of the construction process, we are able to identify possible failure modes to identify hazardous conditions.

PERFORMANCE STANDARD

A well-known tenet of safety engineering states, "Any hazard that has the potential for serious injury or death is always unreasonable and always unacceptable if reasonable design features and/or the use of safety appliances are available to prevent the hazard." The key to successful safety engineering is to identify and design out as many hazards as possible. When this tenet is applied as a design standard, it becomes a routine expectation to design out hazards, thus changing an inherently dangerous facility, product, or service into an inherently safe one.

Achieving Performance

The identification of hazards is the basic building block of ensuring an inherently safe construction project. To many, it is like Lewis Carroll's *Through the Looking Glass*, when Alice remarked, "I can't remember things before they happen," and the Queen described the advantage of living backwards: "Your memory can work both ways!" Often the same hazard that has been causing injury, damage, or downtime, surfaces uncontrolled on multiple occasions. Falling loads due to two-blocking were recurring hazards on construction sites for many years. This trend stopped when anti-two-blocking devices were installed by manufacturers on all new cranes, and retrofitted onto most cranes in the field. Reliance on past experiences can be called "remembering backwards," and as such, it is not all that difficult to begin to control hazards.

Supporting Doctrine

The term "risk" is often used to establish the estimated cost of injuries and damage during the life of the construction process. This terminology is heavily used for insurance purposes. The typical use of the term has no place in a hazard analysis, as it is replaced in system safety with reliability calculations, which will be discussed in Chapter 6. The term "risk" should not be used, because it is a nondefinitive and speculative term that fails to describe peril in terms of injury or damage. This term allows management and engineers to consider risk as an unavoidable cost of business, and fails to focus on specific hazards.

Such terminology can be sidestepped. The National Safety Council has created this general statement of a standard of care: "Needless destruction of life and health is a moral evil. Failure to take necessary precautions against predictable accidents and occupational illnesses involves a moral responsibility for those accidents and occupational illnesses"[1]. Explicit terms stating the peril that the hazard can cause or has repeatedly caused in the past is necessary in order to convey the magnitude of danger to the decision maker, reader, or others. Our overly polite society fails to realistically address life-threatening hazards in terms that command attention and action to eliminate the hazard. Without the use of explicit terms, there is little moral compulsion to ensure the priority of engineering controls and hazard-prevention measures. Mild terms such as "risk analysis," which offer no reference to a specific hazard, allowed behavior-based safety to become a cure-all without ever addressing the need to prevent hazards. The recently published book, *Human Factors: Causes and Control*[2] places anticipated responses to potential hazards into four basic reactions:

- No response was made, as the hazard was not perceived and no one was aware of potential danger
- Parties in charge were aware of the hazard but made no response because they arrived at one of the following three conclusions
  - Assumption that the hazard posed no danger
  - Assumption that the hazard was open and obvious and easily overcome
  - Belief that the user's personal skill would overcome potential danger
- Parties in charge were influenced to accept the hazard as inherent to the activity
- *Parties in charge failed to recognize that a hazard must exist for an unsafe act to occur*

The engineer's talents are overlooked when management adopts programs that examine only the worker's or user's behavior. Management needs to adopt stringent and forceful hazard-prevention policies to look upstream to identify the conditions that create hazardous circumstances, so the hazards are controlled prior to the workers' arrival at the worksite.

---

[1]NSC, Accident Prevention Manual, 6th ed., 1969, p. 3 (remains in subsequent editions)

[2]George Peters and Barbara Peters, *Human Factors: Casused and Control*, CRC Press/Taylor and Francis Group, February 2006.

When establishing a standard of care with engineering controls for each hazard in the construction process, a reasonable projection of safety performance can be calculated. Specific calculation is far more effective than reliance upon underwriters' broad risk assessments based upon occurrences or similar construction experiences in previous years. The most puzzling process is how an engineer can initially identify a hazard and *how a potential hazard is recognized.* Chapter 3, outlining "Principle Three," provides a checklist covering seven different categories, which will aid the engineer in design and planning to identify hazards that are often difficult to foresee. With a listing of numerous hazards within each category, hazard identification at the time of design or construction planning becomes a reasonable process.

3 Principle Three: Categories of Hazards

This chapter provides an easy method for seeking out and finding hazards. The third step in hazard identification is to determine into which of the following seven categories the source of the hazard can be placed.

DETERMINING THE TYPE OF HAZARD

Hazard Sources

Natural environment

Structural/mechanical

Electrical

Chemical

Radiant energy

Biological

Automated systems/artificial intelligence

Each of the following lists contains just a few examples that serve as a starting point for the engineer when determining the nature of the hazard. A prominent source of the identification of potential hazards is found in Willie Hammer's book, *Product Safety Management and Engineering* (Published by ASSE, 1993). In his monthly column in the *Journal of System Safety* (First quarter, 2001), Charles Hoes developed a thought-provoking system of generic hazard categories. *Safety Through Design* (Christensen and Manuele, eds., National Safety Council, 1999.) also refers to similar groupings of hazards. Though arranged differently, some of these categories are included in the following lists. These topics are by no means complete, but are meant to give the reader an idea of how to categorize a new hazard. The featured list should be a starting point for developing additional listings for failure modes.

It is important to note that hazard categories may overlap. It is common to encounter a hazard that contains simultaneous natural, mechanical, and chemical properties. In these cases, specific hazards should be broken down into as many individual properties as possible.

Natural Hazards

The laws of gravity cannot be repealed, nor can the weather be programmed or the ocean drained. The following are some of the possible sources of hazards that the design engineer must contend with in the natural environment.

Natural/Environmental Hazards

A. Gravity
   1. Falls, same level
   2. Fall from elevation
   3. Falling objects
   4. Impact
   5. Acceleration
      (a) Sloshing of liquids
      (b) Inadvertent motion
      (c) Movement of loose objects

B. Slopes
   1. Upset
   2. Rollover
   3. Sliding
   4. Unstable surfaces
      (a) Earthquakes
      (b) Avalanche

C. Water
   1. Floating
   2. Sinking
   3. Drowning
   4. Tides
   5. Floods
   6. Oceanic disturbances

D. Atmosphere
   1. Change in altitude
   2. Humidity
      (a) Excessive moisture (high humidity)
      (b) Excessive dryness (low humidity)
      (c) Condensation
   3. Wind
      (a) Wind chill
      (b) Structural pressure
   4. Visibility (fog, etc.)
      (a) Daylight
      (b) Darkness
      (c) Glare

   5. Dust
   6. Temperature

E. Limitations on human performance
   1. Fatigue
   2. Error
   3. Distraction
   4. Anthropometric
   5. Ergonomic

Structural/Mechanical Hazards

As engineers, we must identify mechanical hazards while considering their mechanical advantage, but also their possible danger.

Structural/Mechanical Hazards

A. Surfaces
   1. Lack of traction
   2. Instability
   3. Protruding obstacles
   4. Incline
      (a) Steps
      (b) Ladders

B. Lever

C. Rotation
   1. Wheels
   2. Gears
   3. Pulley
   4. Screw
   5. Auger
   6. Cams
   7. Pinch point
   8. Friction

D. Reciprocation

E. Compression
   1. Shearing
   2. Puncture
   3. Structural failure
   4. Ejected fragments

F. Causes of vibration
   1. Noise
   2. Dislocation
   3. Parts failure

G. Pneumatic (pressure) hazards
   1. Compressed gasses
   2. Unintended release of gasses
   3. Blown objects

4. Water/liquid hammer
5. Container, hose, pipe, or vessel rupture
6. Overpressure

H. Metal fatigue

I. Bending/hinge

J. Tension/spring

K. Hydraulic forces
1. Liquid jet
2. Rupture of pipe, hose, or vessel
3. Overpressure

L. Vacuum/negative pressure effects

M. Entanglement
1. Noose
2. Snagging
3. Entrapment

N. Impact

O. Velocity

P. Airborne

Q. Blind zone

R. Confined space

S. Waste disposal

T. Access
1. Lack of access
2. Unguarded, elevated location
3. Low overhead
4. Exposure to adjacent and/or proximity hazards

Electrical Hazards

For all its advantages, electricity is a power source that is silently conveyed—and deadly.

Electrical Hazards

A. Voltage, amperage (causing shock, burn, fibrillation of the heart)

B. Alternating current

C. Direct current

D. Spark/arcs

E. Electrostatic

F. Source of dangerous heat

G. Ground

H. Capacitance

I. Sneak circuits

Chemical Hazards

Chemical hazards are a real Pandora's Box of toxic substances that have many potential dangers in a number of forms. To begin this analysis, the following clues should be a helpful approach.

Chemical Hazards

A. Combustion/fire

B. Corrosive/corrosion

C. Toxic Substance
   1. Liquids
   2. Fumes/vapors
   3. Dust

D. Degradation

E. Exothermic (hot)

F. Endothermic (cold)

G. Decomposition

H. Hydrogen embrittlement

I. Disassociation

J. Combination

K. Replacement

Radiant Energy Hazards

Though a major building block of our civilization, radiant energy can create many perils if improperly used. This short list is a starting point:

Radiant Energy Hazards

A. Sound

B. Heat

C. Light
   1. Ultraviolet
   2. Infrared

D. Radio frequency

E. X-ray

F. Nuclear

Biological Hazards

Biological hazards can threaten our health and be potentially fatal. These can be classified in six categories:

Biological Hazards

A. Allergens
   1. Mold
   2. Pollen

B. Organic Carcinogens

C. Infectious agents
   1. Bacteria
   2. Virus
   3. Fungi

D. Agents known to cause disease in humans

E. Venom

F. Conditions that produce sustained mental or physical stress in humans

Automated Systems Hazards

Automated systems hazards are caused by faulty computer hardware or software. The advantages of computers are immeasurable in any sound safety program. An excellent example of technology working for safety comes in the form of computer programs that test load-moment devices on cranes to prevent overload and crane upset. Such programs can be used in conjunction with advanced methods of testing computer firmware circuits for possible failure modes without destructive or competitive cycling, such as that referenced by Todd Isaac, Robert Konkle, and Juan Fernandez in a pilot study by Raytheon. Yet for all their usefulness, automated systems can and do fail on occasion. Parties who do not perform a thorough evaluation of high-risk software programs are at risk for a serious error that could cause injury, damage, or death.

Automated Systems Hazards

A. Program error

B. Technical malfunction

The seven categories of hazards are intended to spur the engineer, safety professional, or anyone else for that matter to fully realize that the nature of hazards is easily understandable and therefore manageable. Once a hazard is isolated, it becomes easier to begin a systematic evaluation of possible controls.

4 Principle Four: The Safe Design Hierarchy to Physically Control Hazards

The following hierarchy of engineering control has become the accepted sequence for evaluating design to best prevent hazards:

1. Elimination of the hazard
2. Guarding to prevent the hazard from causing harm
3. Including safety factors to minimize the hazard
4. Using redundancy for a group of parallel safeguards to require them all to be breached before a harm-causing failure mode occurs
5. Using reliability to mathematically calculate the qualitative numerical probability of eliminating or minimizing a harm-causing failure mode (see Chapter 6)

As construction projects become more complex and sophisticated, safety must be addressed with the same attention to technical detail as is applied to the engineering of these projects themselves. The project critical path (*critical path* is a generic term for the entire construction planning schedule, including site preparation, procurement, and the whole erection cycle) should be highlighted at those points where hazards have been identified in order to recognize potential problem areas. For effective hazard elimination, the entire construction process needs to be examined in this fashion. Listing hazards in the critical path forces the planner to consider itemized alternatives. This leads to the application of a systems safety approach, the same approach that has become the backbone of aerospace and nuclear energy design.

ADDITIONAL CONSIDERATIONS

System safety relies heavily upon the provision for *safety factors* and *redundancy*, in addition to hazard elimination and guarding. It is in this manner that foreseeable error is prevented. To achieve zero-injury, damage, and loss of completion scheduling, reliance on behavior modification to ensure error-free human performance becomes unrealistic. A paraphrase of the age-old saying changed to read, "To err is human, to forgive design" has proven time and again to be a sound philosophy supporting the concept that the elimination of error-provocative circumstances is the basis of system safety. In *Human Error: Causes and Control*[1] (see Chapter 2), human-factors specialist George Peters asserts that construction always presents a complex set of enacting priorities. Large projects may require a macrosystem hazard analysis, which requires encompassing a wide range of skills assigned different tasks, all with different opportunities for error. To support the prevention of error-provocative circumstances, a look at frequency, opportunities for, and severity of injury is necessary.

Safety Factors

Safety factors can be easily explained by the example of a bridge with a posted ten-ton load limit, which is designed to sustain up to 30 tons, thus allowing for foreseeable misuse. Closer to the topic of safety of construction equipment is an example of a questionable safety factor. Cranes are generally rated at a capability that is 85% of the tipping load at any radius. By industrial standards, this is a rather thin margin. In some cranes, rated capacity is only 85% of the structural design of the telescoping boom, which is far less than the tipping load. In such a circumstance, the consequences of an overload would not be a crane upset but a structural collapse of the boom.

Redundancy

Redundancy encompasses a series of safeguards, each of which must fail before the system experiences actual failure mode. A good example is the fuel system on a military helicopter, which has several fuel tanks and a number of fuel lines. To prevent leaks in the event of penetration by enemy bullets, the fuel tank is self-sealing. Both ends of all fuel lines have automatic shutoffs in case one is broken, as fuel has several other routes through different lines to the engine.

Reliability

Reliability is no more than a numerical confidence rating, such as a failure mode that may fail 1 time in 1,000 cycles. The big guess is *when* it will fail. If it fails on the first cycle, it is chancy that 999 successes will follow. Reliability is the judgment to quantify a system's ability to succeed and is not a method of control. This function attempts to take the guesswork out of the hazard-prevention methods of an entire project. The subject of reliability is usually integrated with the use of a fault-tree analysis, addressed further in Chapter 6.

[1]George Peters and Barbara Peters, *Human Error: Causes and Control,* Taylor and Francis, 2006.

COMPONENTS AND APPLICATION OF THE DESIGN HIERARCHY

Each of the following four categories of engineering controls briefly addresses various design choices to achieve an inherently safe design with an expectation of a near-zero harm-causing failure mode. The engineer is encouraged to expand the listing in each of the four headings to accommodate a specific circumstance.

Hazard Elimination

Some safety appliances, such as an overpressure relief valve on a pressure vessel or an air compressor, can entirely eliminate the hazard if they work reliably. The following are some other ways to eliminate hazards:

- Avoid the hazard with alternate safer design and planning. Conducting a prework evaluation of construction methods and processes is an effective and appropriate time to eliminate hazards.
- Substitute with safer construction machinery.
- Relocate any dangerous facilities (such as powerlines or other utilities) away from the construction site.
- Provide design criteria to suppliers of structural components to ensure safe assembly at the construction site.

Guarding the Hazard

This category includes the use of safety appliances to overcome foreseeable operator/user error. Examples of these include anti-two-blocking devices and load measuring indicators, which are designed to intercede; safe-space clearance devices; and insulated links for cranes.

- Establish barricades around any danger zones to eliminate hazardous conflict between equipment and existing facilities. For instance, safe-access provision with staging and guardrails guard against fall hazards.
- Provide automatic interlocks that will disarm the hazard for service and maintenance functions.
- Provide detection systems that audibly and visually warn of a changing circumstance and will intercede before the hazard becomes active and produces a harm-causing failure mode.

Safety Factors

- Raise the structural strengths above the foreseeable misuse and wear limits to reduce failure mode occurrences.
- Reduce exposure to toxic materials.
- Ensure that the structural design is well above the rated capacity in the event of an unintended overload. (Bridges, even those with posted weight limits for autos, should be able to withstand foreseeable exposure to excessively heavy vehicles, such as ready-mix trucks and other load-bearing vehicles.)

- Ensure that cable-tension loading is sufficient to overcome foreseeable wear, and that the sheave diameters will not accelerate wear.
- Ensure that limits for toxic radiation, gas, vapors, and dust are well below health hazards.

Redundancy

A combination of safeguards will collaborate to achieve an effective hazard control network.

- Install design barriers in parallel so that each one must fail sequentially, like a row of dominoes, before the hazard can cause a harm-producing failure mode. For example, an insulated link of a crane's hoist will protect the person guiding or touching the load (such as a steel beam), but will not protect the individual from touching the crane's outrigger. A proximity warning device can audibly warn of an adjacent powerline and alert the crane operator to stop boom movement and avoid touching the powerline. The proximity alarm is a redundant safeguard. Additionally, workers should be trained to avoid touching the load or crane upon hearing the alarm. The combination of the proximity alarm, insulated link, and a designated spotter provides redundancy and a reasonable reliability of avoiding unintentional crane/powerline contact.
- Ensure that each barrier in concert with other barriers covers the entire spectrum of failure modes inherent in the specific equipment, as well as structural weakness and construction methods used at the work site.

These four engineering controls are the options that the engineer possesses to physically control a hazard from becoming armed or triggered into the action mode as defined in Principle One. Engineering controls often take ingenuity to design, and usually require an initial cost. When initial cost becomes the basis for rejection, while failing to address the earning power of the engineering control, multiple injuries are usually the result. In the 1950s, specifications required staging on the slip forms used for concrete placement on dams. Because the forms were reusable for a number of applications, the savings in labor far exceeded the cost of the staging[2]. Chapter 15 will discuss in greater detail the economic value and earning capacity of engineering controls.

[2]David McCollum, "How Proper Scaffolding Cut Costs," *Western Construction,* September, 1957.

5 Principle Five: Control the Hazard with the Appropriate Design Improvement or Appliance

The concepts originally developed by the chemical industry for production processes and system safety innovations for aerospace are remarkably similar to the current principles of inherently safe design. When applied to the construction industry, these concepts can promote safety for construction processes specific to that field. Initially, the contractor's role starts when the project is advertised for bid. At that time a rudimentary construction plan is developed, primarily to determine costs; however, the assessment of the inherent hazards must also be performed and figured into the costs. Once the successful bidder is selected, site-specific construction planning affords the opportunity to screen the use of construction equipment to ensure that it is safe for its intended purpose. This two-phase approach includes:

- Safety in the construction sequence plan:
  - Outline specific phases of the project
  - List of all possible hazards and ways to prevent them
- Ensure that the construction equipment used on the site is safe for its intended use by creating a listing for each piece of equipment that includes:
  - Anticipated hazards
  - Ways that design or use of appliances can be achieved to ensure an inherently safe construction site

VISUALIZE A PATH TO SAFETY

A critical path or other master construction schedule provides a visual aid that highlights any potential hazards and assures that everyone associated with the project receives notice and begins to consider the necessary safety measures to achieve inherently safe construction. When creating such a plan, one must closely examine the hierarchy of design in conjunction with the identified hazards. The most efficient way to accomplish this is to marry the hazard to the appropriate engineering control. To assist the engineer, a simple worksheet matrix has been developed to analyze the hazard in order to determine the appropriate control.

Creating a Matrix

Use of a Hazard Identification and Prevention Matrix, shown in below, can be a useful approach to a design and construction planning guide. This matrix is an innovative tool for engineers to quickly chart each hazard, define the necessary safety engineering, and arrive at a reliability evaluation. The horizontal categories at the top list safety controls and provide space to note specific hazards and prevention measures. The seven vertical categories list likely hazard types; how they can be made inherently safe can then be listed to the right of each.

This matrix allows the design engineer and construction manager to graphically identify the hazard and focus on the necessary design features or appliances that prevent the hazard from becoming armed or active. If the engineer desires to establish numerical reliability values to determine the increased safety of a specific design improvement, the column on the far right provides a space for this value. Chapter 6 gives an overview of reliability incorporated into system safety. This methodology gives management a comprehensive safety appraisal of new products, facilities, and systems.

	Eliminate the Hazard		Guard the Hazard		Provide a Safety Factor		Provide Redundancy		Provide Reliability
	Hazard	Safety	Hazard	Safety	Hazard	Safety	Hazard	Safety	
Natural									
Structural/ Mechanical									
Electrical									
Chemical									
Radiant Energy									
Biological									
Artificial Intelligence									

MAKING THE METHODOLOGY WORK

The question now becomes, "How can we use the information on the matrix to transition from identification to implementation?" The answer is obvious: we need to expand the knowledge of all engineers in systems safety and apply this knowledge to the development of system studies for complex construction sites and the machines on these sites. The system safety engineer must have a flair for mathematics and statistical processes. (In electronics, operational reliability is generally computed with exponential expansions; maintainability is computed in the lognormal; availability [probability of readiness for use when needed] follows the F distribution; static function of storage, etc. is a binomial factor; and when only a few test samples are available, an applied binomial to exponential results is a common approach.) The course we must take to fulfill today's requirements for technical safety will be arduous; the mathematical processes are involved and will be different for each discipline of engineering.

A New Tool for Safe Design

The design engineer must be proficient in his specialty, but must also become knowledgeable in new engineering tools of system safety. One way to build this complex skill set is to seek the assistance of a systems safety engineer. Most engineers' talents are directed toward designing high-performance systems. Their safety knowledge is usually limited to a specific subsystem and perhaps a safe interface to adjoining parts. Such specialization leads to a limited safety overview, particularly where many engineering disciplines are involved in the entire system. Therefore, the design engineer needs the help of a special type of professional engineer: one with a thorough knowledge of system safety, who can participate as a member of the design team, and can systematically analyze the system for unsafe conditions.

The most valid and authoritative proof of what is accepted as inherently safe design is a record of injury-free performance. Once a new design feature of a safety appliance is adopted, it is necessary to develop a record of performance. The easiest system to use is to record the injuries in the number of units multiplied by the number of years of use. From there, a more refined analysis of how the exposure to a hazard can be overcome by design rather than reliance on human performance can be determined. The design and construction of a facility is really a system of many engineering disciplines that work collectively to design multiple components and assemble the resources to erect the facility. Illustrated on page 28 is a system-safety flow chart for examining a master construction safety plan. The master plan is integrated with the critical path or other construction management schedules (often known as the critical path), and provides the opportunity to address each hazard that will arise during the phases of the construction process. The figure on page 29 shows how a piece of equipment can also be evaluated at the time of design and manufacture to identify hazardous conditions. Many pieces of equipment brought onto the worksite have never been evaluated for hazards. When selecting equipment for construction planning, every piece needs to be evaluated for hazards to ensure that only inherently safe equipment is brought onto the project. These charts provide insight into how the design of construction equipment can be analyzed with system safety in mind. Prevention of construction hazards always needs to address project planning and the equipment to be used.

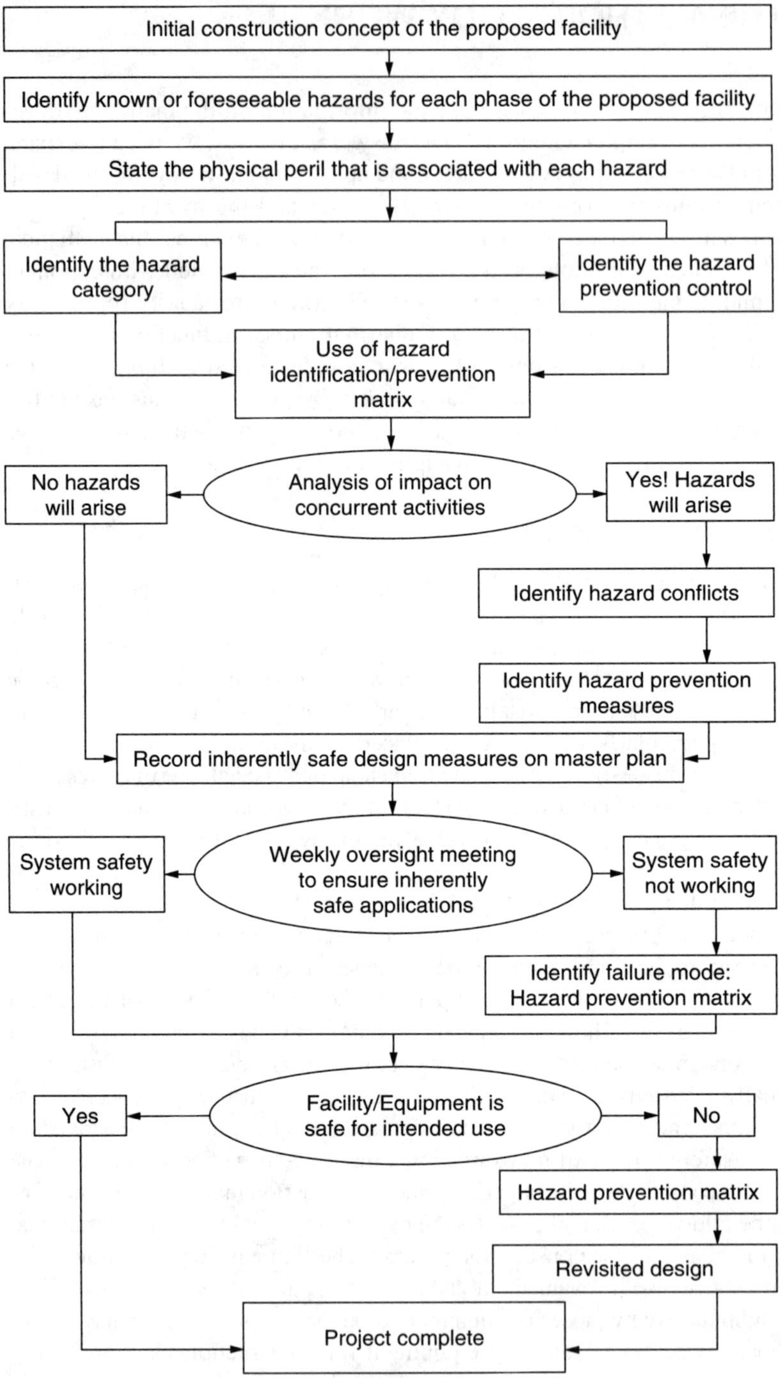
Initial construction concept of the proposed facility
Identify known or foreseeable hazards for each phase of the proposed facility
State the physical peril that is associated with each hazard
Identify the hazard category
Identify the hazard prevention control
Use of hazard identification/prevention matrix
No hazards will arise
Analysis of impact on concurrent activities
Yes! Hazards will arise
Identify hazard conflicts
Identify hazard prevention measures
Record inherently safe design measures on master plan
System safety working
Weekly oversight meeting to ensure inherently safe applications
System safety not working
Identify failure mode: Hazard prevention matrix
Yes
Facility/Equipment is safe for intended use
No
Hazard prevention matrix
Revisited design
Project complete

Today's construction is done with many types of machines, which may or may not be entirely safe for their intended use. The diagram shown below provides a brief outline of how each hazard should be analyzed. This overview of system safety shows the equipment designer how to ensure inherently safe construction equipment.

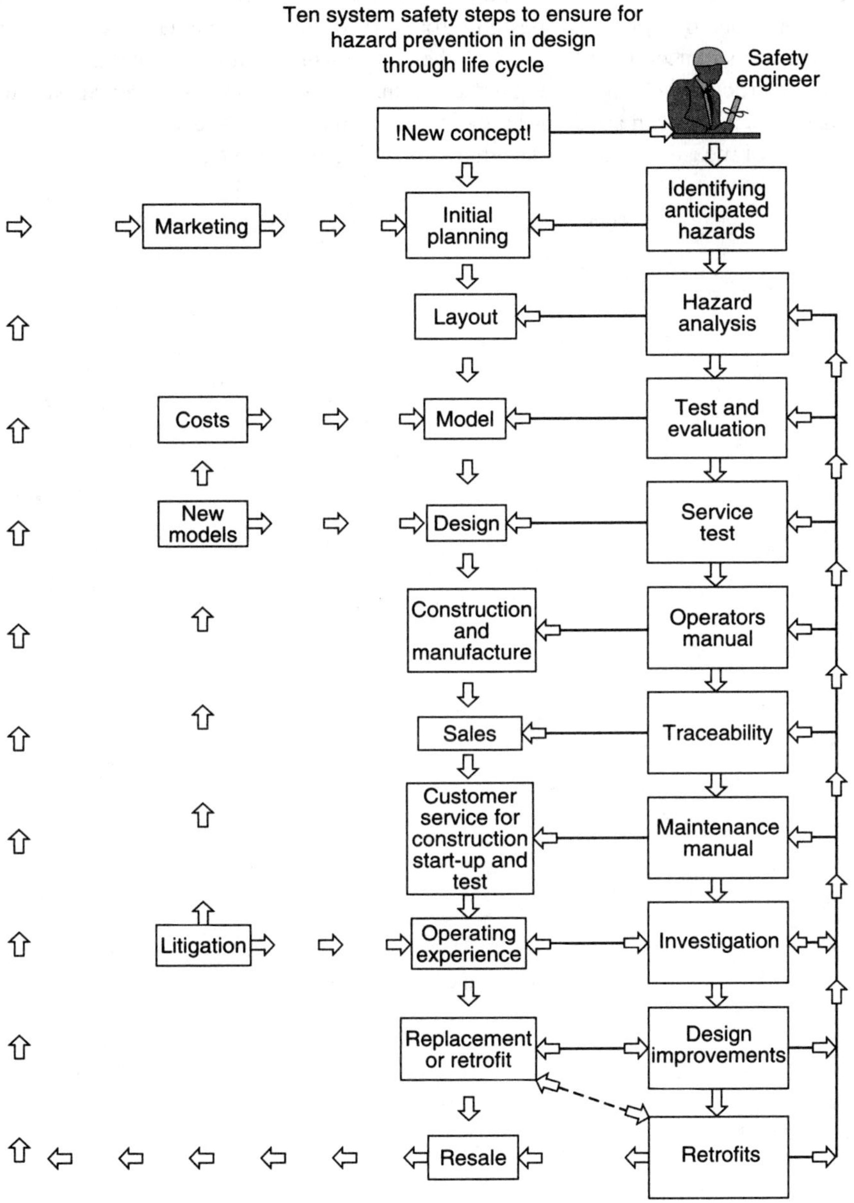

Specify Needs and Tools

Once the designer or the construction manager has completed the matrix for each hazard, they have a serial list of all the hazards that need to be accounted for within the design or concept by means of the critical-path construction schedule. By providing a matrix showing both the hazard and the means of prevention for each hazard in a new design, the designer has the tools to improve the design through recognition of the hazards and identification of the ways to accommodate them. When the construction manager is developing a master construction schedule, he can use the matrix as a basis for a critical path to prevent each hazard that can arise during the project. Chapters 14 and 15 discuss this integration of control measures into the design and planning documents.

6 Reliability: A Method to Evaluate Probable Safe Performance

The concept that safety is everybody's business has made it nobody's specific responsibility, and has far too often become the road to product-liability lawsuits. Briefly, system safety engineering must be supported by reliability studies and include the following concepts: a life-cycle concern unaffected by organizational structure, application of appropriate engineering disciplines, and a technical information-gathering function for decision makers. Reliability provides an overview to gauge the efficacy of (1) hazard elimination, (2) guarding, (3) safety factors, and (4) redundancy by making a quantitative assessment of the likelihood of a construction-phase failure mode. After a reliability assessment, the construction manager and safety engineer have the opportunity to list site modifications to ensure inherently safe construction.

PROOF OF SAFER DESIGN

To establish a measure of proof that the above four design options will in fact eliminate or minimize hazardous failure modes, the engineer has the option of conducting a reliability analysis. Though considered to be tedious or abstract, reliability calculations are a necessary part of successful system safety. When completed, the reliability analysis provides an assessment of the accuracy and efficiency of the controls incorporated into the design.

This process is conducted at the end of the engineering hazard-control hierarchy and

- provides probabilities of failure for each of the identified harm-producing failure modes;
- provides a quantitative analysis of how inherently safe the life cycle of a construction project can be made;
- defines the actual peril that can arise from the specific hazard; and,
- recognizes that machine-dependant safeguards, such as warning labels, verbal instructions, and training processes are not fail-safe because of the inherent and behavior-induced error.

An accurate reliability calculation can help define and isolate each specific error-producing factor. With inherently safe design, training remains very important, as it ensures that the operator/user is aware of the hazard, the peril, and how the design feature or appliance has provided for their safety.

The previously mentioned methods of controlling hazards—elimination, guarding, safety factors, and redundancy—are applied to establish a reliable, fail-safe system. The reliability of each hazard control chosen should be analyzed to establish a numerical quantitative rating of success or failure. This analysis will then determine the relative effectiveness of the chosen control, freeing the professional engineer and safety professional from reliance on intuitive processes. The reliance on intuition is really no better than guesswork as to how the machine will be used or how the facility will be constructed.

The application of system safety principles has been largely limited to the aerospace and electronics community, and has been adopted by very few nationally known foreword-thinking designers and builders. In over four decades of experience, the aerospace industry has shown that qualifying risk as a peril stated in terms of reliability has made it easier for management to provide funds for design improvements. Previously, when dealing with unknowns, management has been reluctant to authorize money during the design stage for safety modifications, because some or all potential modifications have not been proven to work. The system safety-analysis approach with reliability factors has shown that hazards can be effectively identified, controlled, and measured without the need for costly accidents to identify and highlight potential hazards. Then, by a process of assigning values to the various failure modes, the engineer can project the likelihood of injury- and damage-producing incidents. With a reliability value, a cost analysis can be developed to determine the actual life-saving and earning power of the proposed hazard controls. Additional information on this topic is presented in Chapter 15. This method presupposes that allocating funds for a systematic approach that identifies all possible hazards during the design and analysis stages can substantially reduce the overall cost the consumer would have to bear as a result of accidents. System safety supported by reliability computations provides a positive demonstration by engineers that will aid in gaining full acceptance and support by the corporate board and executive leadership.

OVERVIEW OF RELIABILITY ASSESSMENTS

With a system analysis of a construction project with reliability values stated, the system safety engineer is able to provide an estimate of probable injury-free performance. This engineering design aid will also be useful during the design review of new facilities. It will provide a means of ensuring that all hazards that may surface during the construction and operation cycle of the facility have been identified and controlled.[1]

[1]David V. MacCollum, "Reliability as a Quantitative Safety Factor," first published in the *ASSE Journal,* May 1969. The article was selected for the ASSE/Veterans of Safety first place award in 1969 and was reprinted in *Selected Readings* in the Safety Academy Press, 1973 and *Directions in Safety,* Charles C. Thomas (pub), 1976.

Developing Numerical Values

A mathematical reliability approach can provide insight into replacement cycles of critical parts that are subject to failure. For example, spurious x-ray emission can occur in radio tubes operating with high-plate voltages in radar transmitters. This phenomenon often does not become apparent until the tube has been in service for several thousand hours. Testing will show which tubes will fail and when, and which will cause radiation. The reliability approach can then predict a safe tube-change cycle to preclude this type of failure. It is easy to see how a change in maintenance instructions to replace the potentially dangerous tube every 1000 hours rather than expending its entire 5000-hour life can eliminate a hazard and avoid costly redesign of an expensive radar transmitter. Using reliability to provide safety insight has almost unlimited possibilities in the extension of testing and for predicting environmental changes, human performance, and even undesirable human behavior.

Traps of the Numbers Game

There are many pitfalls in the use of mathematical reliability techniques to determine the probability of accident-producing failures. Inadequate mathematical foresight can lead to the conclusion that a machine with 100 components in series, each 99% reliable, is really only 36% reliable. If the machine has 400 components in series, it is only 2% reliable; if it ever works at all, it is more luck than anything else. In practice, the eventuality of a catastrophic incident is not as inevitably hopeless as the example of a machine with 400 series components. The assumptions applied to this circumstance were that failure of any one component will cause the system to fail; and that failure of any component is independent of any other components' failure. Yet a machine with many components (parts) is not likely to fail every time any one part fails. More often than not, failure of any or some of the parts will cause only a reduced level of performance rather than complete breakdown. So when applying theory to the individual failure of a part in a machine, an unknowledgeable safety specialist cannot analyze the impact.

Interpreting figures is somewhat commonplace, but knowledge of statistics is needed to avoid hazards in the following scenario: A manufacturer does not desire to comply with the reliability requirements of his contract for some specialized electronic equipment. He reluctantly sets aside 10 samples of his product for testing. The test is run, and five samples fail. To avoid trouble with acceptance tests, he hires a statistician to compile results into a favorable report. The statistician uses a table of 50% confidence limits for the binomial distribution, showing a reliability of the equipment of at least 45%. His report, minus actual test data, is sent to the buyer's purchasing agent, Mr. Jones. It states only that the item is 45% reliable, with 50% confidence. Mr. Jones has this conversation with his secretary:

Mr. Jones: "Sally, this equipment is apparently 45% reliable, but I don't know what they mean by 50% confidence."

Sally: "I think it means he is in error 50% of the time."

Mr. Jones: "If their statistician does not trust his work 50% of the time, they should hire someone who doesn't make as many mistakes."

Sally: "Maybe we should send it back and ask for more tests, and maybe they won't make so many mistakes."

Mr. Jones: "Write a letter to that effect."

When he received the letter, the manufacturer decided not to give up and hired a reliability expert, who had the samples repaired and retested. Two failed, and the other eight, with a great stretch of the imagination and interpretation, passed. The expert reasoned that he had 20 samples with 13 successes and 7 failures. He also referred to the same table of 50 percentage points for binomial distribution and found his reliability was at least 48%, with 50% confidence. He reported reliability as 48%, not mentioning either confidence level or test data.

The false logic here is that reliability studies cannot be made without failures, so the more failures the better. If the initial test had 10 successes, reliability could have been stated as 94% for this binomial distribution table, with 50% confidence. With such mathematical juggling, it is imperative that the engineer have a sound knowledge of statistical methods before entering this field.

Available Reliability Models

Accuracy of reliability predictions depends on the robustness of mathematical models and available data, the latter most often being the limiting factor. When inadequacies exist in the data, uncertainties will result in the predictions. When data is unavailable, a Bayesian approach is often taken by applying judgment, beliefs, knowledge, and so on into the calculations. This should warn that Bayes' doctrine of chance should be applied to the solution of safety problems only with extreme caution.[2]

Safety standards should also be applied with caution. Compliance with standards will preclude a failure mode, and most accidents will be avoided. Far too often industry standards are merely guides to assure the particular industry the maximum return for the minimum capital outlay, without consideration of all possible accident causes. The National Electrical Safety Code is a specific example. It provides special requirements for the safety of the utility lineman, but has only "motherhood" statements concerning public safety. Standards are nonexistent for guarding a transformer bank mounted on an H-frame to, for example, prevent a child from attempting to retrieve a kite. Another example is the manufacture and sale of unvented gas heaters of less than 50,000 BTU prior to the 1970s. It was inferred that they were safe; however, it became known that these gas heaters caused asphyxiation in a closed room. These two examples illustrate that "safety standards" are an invalid basis for total safety concepts. The engineer using the reliability technique to support a systems analysis must consider that established standards might be inadequate when considering an entire system's safety effectiveness.

---

[2] Thomas Bayes published a paper on solving a problem in the *Doctrine of Chance,* (1763), a method for estimating unknown conditional probabilities.

Balancing Hazards and Safeguards

Next is the trap we fall into when making determinations of acceptable hazards, which are variable in themselves. For example, a picnic is planned for next Sunday, if it doesn't rain. The weatherman predicts a chance of 1 in 100 that rain will fall (1%). The normal planner sees this as an acceptable risk and proceeds with the picnic. But if we are designing nuclear warheads, and physicists tell us a design weakness will cause premature detonation at the rate of 1 in 1,000, we all view this as an unacceptable risk because of the severe consequences of a nuclear blast. The mathematical and statistical risk-rating function of reliability techniques makes systems safety analysis workable. Reliability is a function of time, conditions of use, and environment. This data is combined with the systems analysis, which considers the worst possible event and makes a comprehensive examination to determine if it can occur under any combination of circumstances. Often this technique is referred to as the fault-tree analysis, which is only a graphic representation of parallel and sequential combinations of component states that will result in the worst possible event, or any other undesired events.

We also need to examine those events considered to be undesirable. Unfortunately, our initial thoughts are often on the spectacular, such as the death of an astronaut, a bridge collapse, or other newsworthy accidents considered to be the result of a gross hazard; while many injuries and deaths of construction, maintenance, and operational personnel over a long period are often indifferently viewed as the inconsequential cost of doing business. Though some of the examples in Part II depict spectacular failure modes, many construction hazards are ongoing precisely because they are perceived to be minor and not worth correction. Correct identification and reliability analysis of a hazard will help ensure that it is controlled in the most cost-effective way possible.

"THUMBNAIL" EXAMPLE OF A SYSTEM SAFETY ANALYSIS

There is a lot that can be learned about the system safety hazard analysis by a transfer of the knowledge generated in the development of previously marketed construction equipment or a similar previously built construction project. It is important to avoid a defensive attitude when using examples too close to home that may be clouded with bias. Let us consider the sequence of undesirable events that in years past might have surrounded the loss of a large passenger aircraft as an example of how systems safety is made to work with reliability mathematics. Included in the design of a new aircraft is a horizontal stabilizer taken from the design of an older craft of comparable size. It is operated by an electric switch that controls a reversible electric motor that turns a worm gear, that in turn moves a screw jack that forces the stabilizer down or back to a level position, depending on the direction in which the worm gear rotates. There is also a manual trim wheel between the pilot and copilot's seats, which accomplishes the trim via the horizontal stabilizer. The pilot cannot override the electric motor to the worm gear with the manual trim wheel, or slow the movement, by grasping it.

Logic analysis showed that a critical hazard existed. It was reasoned that if the electric switch to the motor were to fail in the depress mode, the craft would be placed in an outside loop and would crash. This would take only one or two minutes. The pilots would have almost no time to save themselves from catastrophe by finding the circuit breaker and throwing it open. The mechanics of this hazard are simple: The surface of the horizontal stabilizer is very large as compared with the elevator surface. The pilots have only a short time to control the aircraft, because the authority of the stabilizer soon exceeds the correction that can be made by the elevator. Therefore, the aircraft is placed quickly in an irreversible loop, which quickly leads to destruction. The switch is a good example of inadequate safety design.

The designers of the horizontal stabilizer on the earlier aircraft considered only the failure of the electric motor by providing a manual trim as part of the stabilizer. They did not recognize the consequences of the switch failing in a closed position, which would depress the stabilizer—in reality, the worst possible event. Suppose the accident experience of the earlier aircraft included the loss of two planes over the ocean shortly after changing elevation. No investigation would be possible to reveal switch failure, because the wreckage could never be recovered. This failure of the switch in the closed position could have been the cause.

Fault Trees and Symbols

With an example of a probable accident analysis that could result from an undesired event (such as a switch failing in a closed position), we can apply some terminology and develop a fault-tree analysis. Logic symbols are used to adapt the analysis to a graphic presentation. These basic symbols are shown below. They are used in the next two illustrations to show how the hazards result in the loss of the aircraft by depicting a critical path (fault tree) and the chance of failure.

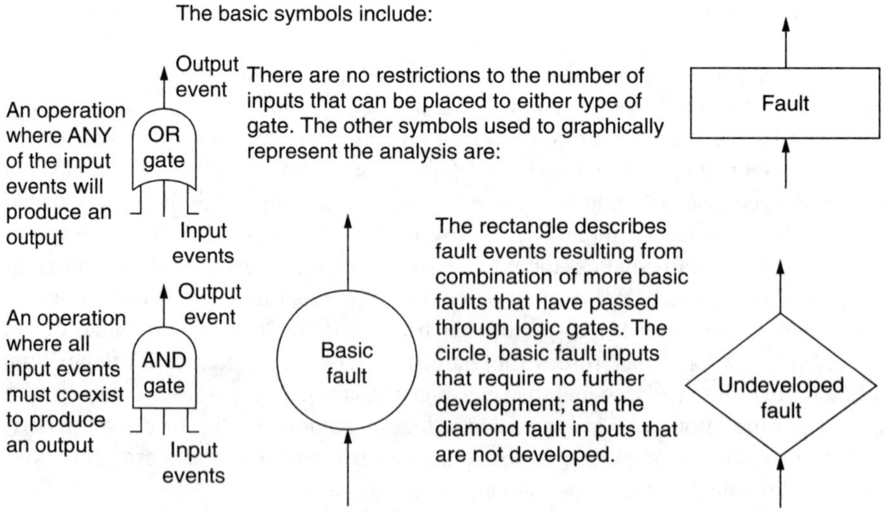

The fault tree graphically charts the route that an undesirable event can take to cause catastrophe.

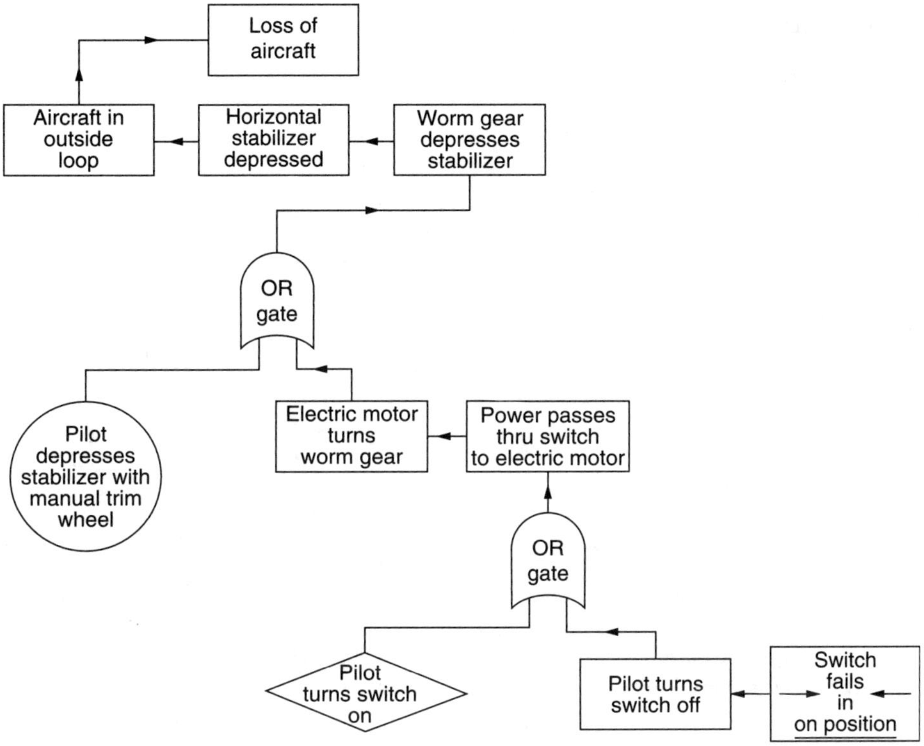

The fault tree clearly presents to the designer a number of choices wherein the circle can be interrupted for the insertion of a safety appliance that can make the system fail-safe. For instance, if the possibility of a tractor rollover is the basic fault, appliances can be added to the machine, such as a ROPS and automatic seat belts, in order to change the outcome.

Caution dictates that a new fault tree should be graphed to show the added safety appliance to ensure that no new hazards are introduced into the system.

With the sequence of undesirable events shown schematically, risk of accidents can be assessed in meaningful values, such as 1 in 100, 1 in 1,000,000, and so on.

The likelihood of an accident must be synthesized to determine if management can live with this risk of failure, or can support the cost of design changes to overcome the hazard. With such information available, decisions can be made based on a knowledge of the possible mode of the undesired event, the likelihood of it occurring, and the cost of eliminating it.

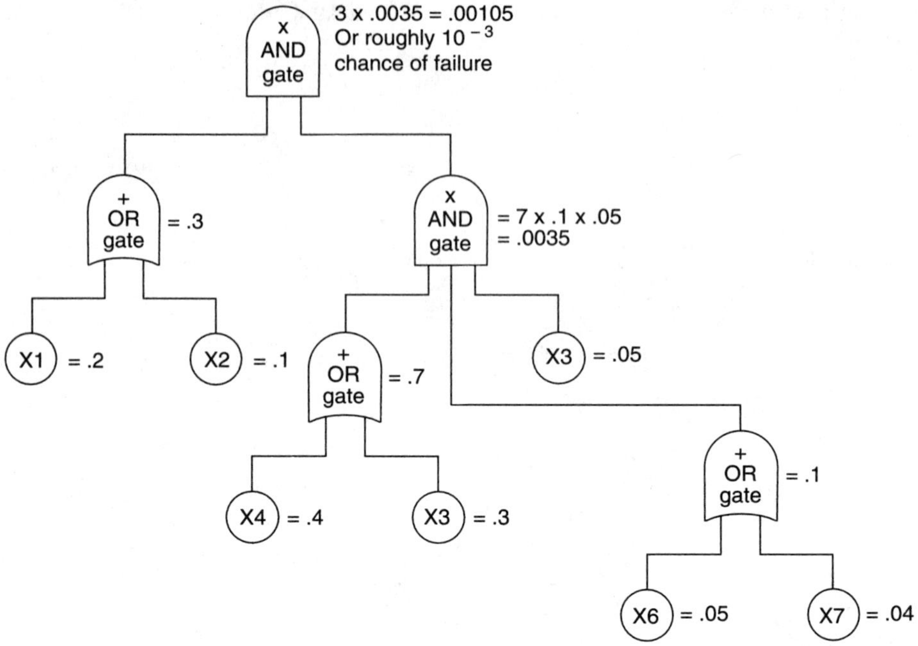

Engineers with a basic knowledge of system safety concepts can easily interpret the construction process as a system to identify and prevent hazards in the design and planning stages. Chapters 7 through 13 provide 50 examples of how the principles of safer design can look upstream and prevent lost time, damage, injuries, and death.

Part II Examples of Engineering Control

Chapters 7 through 12 contain forty-eight examples of how hazards can be identified and overcome with engineering controls. These chapters group examples of hazards by sources where they frequently occur:

- Chapter 7: Crane Hazards
- Chapter 8: Other Equipment Hazards
- Chapter 9: Universal Hazards
- Chapter 10: Access Hazards
- Chapter 11: Construction Methods Hazards
- Chapter 12: Operation and Maintenance Hazards

APPLICATION OF METHODOLOGY

Part II of this book illustrates the application and uses of the "Five Principles" of system safety for engineers. Each example in these chapters will identify the hazard syndrome in its dormant stage, how it is armed, and the peril it presents in the action mode. A specific summary of a typical occurrence of the hazard and its history is reviewed in order to provide background information on how and where it occurs in an effort to identify this syndrome. The matrix will illustrate how to define the hazard and appropriate engineering controls. A summary of standards and literature, with commentary and recommendations for appropriate engineering action, will accompany each hazard analysis. These examples allow the engineer to become aware of

- critical hazards that plague construction;
- how hazards can be identified at the time of design or construction planning, giving the opportunity for correction before anyone is endangered by the hazard;

- actual peril in a given hazard in terms that succinctly describe the defects that cause failure modes; and
- design remedies that eliminate or minimize the hazard's ability to cause injury or damage.

Part II is the most vital part of this book, as it provides real-life examples of how engineering can provide safer design and construction planning to eliminate hazards.

7 Crane Hazards

INTRODUCTION

The myriad hazards associated with crane design are addressed at length in other publications.[1] Six of the most frequent life-taking hazards are discussed in this chapter:

1. Two-blocking
2. Crane upset
3. Jib-boom stowage on hydraulic telescoping booms
4. Latticework boom disassembly
5. Crane operator protection
6. Powerline contact

As cranes are used on most construction projects, and are a source of a significant percentage of injuries, an overview of how crane hazards can be identified is needed by construction managers and those responsible for construction safety.

THE ROLE OF CRANES IN CONSTRUCTION

Engineering improvements play a vital role in the rapid development of sophisticated crane design. Cranes and aircraft have a lot in common when their rapid development and versatile use is compared. The design of both aircraft and cranes has become more complex as advancements in technology are made. There has been increased focus on providing design aids for pilots. Right now, cranes are in a phase of development as dramatic as the change aircraft underwent several decades ago. The incorporation of electronics to make cranes an integrated system of mechanical, hydraulic, and electrical

[1]David V. MacCollum, PE, CSP, *Crane Hazards and Their Prevention,* American Society of Safety Engineers, 1993, 2004.

components that depend upon computer software, has created a new priority for their safe design. No longer can crane designers consider available technology as frivolous bells and whistles that give a false sense of security to the crane operator. Just as World War I flying aces and the stunt barnstormers of the 1920s have been replaced by highly-trained and licensed pilots for particular aircraft, so must today's crane operators be held to the same level of performance through training and licensing. Located in Fairfax, VA, the National Commission for Certification of Crane Operators is considered to be the largest and most authoritative crane operator licensing organization in the United States.[2]

Operational Challenges

Crane operators can no longer judge the lifting capacity of a crane by the "seat of their pants" in cases without a load-moment indicator. To assume a crane is overloaded when it starts to become light and raises off its outriggers is very dangerous, as some newer cranes carry extraheavy counterweights that increase the tipping load far beyond the designed structural strength of the boom. The use of a load-moment indicator (LMI) is a must to limit the crane's lifting capacity to within its rated capacity, which is based upon both the structural strength of the boom and the stability of the crane. Cranes today come in a wide variety of models with many applications. Crane design now meets the many needs of construction with telescoping, latticework, articulating boom cranes, knuckle boom cranes, tower cranes, and straddle cranes; as well as excavators, backhoes, and many other devices used for mobile lifting; all of which come in a variety of carrier bodies, which range from crawler treads to rough-terrain and conventional wheeled transport. Cranes now have a number of attachments: draglines, clamshells, buckets, vibratory pile drivers, aerial lifts, and pumpcrete machines with articulated booms. It is well-established in the rules for most operator's manuals that cranes should lift only freely suspended loads. A crane using a vibratory pile driver to pull pilings does not have a freely suspended load, so a load-moment device is necessary to prevent overloading. These complex parts and other hazards inherent in cranes make knowledge of crane design components and how to eliminate hazards through engineering controls necessary on any construction project.

SECTION 1

Two-Blocking

The Hazard Syndrome

The hazard of two-blocking is *dormant* when the crane is not in use. The hazard becomes *armed* when the operator is using the crane, as use may allow the headache

[2]2750 Prosperity Ave, Suite 620-CCO; Fairfax, VA, 22301-4312, (703) 560-2391

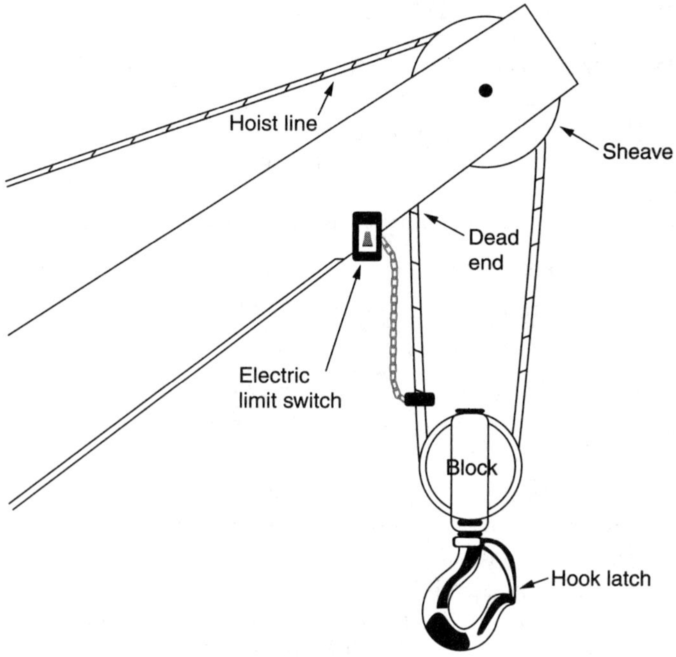

ball to approach the vicinity of the tip of the crane boom. The hazard is *active* when two-blocking occurs and the load hook assembly or lower load block contacts the sheave on the boom tip or upper load block, causing the hoist line to part and break, or pull the headache ball and hook up over the top of the sheave at the tip of the boom. The load straps become disconnected, and the load is lost, becoming a falling object and imperiling the lives of those working below. The illustration above shows a boom tip unprotected and able to two-block.

Scope

Current OSHA requirements do not require anti-two-blocking devices on all cranes, but these safety features came to pass as standard equipment on cranes from repeated liability lawsuits and heavy payments. Justice systems believed in the value of inherently safer design, and helped to ensure improved design on all new cranes. Two-blocking provides an example of the value of the justice system's potential to influence industrial safety.

Today, anti-two-blocking devices are found on all new cranes despite the lack of government standards. The standardization of anti-two-blocking devices on cranes is an important triumph for safety. The use of such a device has made an inherently dangerous aspect of crane use into an inherently safe one. By looking at ways the hazard of two-blocking has been overcome by liability litigation and industry acceptance, we hope to depict similar patterns in the next 49 examples and show how life-threatening hazards can be overcome by voluntary industry acceptance of safe design or process. A proactive approach to design safety would be more effective and beneficial to construction than the currently practiced approach by punitive government regulation or liability lawsuits.

Peril from the Failure Mode

Death and crippling injury often result when a line two-blocks. If the load line breaks under tension, and it happens to be supporting a worker in a personnel basket or several workers on a floating scaffold, it causes people to fall. Other times, people are hit by the falling load, headache ball, or straps on the hook.

On latticework booms, the following stresses in combination can break the hoist cable:

- The weight of the load
- The weight of the headache ball and rigging
- The tension caused by the hoist drum
- The weight of the boom when the pendant goes slack
- Whip caused by a long latticework boom

A crawler crane walking to a location with a raised boom and the headache ball snug against the boom tip sheave can snap the hoist line. On hydraulic boom cranes, the power of the hydraulic rams that extend the boom often create sufficient tension to break the hoist cable if the line two-blocks. Both crane types will two-block when the hook is near the tip and the boom is lowered. An operator can forget to release or pay out the load line when extending the boom, and the hoist cable can be broken. This hazard fits in the category of *structural/mechanical* hazards under tension. However, the peril also results from the *natural* source of gravity.

The consequences of two-blocking are devastating to all parties involved and include

- the loss of life;
- OSHA fines;
- compensation expense;
- the employer's embarrassment over local TV and news coverage; and,
- the crane owner's and the manufacturer's liability loss for not providing an anti-two-blocking device.

Brief Summary of the Occurrence

A falling headache ball struck the deceased when the crane two-blocked and the hoist line broke. The anti-two-block electrical switch had been broken by the previous renter, and the crane repair was delayed until a replacement switch arrived. The deceased repairmen had been directed to use the crane to assist in lifting other crane parts to be repaired. The manufacturer's operator's manual specified that the crane was not to be in use when the anti-two-blocking device was nonfunctional.

The case against the manufacturer was dropped, as the crane was equipped with an anti-two-blocking device. When the manufacturer started to equip cranes with this device and specify its proper use in their service manual, the responsibility for the use of anti-two-blocking devices was transferred to the user. The case against the employer was reopened for intentional negligence.

Date of Occurrence

February 26, 1990

Source of Information

234th Judicial District, Harris County, TX, # 90-54322

Contributory Factors

A crane equipped with a jib or flyline on the boom is even *more* prone to two-blocking, since the operator must watch two separate hooks in addition to the signaler and the load. An operator can easily be unaware of a two-blocking situation that may be occurring on the line not being used. Human-factors logic tells us that when an operator must use two controls—one for the hoist cable and one for the hydraulic boom extension—the chance of error is increased. With a second hook to watch, the operator is clearly overtasked.

It took years for the manufacturers, rental firms, and crane owners to adopt the use of anti-two-blocking devices. When confronted with liability claims for failure to provide these safety appliances, the defendants introduced many myths that decried the reliability of anti-two-blocking devices. Indeed, the introduction of accessories to enhance operator performance always seems to solicit negative pronouncements by the "Doubting Thomases." This, coupled with courtroom testimony by purported "safety experts," who denounce the effectiveness of safety devices such as the anti-two-block, dramatically hinders industry acceptance of these lifesaving appliances. These positions have endangered the lives of many. Here are some of the oft-repeated untruths regarding the anti-two-blocking device:

- *The anti-two-blocking device often does not function, as the crew frequently disables the device because they say they don't like to hear it ringing while roading the crane short distances when the hook is pulled to the boom tip*. If supervisors are not able to convey to the crew that it is "road day" (termination) for those who disable valuable safety appliances, the supervisor should be dismissed as well. As for the alarm ringing when roading the crane with its boom up (hopefully not high enough to touch powerlines), the hook only needs to be secured near the base of the boom with a clevis or other securing device to avoid this.
- *The operator or signaler cannot watch the load and the boom tip simultaneously*! "Human factors scientists maintain that requiring an equipment operator to perform multiple tasks diminishes performance."[3] If the crane has a jib extension, there are now three things to watch: two boom tips and the load.
- *When a crane is equipped with an anti-two-blocking device, crane operators feel they don't have to watch the hook anymore*. The truth is, operators should be watching the load, not the hook, as the anti-two-blocking device relieves them of this task.

[3]Lehman and Gage, "How Much is Safety Really Worth? Countering False Hypotheses," *ASSE Professional Safety*, May 1995.

In order to combat similar unfair pronouncements regarding other safety appliances, we must raise awareness about the effectiveness of system safety and the fallacy of relying solely on human performance. When used properly, the anti-two-blocking device continues to save lives. This fact is proven by statistical information and the virtual extinction of anti-two-block liability lawsuits, which saves the manufacturers millions of dollars in reduced claims of liability from two-blocking cases.

History of the Hazard

Early anti-two-blocking devices worked the same as they do today. When the hook or lifting block approaches a suspended bar, the limit switch or interlock will close the circuit, sound the alarm, and stop the hoist motor. Over the years, there have probably been thousands of two-blocking occurrences resulting in a broken hoist cable, with most of them going unrecorded because no one, luckily, was injured when the hook or load dropped. The cost of preventing this hazard would have been miniscule when compared with the physical and monetary catastrophe that resulted.

Early 1900s: Limit switches on bridge cranes were made available to prevent over-reeling of the hook up and over a hoist drum.

1906: The first patent of an anti-two-blocking device was issued in 1906. HIFI has copies in its files of over 53 patents issued from 1906 to 1978 for anti-two-blocking devices and limit switches. For further information, contact HIFI.

Post WWII: Hydraulic cranes with telescoping booms were introduced into the marketplace.

1960s: Galion Crane Company was the first to file for a patent on an anti-two-blocking device for hydraulic telescoping booms. This was a hydraulic valve arrangement that allowed the hoist drum to release the hoist cable when the telescoping boom was being extended, effectively preventing two-blocking. Drott filed patents for circuitry that provided an alarm and a mechanism that would intercede to prevent further extension of the telescoping boom.

1965: The first military specifications for the purchase of cranes requiring anti-two-blocking devices were written.

1966–1970: Galion installed an anti-two-blocking system as standard equipment on its cranes.

1967–1968: Austin-Western, a division of Clark Equipment Company, offered an electric anti-two-blocking device.

1969: Drott offered its first anti-two-block control as an option. The Bantam division of Koehring offered these devices on some of its rough-terrain cranes. Grove, the largest manufacturer of hydraulic cranes in the United States, made available an anti-two-blocking device for truck-mounted cranes.

1969–1970: Harnischfeger (P&H) made available an anti-two-blocking device in this period for rough-terrain and small-truck hydraulic cranes.

1970s–1980s: Specialty crane manufacturers, such as Ben-Ju and Broderson, offered anti-two-block warning and control lockout devices.

1975: Bucyrus Erie Company introduced a hydraulic anti-two-blocking system for its marine cranes.

1978: American Hoist and Derrick (AmHoist) installed an anti-two-blocking system as standard equipment on its 18-ton hydraulic crane.

1979–1980: Drott adopted the anti-two-blocking device as standard equipment.

1981: Grove standardized the Krueger Anti-Two-block Warning System for all cranes. By the end of 1988, Grove had installed the Pat Load Indicator System as standard equipment. Beginning in 1989 all new Grove products were equipped with Load Moment Indicator Systems that included anti-two-block capability.

1986: National Crane Corporation was the first to provide an anti-two-blocking device for pedestal cranes mounted on flatbed trucks.

1990-present: Most new cranes are equipped with some form of anti-two-blocking device.

When anti-two-blocking devices are used and properly maintained, their reliability record is perfect. There are no known failures of a properly maintained anti-two-blocking device.

Hazard Identification and Prevention Matrix

Two-blocking presents the natural hazard of a falling object caused by gravity, and can be effectively guarded against by an anti-two-blocking device. It is caused by the hazard of cable tension.

	Eliminate the Hazard		Guard the Hazard		Provide a Safety Factor		Provide Redundancy		Provide Reliability
	Hazard	Safety	Hazard	Safety	Hazard	Safety	Hazard	Safety	
Natural	Gravity falling object								
Mechanical	Cable tension	Anti-two-block stops source tension							100%
Electrical				Anti-two-block alarm					100%
Chemical									
Radiant Energy									
Biological									
Automated Systems									

Applicable Standards and Literature

Anti-two-blocking devices, consistent with available state-of-the-art technology and current custom and practice, are not currently required by OSHA on all cranes. When operators lift personnel with a crane, anti-two-blocking devices are required for this limited exposure.

OSHA

1926.550(g) (3)(ii)(c) only requires an anti-two-blocking device on cranes being used to suspend a personnel platform.

> (g) Crane or derrick suspended personnel platforms (1) Scope, application, and definitions- (i) Scope and application: This standard applies to the design, construction, testing, use and maintenance of personnel platforms, and the hoisting of personnel platforms on the load lines of cranes or derricks.
>
> (3) *Cranes and Derricks*—(i) *Operational Criteria* (A) Hoisting of the personnel platform shall be performed in a slow, controlled, cautions manner with no sudden movements of the crane or derrick, or the platform.
>
> (ii) *Instruments and Components*. (A) Cranes and derricks with variable angle booms shall be equipped with a boom angle indicator, readily visible to the operator. (C) A positive acting device shall be used which prevents contact between the load block or overhaul ball and the boom tip (anti-two-blocking device), or a system shall be used which deactivates the hoisting action before damage occurs in the event of a two-blocking situation (two-block damage prevention feature).

Interesting points to note about the OSHA standards include:

(a) It is interesting that OSHA 1910.179(g)(5)(iv) for overhead bridge and gantry cranes requires that "The hoisting motion of all electric traveling cranes shall be provided with an overtravel limit switch in the hoisting direction." This is another way of saying "an anti-two-blocking device system."

(b) OSHA 1910.179(n)(4)(i), "Hoist Limit Switch," requires that "At the beginning of each operator's shift, the upper limit switch of each hoist shall be tried out under no load . . ." This requirement shows that anti-two-blocking concepts have been applied on overhead and gantry cranes.

(c) OSHA subpart N, 1926.550(a)(1) requires that "The employer shall comply with the manufacturer's specifications and limitations applicable to the operation of any and all cranes and derricks." Occasionally, crane owners have failed to provide functioning anti-two-blocking devices as recommended by the manufacturers.

(d) OSHA CFR 1926.550(4)(ii)(F) states the following concerning the design of the personnel basket to be used: "In addition to the use of hard hats, employees shall be protected by overhead protection on the personnel platform when employees are exposed to falling objects."
Comment: If the hoist cable is two-blocked and the personnel basket is without overhead protection, the falling headache ball or block will surely strike the occupants, and hard hats will provide little protection.

ANSI

(a) The old ANSI B30.15-1973, Mobile Hydraulic Cranes, Section 15-1.3.2(d), Two-blocking Damage Preventive Feature states, "On a telescoping-boom crane, with less than 60 feet of extended boom, a two-blocking damage preventive feature shall be provided capable of preventing damage to the hoist rope, and/or other machine components, when hoisting the load, extending the boom, or lowering a boom on a machine having a stationary winch mounted to the rear of the boom hinge."

Comment: Considering the fact that the longer the boom, the further the boom tip is from the observing operator, it seems that an anti-two-blocking device requirement limited to booms less than 60 feet is insufficient.

(b) ASME B30.5-1994 (Revision of ASME/ANSI B30.5-1989) Mobile and Locomotive Cranes, Section 5-1.9.9 Two–Blocking Features, still recommends an anti-two-blocking device as a damage-preventive feature for the boom tip. This new standard now requires the anti-two–block device on all cranes as follows:

(i) Telescopic boom cranes manufactured after February 28, 1992, shall be equipped with an anti-two–block device or a two–block damage prevention feature for all points of two-blocking. Telescopic boom cranes manufactured before February 28, 1992 should be equipped with a two–block warning feature(s), a two–block damage prevention feature, or an anti-two–block device for all points of two-blocking (i.e. jibs, extensions, etc.).

(ii) Lattice boom cranes manufactured after February 28, 1992, shall be equipped with a two–block warning feature which functions for all points of two-blocking. Lattice boom cranes manufactured before February 28, 1992, should be equipped with a two–block warning feature which functions for all points of two-blocking."

Comment: This standard fails to identify two-blocking as a life-threatening hazard. From a human-factors standpoint, such devices should be standard equipment on all cranes because of the dire consequences associated with this hazard.

(c) Holland, Belgium, France, Germany, Finland, Sweden, Switzerland, the Czech and Slovak Republics, the former Soviet Union, and Japan all require anti-two-blocking devices on cranes. The United Kingdom has no specific requirement for an anti-two-blocking device, but their rules achieve the same results by limiting the hoisting capacity to the rated capacity of the cable. As a result, no load is imposed upon a hoist cable that exceeds the rated capacity, which is well below the breaking strength of the hoist cable. In short, most European countries have arrived at methods to prevent the two-blocking failure mode.

U.S. Army Corps of Engineers

Safety and Health Requirements Manual EM 385-1-1 (November 3, 2003) Section 16.D.01e requires anti-two-blocking devices as follows:

(a) Lattice boom cranes shall be equipped with an anti-two–block device to stop the load hoisting function before the load block or load contacts the boom tip. Cranes that are used exclusively for duty cycle operations (such as clamshell, dragline, grapple, or pile driving) are exempt from anti-two–block equipment requirements.

(b) Telescopic boom cranes shall be equipped with an anti-two–block device to stop the load hoisting function before the load block or load contacts the boom tip and to prevent damage to the hoist rope or other machine components when extending the boom.

(c) Telescopic boom cranes which are used exclusively for duty cycle operations (such as clamshell, dragline, grapple, or pile driving) shall be equipped with a two-blocking damage prevention feature or warning device to prevent damage to the hoist rope or other machine components when extending the boom.

Society of Automotive Engineers (SAE)

Standard "Two-Blocking Warning and Limit Systems in Lifting Crane Service" J1305, June 1987, provides a recommended practice for minimum performance requirements of devices that signal or automatically prevent two-blocking.

Department of Defense (DOD)

MIL-C-29918B (USAF), "Crane, Truck-Mounted, A/S32A-16," Paragraph 3.7.8, August 27, 1965

MIL-C-27840C (USAF), "Crane, Truck-Mounted, A/S32H11, 4,000 lb. Capacity, 4 × 4," Paragraph 3.7.6, April 12, 1966

MIL-C-27718C (USAF), "Crane, Hydraulic, Truck-Mounted, AS32A-16, 16,500 lb. Capacity, 6 × 4, GED"

MIL-C-28616 (YD), "Crane, Shovel, Convertible, Crawler-Mounted, Commercial DED," Paragraph 3.9.1, April 30, 1971

MIL-C-28622 (YD), "Crane, Hydraulic, Wheel-Mounted, 4 × 4, Full Revolving (Shipboard Aircraft Crash Salvage)," Paragraph 3.6.2, September 30, 1971

MIL-C-23877B (YD), "Cranes, Hydraulic, Wheel-Mounted, 5 & $12^1/_2$ Ton Capacities," Paragraph 3.6.4, May 1, 1972

MIL-C-28614A (YD), "Crane, Hydraulic, Full-Revolving, Wheel Mounted, 4 × 4, DED," Paragraph 3.10.2, June 12, 1975

Other References

- David V. MacCollum, *Crane Hazards and their Prevention,* ASSE, 1993.
- D.E. Dickie, *Crane Handbook*, Canadian Construction Association of Ontario, 1972: P 130, Figure 4.62 recognizes that the hazard of two-blocking will cause the hoist cable to break.
- "Crane Two-Blocking" *Hazard Information Newsletter*, Volume 1, Issue 3, June 1996.

Commentary

The slow universal acceptance of the anti-two-blocking device has been similar to the slow development and reluctant acceptance of rollover protective structures (ROPS). Repeated costly litigation finally proved to be the instrument necessary to move the crane companies, equipment rental firms, and contractors towards safety. Liability litigation successfully prompted the near-universal adoption of anti-two-blocking devices after decades of industry reluctance to embrace this lifesaving device. After the industry became swamped with lawsuits for failure to provide the anti-two-blocking devices, all crane manufacturers began to provide them as standard equipment after delaying their use for years.

Recommendations

Industry acceptance of this safety appliance now decrees that all new and old cranes be equipped with anti-two-blocking devices, surpassing OSHA regulations. No substantial design improvements had been made prior to industry adoption of this device. Liability awards seem directly responsible for prompting this change. The path of acceptance of the anti-two-blocking device can be used as a model for the subsequent 49 hazards outlined in this book, as we enlist the engineering profession in developing more safety appliances to overcome inherent hazards with inherently safe design.

SECTION 2

Crane Upset from Overload

The Hazard Syndrome

A crane can upset from overloading. Any crane without a load-moment indicator (LMI) presents a *dormant* hazard that can become *armed* at any time when the weight of the load on a crane exceeds its rated capacity and causes it to destabilize. Analysis shows that over half of the incidents involving this hazard occur when the crane operator is either sluing the crane cab, or extending or lowering the boom with the outriggers retracted. Both these actions can *activate* the hazard to upset by increasing the lifting radius. At that point upset occurs quickly.

Scope

Retracted outriggers and improper boom use, such as extending the boom while lifting a load, often result in upset. The rated capacity for most cranes is calculated at 85% of the weight that will cause the crane to tip from overload. This leaves a 15% safety factor, which is a very thin margin in terms of overload. Snagged loads, soil conditions, and other outside factors can cause a crane to exceed that 15% margin in the blink of an eye. Hydraulic telescoping cranes can easily upset with or without a load on the boom, threatening the lives of the operator and bystanders. Upset occurs in a number of ways. From review of over one thousand crane upsets that occurred over a twenty-five-year period, it was calculated that one upset occured in about every ten-thousand hours of crane use. Nearly 75% of these upsets were the result of error-provocative circumstances that caused the operator to inadvertently exceed the crane's tilting load capacity. The following breakdown was made:

%	Circumstance
15%	In the travel mode with a pick-and-carry load
39%	Making a swing with outriggers retracted
15%	Making a pick with outriggers retracted
14%	Making an overload pick or swing with outriggers extended
6%	Making a pick or swing (use of outriggers unknown)
7%	Outrigger failure
4%	Load not freely suspended or was pulled to the side
100%	Total

The first step in designing out this hazard is to decrease the point of rated capacity and increase the margin of error for the weight needed to tip a load. The construction manager can also prevent this hazard by premeasuring loads so as to not exceed the rated capacity.

Peril from the Failure Mode

The significant consequences of crane upset are a mixed bag of either one or a combination of the following:

%	Consequence
3%	Fatalities
8%	Lost-time injuries
24%	Lost-production time
99%	Damage to the crane
20%	Serious damage to other property

This hazard is both natural and structural/mechanical, because it is caused by gravity and instability applied by outside factors.

Brief Summary of the Occurrence

A truck driver was crushed to death by an overturning large truck-mounted hydraulic telescoping-boom crane. The crane was offloading heavy concrete barriers on a freeway when it upset. The driver assumed that he was in a safe place, standing by his truck cab next to the crane. The general contractor's work plan included using the crane with its outrigger retracted and keeping the crane inside the freeway's inside lane. The middle lane was used to offload the concrete barriers from rental trucks, and the outside lane was reserved for southbound traffic. The crane's load-moment indicator (LMI) had been disconnected. Had the LMI been in use, the interlocks on the outriggers would have not allowed use of the crane's boom with the outriggers retracted. The key to disconnect the LMI and allow unsafe use of the crane was provided by the rental company. The work was being done at night, and the upset occurred around midnight. Mere moments before the crane upset, a station wagon with several persons in it passed the crane on the southbound lane. The vehicle just missed being struck and its passengers possibly killed by the tipping crane's boom. The freeway was flanked on both sides with access roads, and southbound traffic could have been easily routed onto these roads, affording sufficient space for the crane's outriggers to be extended. The general contractor and the rental firm were found to be negligent in crane usage. Construction management at the time of planning failed to look upstream and utilize the rerouting of public traffic to the access roads, which would have allowed the use of the crane's outriggers and eliminated an *armed* hazard to public travel in a construction worksite.

Date of Occurrence

July 13, 1988

Source of Information

45th Judicial District Court of Bexar County, TX, # 98-CI-10315

Contributory Factors

In the aforementioned case study, the lack of construction safety planning to reroute public traffic and allocate space to extend the crane's outriggers was the principle deficiency. In many cases of upset, different factors combine to cause imbalance that ultimately leads to the upset. The following fault-tree analysis lists events that can create a change of circumstances to trigger the hazard into the active mode from the armed mode.

The next image uses a similar fault tree to list the factors that can prevent or control the change of circumstances that activate the hazard. When used in conjunction with each other, these features make crane upset a virtual impossibility.

The maximum load that a crane is capable of lifting safely varies depending on the circumstance. Retracted outriggers, extended booms, and the arc of the slue (rotation) are all factors that must be considered before a lift can be made safely. For many years, the necessary calculation for safe loads was a time-consuming and error-prone process performed

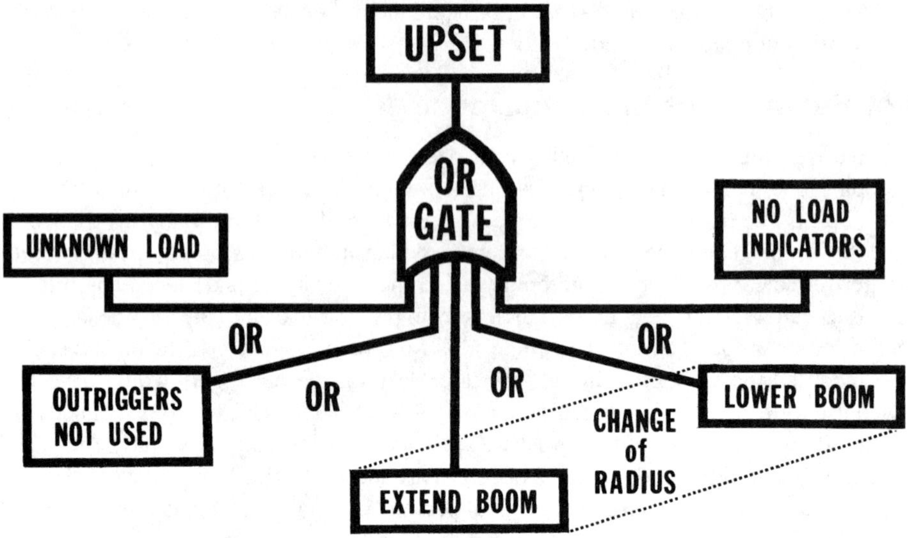

AS NO SAFEGUARD EXISTS TO PREVENT UPSET

by operators. Load-moment indicators (LMIs) are a technological innovation that can eliminate the potential for such human miscalculation of a crane's lifting capacity, but only when they are used correctly. Unfortunately, many crane operators are still not familiar with the use of LMIs. In addition, there is no uniform LMI, just as there is no uniform type of crane. Each brand comes with variances that the operator must quickly acquaint him or herself with.

The danger from an operator estimating the lifting capacity of a crane varies depending on crane type. On flatbed-mounted pedestal cranes, the hydraulic boom is located

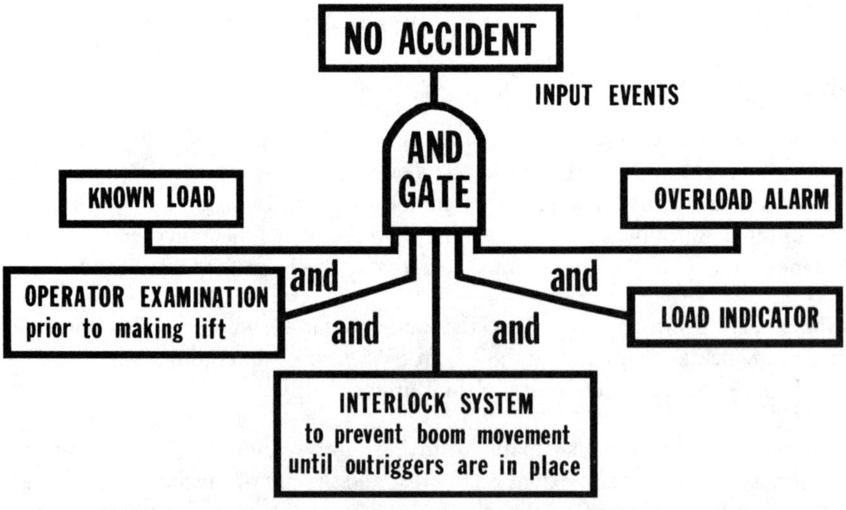

ALL SAFEGUARDS MUST BE REMOVED BEFORE AN UPSET CAN OCCUR

directly behind the truck cab, and the over-the-rear or over-the-cab tipping load can be as much as twice the rated capacity. The crane can easily pick a load over the rear or cab that will cause upset when the load is slued to either side. The LMI should be considered as a necessary aid for all crane operators.

On some high-reach hydraulic cranes, the weight of the boom itself—without a load—can create an imbalance and cause upset when the boom is positioned at a low angle. A crane can be upset when outriggers cannot be extended or lowered on one side to work in a confined location or are not repositioned after having been momentarily retracted or raised to let traffic pass. On some makes this can occur even with outriggers extended. On some of the newer cranes, the counterweights are heavy enough to raise the crane's lifting capacity in a few of its configurations. As rated capacity is 85% of the structural design, the boom can collapse at 100% of its rated capacity before it even reaches tipping load.

Today's crane operator is confronted with a number of error-provocative circumstances that can cause upset or boom failure:

- The ability to extend the hydraulic boom while lowering it
- The choice of extending or retracting outriggers
- Soil that is unable to support the outriggers
- Outrigger failure
- The changing lifting capacities as the boom is rotated to various positions
- An unknown load weight
- Incorrect labeling of load weight
- Loads that are not freely suspended

Modern cranes have controls that respond very quickly. On hydraulic cranes, if the boom is extended while it is lowered, the lifting capacity will be reduced very swiftly.

Cranes are vulnerable to upset largely due to management indifference to the installation of LMIs on older cranes. Cranes are easy to operate and thus attractive to the untrained, tempting them to try their "luck" with self-experimentation in learning how to operate them. This foreseeable misuse is always with us and can be controlled by electronic safeguards. Management is responsible for ensuring the proper installation and training associated with LMIs. An operator should be familiar with the correct use of an LMI before becoming experienced with a crane.

Management often assumes working without safety appliances is an option when operators are experienced; however, even the most experienced and well-trained crane operator without an LMI can be fooled. The radius of a telescoping hydraulic crane can be increased rapidly when lowering and extending the boom, and overload can occur in a flash. Just lowering the boom, extending a boom, or doing both at the same time can quickly cause a crane to exceed its lifting capacity. Load-moment devices are designed to keep boom movement within the rated capacity and assist the operator in error-free operation.

Though the proper use of an LMI remedies all of these potentially lethal situations, many crane operators have not been trained on how to use them; and without the proper training, operators typically avoid these lifesaving devices.

History of the Hazard

Most Western European countries have required load-sensing devices or LMIs since 1960. Today, most manufacturers of large cranes offer state-of-the-art LMIs on new cranes. These little black electronic boxes prevent overloads for any boom radius. The failure to equip older cranes with, or to use, load-measuring systems is almost as frequent and probably as costly as not providing proximity alarms to avoid powerline contact. America's crane safety requirements for load-measuring systems are almost four decades behind those of Western Europe.

Hazard Identification and Prevention Matrix

An upset is the result of unbalance caused by gravity. LMIs with interlocks are reliable methods of control, and can make cranes inherently safe lifts by ensuring that the crane maintains the correct balance. Most LMIs have in their memory all safe

	Eliminate the Hazard		Guard the Hazard		Provide a Safety Factor		Provide Redundancy		Provide Reliability
	Hazard	Safety	Hazard	Safety	Hazard	Safety	Hazard	Safety	
Natural	Gravity/ destabilization			Interlock				Reroute traffic	100%
Structural/ Mechanical	Overload			Use of outriggers					99%
Electrical		LMI detects overload		Alarm on LMI warns				LMI stops boom movement	99%
Chemical									
Radiant Energy									
Biological									
Automated Systems									

rated capacities, and will give warning when that specific crane reaches 90% of rated capacity. Balance may be achieved through outriggers or limited load capacities.

Applicable Standards and Literature

OSHA

OSHA does not require load-measuring or load-moment devices. The Occupational Safety and Health Act of 1970, Section 5(a)(1), General Duties Clause may be used to address this hazard. OSHA 1910.179(b)(5), 1910.180(c), 1910.181(c), and 1926.550(a)(1) require compliance with the manufacturer's rating charts. OSHA 1910.18u0(h)(3)(vii) states, "On truck-mounted cranes, no loads shall be lifted over the front area except as approved by the crane manufacturer." Lifting over the front of a truck-mounted crane greatly increases the tipping load, and the boom and turntable may be overstressed. Then sluing the load to either side would most certainly result in upset. The U.S. Department of Labor's Marine Terminal Standard 1917.46 and Longshoring Standard 1918.74(a)(9) included longshore union involvement and addressed load-indicating devices. Rating charts and overloading are addressed in 1917.45(b) and 1918.74(a)(2). Counterweight restrictions are noted in 1917.45(f)(6).

ANSI

ANSI requirements address compliance with manufacturer's rating charts and currently specify the following:

"All new cranes with a maximum rated load capacity of 3 tons or more should have load indicators." (Section 5-1.9.10© Load Indicators, in B30.5, 1994.)

U.S. Army Corps of Engineers

Safety and Health Requirements, EM 385-1-1, (3 November 2003) Section 16.D.01.a reads, "All lattice-boom and hydraulic mobile cranes shall be equipped with the following:

(a) A boom angle indicator and a load-indicating device, or a load moment indicator (rated capacity indicator): calibration and testing of indicators will be performed in accordance with the manufacturer's recommendations."

This requirement became effective January 1, 1994.

SAE

The Society of Automotive Engineers has developed a number of standards that relate to load-measuring systems and associated operator aids as follows:

(a) Crane Load Stability Test, SAE J765, October 1990

(b) Telescoping Boom Length Indicating System, SAE J1180, October 1980

(c) Radius-of-Load and Boom Angle Measuring System, SAE J375, April 1985

(d) Load Indicating Devices in Lifting Crane Service, SAE J376, April 1985

(e) Crane Load Moment System, SAE J159, April 1985

Department of Defense (DOD)

Numerous U.S. Military Specifications published by the Department of Defense, which require load-measuring and load-moment devices on mobile cranes, show an acceptance and the necessity for load-measuring devices and are listed as follows:

MIL-C-27840B (USAF), "Crane Truck Mounted A/S32H-11," Paragraph 3.5.3.4, November 6, 1964.

MIL-C-27718B (USAF), "Crane, Truck Mounted A/S32A-16," Paragraph 3.5.3.4, August 27, 1965.

MIL-C-27840C (USAF), "Crane, Truck Mounted A/S32H-11, 4,000 Crane Capacity, 4 × 4," Paragraph 3.5.3.4, April 12, 1966.

MIL-T-62089 (AT), "Truck, Maintenance; with Rotating Hydraulic Derrick, Air Transportable, 34,500 pounds, GVW, 6 × 4," Paragraph 3.6.4.5, March 26, 1968.

MIL-C-28564 (YD), "Cranes, Hydraulic, Truck Mounted, 8 × 4, 25-, 40-, and 55-ton Capacities," Paragraph 3.5.4, December 30, 1969.

MIL-C-27718C (USAF), "Crane, Hydraulic, Truck Mounted, A/S32A16, 16,500 Pound Crane Capacity, 6 × 4, GED," Paragraph 3.5.5.4, September 8, 1970.

MIL-C-28614 (YD), "Crane, Hydraulic, Rough-Terrain, 25-ton Capacity, Full Revolving, Wheel-Mounted. 4 × 4. DED." Paragraph 3.6.7, April 8, 1971.

MIL-C-28622 (YD), "Crane, Hydraulic, Wheel-Mounted, 4 × 4, Full Revolving, 45-ton Capacity, DED," Paragraph 3.6.7, September 30, 1971.

MIL-C-28629 (YD), "Crane, Hydraulic, Wheel-Mounted, 4 × 4, Full Revolving, (Shipboard Aircraft Crash-Salvage)," Paragraph 3.6.7, February 29, 1972.

MIL-C-52341E (ME), "Crane, Wheel-Mounted, Diesel Engine-Driven, 20-ton, Rough-Terrain," Paragraph 3.12.13.7, October 24, 1972.

MIL-T-62089A (AT), "Truck, Maintenance; with Rotating Hydraulic Derrick, Air Transportable, 34,500 pounds GVW, 6 × 4," Paragraph 3.6.2.5, December 18, 1973.

MIL-C-52326B, "Crane, Crawler Mounted: 60-ton at 25-foot Radius, Diesel-Engine-Driven," Paragraph 3.15.1.4, April 4, 1974.

MIL-C-29352 (MC), "Crane, Wheel-Mounted: Hydraulic, Rough-Terrain, 30-ton Capacity, Full Revolving, 4 × 4, DED," Paragraph 3.3.2.9, July 17,1974.

MIL-C-28614A (YD), "Cranes, Hydraulic, Full Revolving, Wheel-Mounted, 4 × 4, DED," Paragraph 3.10.7, June 12, 1975.

MIL-C-62253 (AT), "Crane: Material Handling (Rebuild)," Paragraph 3.5, September 16, 1975.

MIL-C-62135A (AT), "Crane; Material Handling," Paragraph 3.5, May 21, 1976.

MIL-C-52991 (ME), "Crane, Truck-Mounted, Hydraulic, 25-ton, Commercial Construction Equipment (CCE)," Paragraph 3.4.10, September 27, 1976.

MIL-T-62089B (AT), "Truck, Maintenance; with Rotating Hydraulic Derrick, Air Transportable, 34,500 pounds GVW, 6 × 4," Paragraph 3.612.5, June 9, 1980.

Other References

- David V. MacCollum, *Construction Safety Planning*, pages 82–83 "LMI", and pages 267–271, "Critical Lift Plan," Van Nostrand Reinhold, 1995.
- David V. MacCollum, *Crane Hazards and Their Prevention*, ASSE, 1993.
- The United Kingdom safety requirements, in the publication *Factories* (1961), No 1581, Part III, "Indication of Safe Working Load of Jib Cranes," para. 30, " . . . requires load-sensing devices for all cranes with a lifting capacity of over 2,000 pounds."
- The Netherlands has the most advanced approach to preventing crane upset from overloading by requiring a Load Moment Limiter (LML) system. As specified by the Netherlands Normalization Institute (NNI), all cranes are required to meet NEN norm standards 2017-2028:
  (a) The LML insures that the crane cannot lift a load that exceeds its rated capacity for that particular radius.
  (b) The LML stops all crane movements leading to an increased radius or to other hazardous circumstances except sluing.
  (c) The crane cannot be started automatically after being switched off by an LML.
- Additional data concerning Dutch crane safety ("Safety the Dutch Way") was published in the March, 1991 issue of Cranes Today, p.31. It is probable that LML will be adopted by all of the European countries that are member states of the, European Economic Community.
- *Mobile Crane Manual,* by D.E. Dickie (1982). Chapters 4, 5, 6, & 7 provide an excellent discussion on how to use load charts. Dickie, of the Canadian Construction Safety Association of Ontario, states in his *Crane Handbook* that 50% of mobile crane failures are caused by improper use of outriggers.
- "Mobile Crane Upset," *Hazard Information Newsletter,* Volume 2, Issue 11, February 1998.

Commentary

On many occasions, construction management promotes the unsafe use of cranes by assigning a crane operator who is unfamiliar with the features of an LMI, or worse, allowing the LMI to be disconnected. As the function of an LMI is to prevent upset by disallowing use of the crane in the event of overload, imbalance, or lack of outrigger stability, use of cranes without them becomes a gross disregard of safety.

The United States is a world leader in electronics, computer technology, smart weapons, and other electronic systems, but we have less than a Third World sophistication when it comes to retrofitting the latest technology on our older cranes. As early as the October

1988 issue, the British magazine *Cranes Today* reported that Wylie Weighload, one of Europe's leading manufacturers of load-measuring systems, was undertaking U.S. expansion that "will give the company direct access to the U.S. market at a time when the indigenous crane manufacturers are starting to fit safe load indicators as standard in response to product liability pressures." Even with the updating of older equipment, the industry still fails to adopt a useful operator aid. LMIs and interlocks are now increasingly accepted as necessary operator aids to prevent undesired movement.

If load-measuring systems on rough-terrain cranes designed for on-rubber and off-rubber use do not function during on-rubber lifts, manufacturers of these cranes should provide interlocks that limit boom use in order to prevent overloading.

Limiting crane operation to only well-qualified, competent, and *licensed* operators who have been certified for their use of LMIs, as well as specific makes and models of cranes is essential to the prevention of mobile crane upset. To meet this need for competent and qualified crane operators, the National Commission for Certification of Crane Operators was established in 1995. This group has taken on a difficult task and is to be commended for their forward thinking and for providing a very necessary service. For further information about this organization, write or call:

National Commission for Certification of

Crane Operators

2750 Prosperity Avenue, Suite 620-CCO

Fairfax, VA 22031-4312

Telephone: (703) 560-2391

Fax: (703) 560-2392

Recommendations

Construction safety planning should include an analysis of all crane lift tasks in order to achieve an inherently safer construction site. Further, as a requirement, all cranes should be equipped with LMIs as well as interlocks to prevent the retraction of the outriggers when the boom is extended. A study should be directed to expand upon the cost benefits of LMIs. Designers should ensure that outrigger pads are of sufficient size to afford support on wet sand, and include sufficient information in the operating manual on how to provide adequate support on different terrains.

Crane operators should receive an extended crane-operator certification that proves they are qualified in the use of LMIs, similar to the licensing that aircraft pilots must undergo to use electronic navigation systems. This certification of qualified operators will provide a basis to give crane operators authority to ensure that the lift is within its rated capacity. The crane operator needs to be set apart, as is the pilot of an aircraft, with a mystique and authority to take charge and not make any lift until the planning meets his or her approval. Requirements should extend to the equipping of cranes with LMIs and ensuring that crane operators are certified in their use. Design of a crane's LMI should always include lifting limits when a crane's outriggers are retracted.

SECTION 3

Jib-Boom Stowage on Hydraulic Cranes

The Hazard Syndrome

A jib-boom stowage system that relies upon manual pin placement is a *dormant* hazard. This hazard becomes *armed* when jib stowage is attempted with the use of a pin to anchor the jib to the side of the main hydraulic boom, as it is often misaligned with the anchor-pin holes. Misalignment results in the *active* falling-object hazard of the jib boom falling free. The images below show a typical jib-boom hinged connection to the outer telescoping boom. Notice the reliance on pins for anchor points and the absence of a connecting pin.

Scope

The correct stowage of this type of boom is dependent upon error-free alignment of the anchoring-pin holes, which can be difficult to visually verify. Whenever misalignment of the pin in the proper hole occurs, the jib boom is in danger of falling from the intended anchor point. Assessment of human visual capability calls for a redesign of the stowage feature, including alignment markers to indicate proper pin connection for safe stowage. Some users of cranes with jib booms have removed the jib booms because they believe it is safer not to rely on an error-provocative pin-anchoring system to secure the jib boom in a stowed position.

Peril from the Failure Mode

When the heavy (approximately 2000 lbs.) jib boom falls due to improper stowage, it may strike someone and cause serious injury or death. Improper stowage is a relatively frequent occurrence on construction sites, and a number of lawsuits have been filed regarding this hazard. It is caused by the *natural* hazards of gravity and foreseeable error by the user. It is compounded by the *mechanical* aspect of a system that fosters unintentional misalignment of the attaching pin.

Brief Summary of the Occurrence

The deceased was in the process of folding the jib boom into a stowed position and believed that the anchor pin was properly placed. The anchor-pin holes on the main boom and jib boom were not aligned, and the deceased was struck by the falling boom and instantly killed. The crane operator and his assistant stated that they had followed the operating instructions, and the pin appeared to be properly inserted.

Date of Occurrence

September 22, 2001

Source of Information

U.S. District Court of South Carolina, Aiken Division, Columbia, SC, # 1 04-21943-27

History of the Hazard

This hazard has maimed and killed people from the time that jib booms began to be stowed on the side of the main boom. The hazard was identified in the late 1970s, and incidents continue to be recorded. Conventional reporting of these occurrences classify them as involving a falling object, but fails to identify the object and the necessary engineering design improvement.

Contributory Factors

The designers of a crane choose a nonautomatic latching system similar to a door or gate latch. The use of a manual pin is an operator- or worker-dependant design, and less effective than a design-dependant system that automatically secures the jib to the main boom. The hazard on some cranes presents additional dangers: To place the securing pin correctly requires someone getting on top of the boom (unsafe access) to insert the anchoring pin. There are no markings on the main boom to show that it has been properly positioned or to align the anchor-pin holes for insertion. Since the anchoring is between the jib boom and main boom, it cannot be easily seen whether the locking pin is in place. A design that automatically latches the boom into place would greatly improve the inherent safety of the jib-stowage feature. A pin-setting marker point visible to the crane operator to indicate proper alignment would ensure for a correct jib-stowage process. The use of pins to attach the jib boom to the outer telescoping boom has also had a record of jib-boom free fall. User error arises as the storage anchor-pin is removed before the jib is secured to the outer telescoping boom. A way to prevent this error is to develop an electronic verification system that will not release the storage latch until the pins for attaching the jib to the telescoping boom are securely in place.

Hazard Identification and Prevention Matrix

The hazard is caused by gravity. It is manifested by a simple mechanical design defect that erroneously relies on perfect user performance. The main defect in the placement of the stowage pin is that in many cases, correct placement of the pin requires perception beyond the reasonable expectation of human performance, because the boom itself creates an obstruction to viewing the pin. Visual obstruction by the boom means the pin cannot be adequately viewed before, during, or after stowage to ensure proper placement and structural integrity. On the matrix, this defect would fall under the category of a mechanical design defect that requires elimination of the hazard with the design of an automatic latching system.

	Eliminate the Hazard		Guard the Hazard		Provide a Safety Factor		Provide Redundancy		Provide Reliability
	Hazard	Safety	Hazard	Safety	Hazard	Safety	Hazard	Safety	
Natural	Gravity/ human performance								
Structural/ Mechanical	Unintentional misalignment (snagging)	Automatic latching system				Backup stowage pin		Visual marks for proper alignment	100%
Electrical				Electronic verification for proper jib stowage					100%
Chemical									
Radiant Energy									
Biological									
Automated Systems									

Applicable Standards and Literature

OSHA

OSHA has not commented on this subject in general, but OSHA 1926.550(a)(1) states, "The employer shall comply with the manufacturer's specifications and limitations applicable to the operation of any and all cranes and derricks." The manufacturer's complicated instructions have been proven to be ineffective.

ANSI

ANSI has not commented on this hazard.

This hazard is the failure to conduct a system safety-hazard analysis at the time of design. Had consideration been given to the following references, the hazard could have been eliminated at the time of design.

Other References

- Amos L. Wood, *The Organization and Utilization of an Aircraft Manufacturer's Air Safety Program;* Fourteenth Annual Meeting of the Institute of Aeronautical Sciences, New York City, January, 1946.
- *System Safety, Design Handbooks Series,* 2nd Edition; U.S. Space Administration, 1969 to Present.
- David V. MacCollum, "Reliability as a Quantitative Safety Factor," *Journal of the American Society of Safety Engineers*, May 1969.
- Willie Hammer, *Handbook of System and Product Safety;* Prentice-Hall, 1972.
- W.G. Johnson and E. Grandjean, *The Management Oversight and Risk Tree-MORT;* Superintendent of Documents, 1973.
- William Johnson and Marcel Dekker, *Mort Safety Assurance System;* National Safety Council, 1980.
- Willie Hammer, *Product Safety Management and Engineering;* Prentice-Hall, 1980.
- Willie Hammer, *Occupational Safety Management and Engineering;* Prentice-Hall, 1981.
- George A. Peters and Barbara J. Peters, *Human Error: Causes and Control;* CRC Press/Taylor and Francis Group, 2006.

Commentary

The design of the pin connection is inherently dangerous because it relies upon user performance beyond ordinary expectations. These design flaws present the need for an automatic locking system to safely anchor the boom in the stowage position. This hazard needs to be widely publicized, with instructions to show how arrows could be placed to ensure a visible check for proper alignment of the latch system. Design criteria should provide a latch that is as simple as a gate latch. An optimum design would also include a retaining bar that requires the latch to be raised before the gate can be opened to release the jib while installing it, and visual markings to show that the boom is properly aligned to ensure for latching.

Recently a new model MTF65G-4 Tadano crane replaced the conventional manual pins used in jib-boom attachment with a "crank into position" device that was featured in *Lift and Access* magazine, July–August, 2006 in an article by Guy Ramsey. Such innovation is the key to the successful elimination of construction-site hazards.

Recommendations

Equipment designers should address this hazard and define a requirement for automatic latching devices to replace the manual pin system. Professional engineer involvement to improve design is the key to stopping inappropriate jib-boom stowage. As a redundant measure, human-factors specialists should be consulted for any task that requires visual verification, as the manual placement of pins always invites the possibility of human error. In this day and age of "miracle" electronics, a monitoring system would ensure proper sequencing of pin placement and removal to avoid an unsecured jib boom. A further challenge to designers would be to "automatically" attach the jib boom and stow it without manual assistance. Such a feature would eliminate the need for performance of this task by the crane operator and the helper, who usually stow the jib boom after use.

In the absence of such advances, the construction manager always has the choice to have written into a contract requirement that the jib booms be removed before the crane is brought to the job site. In this manner, the construction manager does not have to cope with the hazard of unsafe jib-boom stowage.

SECTION 4

Latticework Boom Disassembly

The Hazard Syndrome

Latticework booms are *dormant* hazards that require disassembling for shortening, lengthening, or transporting. When the boom is lowered to a horizontal position for disassembling and is only suspended from the boom tip by its pendant guys, the hazard is *armed.* Boom collapse can occur if each boom section is not completely lowered with the boom tip on the ground or supported by relocating the pennant lines, or stabilized by blocking or cribbing. Collapse can also happen when the tip rests on the ground and the midsection connections are not supported. When the lower pins that connect the boom sections are the first ones knocked out by workers who are under the boom, the hazard is *active,* and the boom collapses downward upon them.

Scope

Telescoping hydraulic crane booms are rapidly becoming the crane of choice, as they are easy to extend and retract and allow for greater speed and mobility. However, a need for cranes with latticework booms remains, and a number are still being manufactured or are in service. When the pins are removed from the lower chord of the boom connectors, the

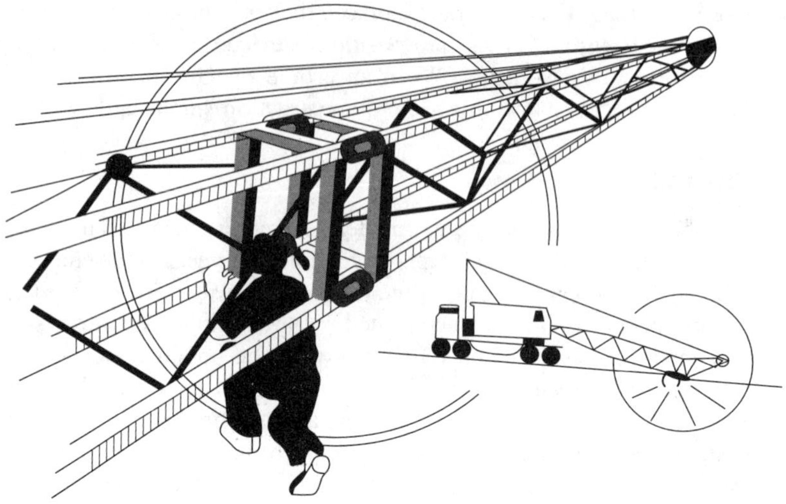

connection point hinges downward and causes the boom sections beyond that point to fall to the ground. To lengthen, shorten, or disassemble for transport a latticework-crane boom the user must rely solely upon error-free worker performance.

Peril from the Failure Mode

When removing the connecting pins, an unsupported latticework-crane boom can fall and crush those underneath. The gravity that causes an object to fall makes it a *natural* hazard, which is compounded by reliance on human performance. *Mechanical* hazards of faulty hinging and boom-section connection with pins are also present during latticework boom disassembly. The boom section displayed below can fall when not braced properly.

There is litigation on record of sixty-two wrongful deaths, three paraplegics, and twenty serious crippling injuries. All cases involve people who were crushed by the falling boom. This number of deaths and crippling injuries shows that this system of disassembling or shortening the boom is an error- provocative task that creates unnecessary danger to the worker. The manufacturers' manuals of instructions, often requiring the use of cribbing, has not met with reasonable success, as cribbing is not always provided and may not be available. Additional manufacturers' instructions for supporting the pinup boom by using the pennant lines to stabilize the section to be disassembled have not been effective.

Brief Summary of the Occurrence

The deceased stepped under the boom, which was supported in a horizontal position by the pinup guys. When he hammered out the second pin on the lower boom chord, the boom hinged downward and crushed him.

Date of Occurrence

1995

Source of Information

U.S. District Court, Western District of New York, # 96CV0855

History of the Hazard

During World War II, as an equipment operator, the author observed personnel attempting to disassemble latticework-crane booms. There were instances when the boom was supported by the pinup guys in a horizontal position, and attempts to separate the sections without cribbing or other support were common. Fortunately the booms were connected with bolts that were parallel to the length of the boom, and when the nuts to the bolts were removed the bolts bound and did not disengage, keeping the boom from falling and forgiving the worker in danger. During the war and immediately afterward, some ingenious construction-worker welders replaced the bolts with pin connections, a practice that was soon adopted by crane manufacturers, and the hazard of boom collapse became rampant, as the pin system is now used on most lattice-boom connections.

Contributory Factors

Rather than develop alternate inherently safe designs, manufacturers assume that instructions are sufficient, though they may be complex. In the operating manuals, the disassembly process generally calls for laying the boom down first so that the boom tip rests on the ground. The next step is to slack and remove the pennant pinup guys at the boom tip and reattach them at the first boom section behind the boom-tip section. The operator must support this hinge joint until the two lower pins are removed and then lower the boom to the ground and then repeat the process until all boom sections are on the ground. This routine never seems to be uniformly followed; neither does the practice of placing cribbing underneath the boom, as cribbing lumber is usually unavailable.

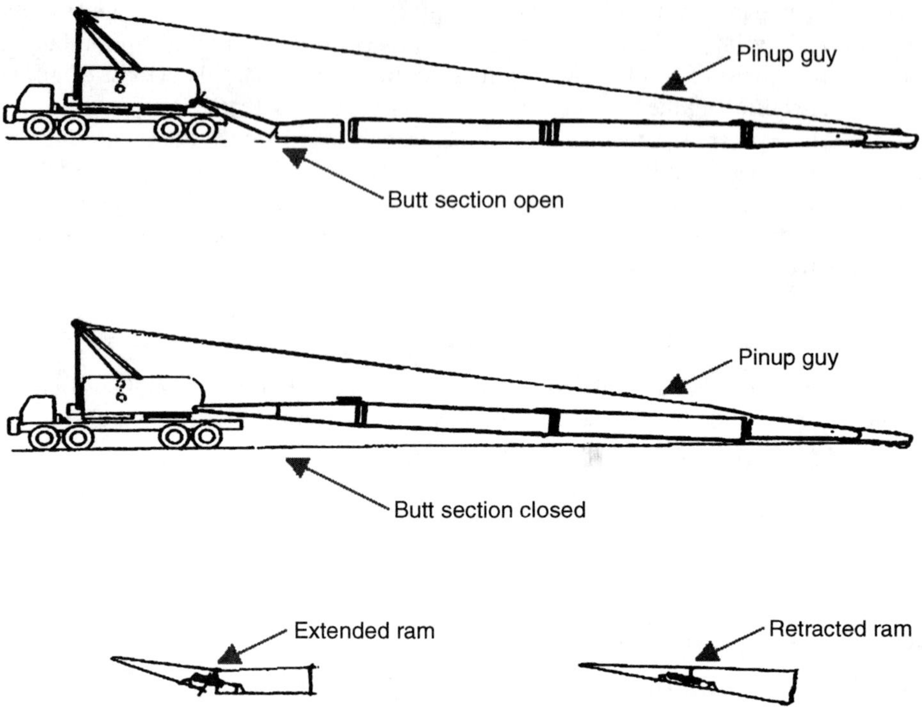

Applicable Standards and Literature

OSHA

1926.00(a)(3)(i) states, "Heavy machinery, equipment, or parts thereof, which are suspended or held aloft by use of slings, hoists, or jacks shall be substantially blocked or cribbed to prevent falling or shifting before employees are permitted to work under or between them."

1910.179(n)(3)(vi) states, "The employer shall require that the operator avoid carrying loads over people." *A suspended boom that is not cribbed is not safe to disassemble.*

1910.179(n)(3)(x) states, "The employer shall insure that the operator does not leave his position at the controls while the load is suspended." This assures that the operator does not leave the cab during disassembly.

1910.180(h)(3)(vi) and OSHA 1910.180(h)(4) contain the same requirements as above for crawler locomotive and truck cranes.

1926.550(a) General Requirements, relies upon the manufacturer's instructions on how to safely disassemble the boom.

Marine Terminal Standard 1917.45(f)(12) discusses assembly and disassembly of crane-boom sections.

ANSI

ANSI standards in general have prohibited work under unsupported suspended loads.

U.S. Army Corps of Engineers

EM 385-1-1 *Safety and Health Requirements Manual* (U.S. Army Corps of Engineers, November 3, 2003) Section 16.D.04 states,

(a) The manufacturer's boom assembly and disassembly procedures shall be followed. The manufacturer's boom assembly and disassembly procedures shall be reviewed by all members of the assembly team prior to assembly and disassembly.
(b) When removing pins or bolts from a boom, workers shall stay out from under the boom.

Other References

- The *Crane Handbook* by D.E. Dickie of Ontario (1975) has detailed discussions on boom disassembly.
- Operator's manuals for various cranes usually have instructions on boom disassembly, but often fail to specify the use of cribbingx.

Hazard Identification and Prevention Matrix

Gravity is a natural equipment hazard that can be eliminated by redesign of the lower boom section. An engineering control that would allow the crane operator to lower the

	Eliminate the Hazard		Guard the Hazard		Provide a Safety Factor		Provide Redundancy		Provide Reliability
	Hazard	Safety	Hazard	Safety	Hazard	Safety	Hazard	Safety	
Natural	Gravity (falling object)		Voluntary use of lumber cribbing						50%
Structural/ Mechanical	Pin removal (falling object)	Redesign to allow boom to be placed flat on ground							95%
Electrical									
Chemical									
Radiant Energy									
Biological									
Automated Systems									

entire boom to the ground is needed, as shown above. The butt section of the boom can then be redesigned as discussed in the commentary.

Commentary

Many operators' manuals have detailed procedures requiring the relocation of the pinup lines as support while disassembling the boom sections. Unfortunately, such operator's manuals rely upon unrealistic expectations of user performance. The instructions require relocation of the pinup guys for each boom section. This time-consuming repetition is an error-provocative routine. The use of cribbing or other systems to support each boom section during pin removal is not considered to be an industry practice, and is not mentioned in the operator's manuals. Use of labels to show how the boom can be disassembled safely when supported by cribbing appear to be ineffective. In addition, many labels quickly become worn or illegible. These labels should be improved to show more detailed reminders of safe disassembly, and to warn of the danger of collapse when unsupported. Complicated instructions requiring detailed actions by crane users have been an open invitation for user error and often result in fatalities. D.E. Dickie, in his *Crane Handbook,* describes the use of step pins that can be inserted only from the inside facing out and that can be removed only by driving them from the outside in; and the use of welded lugs on the outside that prevent pins from being inserted from the outside in and that can be removed only by driving them from the outside. These precautions prevent the worker from standing in a position that could be fatal if the boom were to suddenly fall.

The existing system of instructions and labels does not reliably save lives. New design of the butt boom section is necessary. Preliminary design concepts show that the first section of the boom can be modified so that all other boom sections can be lowered to the ground for disassembly. This approach eliminates the hazard because pin removal can be accomplished only when the latticework-crane-boom sections are on the ground.

Recommendations

Design engineers should develop a prototype design drawing and model of a hydraulic ram for opening and closing hinges on the butt section of the boom. Crane designers should consider designs that allow the first section of the boom to be hydraulically opened from the crane cab controls so the boom can be laid on the ground. In this way the butt section of the boom can be designed with a hydraulic cylinder to open and close this section of the boom, and the operator can lower the other boom sections to the ground for safe pin removal. When the boom is assembled, the hydraulic cylinder can lift the boom up to where just the tip of the boom rests on the ground. From the design drawing, a prototype scale model could be assembled to show feasibility. Currently, Oregon State University has a student participating in this project. With such a study, this design concept should be made available to all latticework-crane manufacturers as a public service.

SECTION 5

Crane Operator Protection from Upset or Falling Objects

The Hazard Syndrome

Most construction equipment, such as tractors, road rollers, graders, scrapers, and earth movers, protect operators with Rollover Protective Structures (ROPS). ROPS standards were first developed by the U.S. Army Corps of Engineers in the 1950s, with specifications of W2 (a structure that resists twice the weight of itself) to withstand structural failure in the event of rollover, and adopted by the Society of Automotive Engineers (SAE) in the mid-1960s. Unfortunately, rough-terrain cranes, crawler-track-mounted cranes, and excavators wherein the operator's cab is mounted on a turntable have no requirement for ROPS. Many crane manufacturers use lightweight sheet metal for the cabs, which presents the *dormant* hazard of crushing. Without a standard for operator protection in the event of severe cab damage, crane operators can be in jeopardy. Cab damage most commonly occurs from upset or being struck by a heavy falling object. The hazard is *armed* when an operator enters the cab. The hazard is *active* when the safety of the cab is compromised by a structural failure in the event of upset or a falling object.

Scope

Some track-mounted excavators are converted into log loaders and cranes used in construction, and in many conversions the operator's cab is raised approximately four feet to improve operator visibility. These track-mounted cranes are chosen on some construction projects because of their ability to traverse very rough and steep terrain. Cranes used in these conditions should have strong cabs, as the risk of upset or rollover is greater. Failure to address the mechanical soundness of a protective cab is a violation of the second fundamental principle of safer design, as they are unsafe for their intended use.

Peril from the Failure Mode

When rough-terrain hydraulic cranes and track-mounted cranes upset, the cab is crushed, usually resulting in serious injury or death to the operator. In some instances the cab is penetrated by a heavy falling object, also resulting in operator injury. This hazard results from *natural* circumstances such as overturning or falling objects and destabilization due to gravity, but remains unguarded by the *structural/mechanical* defect of structural failure.

Brief Summary of the Occurrence

The deceased attempted to travel downhill. To view his path of travel, he raised the boom about 20 degrees and positioned it at the "two o'clock" angle from forward travel. The track vehicle upset in a forward direction, crushing the raised cab and trapping the operator, who suffocated because his chest was crushed by the collapsed cab. Examination of the cab revealed that the approximately 5/32" steel plate had been poorly welded and had failed. The upper corners were without triangular haunch braces for each corner post to support the steel cab roof. The corners collapsed away from the roof and caused the frame to collapse.

Date of Occurrence

September 9, 1988

Source of Information

Circuit Court, Multnomah County, OR, # 0009-09139

History of the Hazard

1919: Edward Hewitt, an agricultural engineer, published a study on the hazard of rollover or wheel tractors.

Late 1920s: Studies detailing the mechanics and forces that create wheel-tractor rollover were produced by an engineer named McKibben.

1930s: Some tractor owners developed their own protective canopies for both wheel- and track-mounted tractors.

1940s: The states of California, Oregon, and Washington required substantial brush guards on a tractor used in logging. These guards were generally strong enough to resist the forces of rollover.

Mid-1950s: The University of California developed a roll bar for small-wheel tractors.

1957: The Portland District U.S. Army Corps of Engineers required a protective rollover canopy designed to resist the forces of rollover, calculated as twice the weight of the tractor (W2) applied laterally at the uppermost top member of the steel canopy.

1970s: OSHA required ROPS by the standards of either the Army Corps of Engineers or the SAE.

1980s: John Sevart, a consulting engineer in Topeka, Kansas, developed a design for ROPS on a rough-terrain crane. It was established by tests that the design of ROPS mounted on a turntable for rough-terrain and track-mounted cranes is reasonably feasible with a strong steel structure. Currently, Olaf Knezivac, an engineer with the British Columbia Workers' Compensation Board in Canada, has developed standards for operator protection that have proven to be effective in an actual crane upset, thus rendering the cabs reasonably protective.

The force of upset is greatly diminished when the cab strikes soil and the low density of the soil allows it to be depressed. The cab slides when it falls on a rocky surface. Both of these actions serve to dissipate the latent force on the side and top of the cab, allowing for the safe design of a crane cab. This design would also provide reasonable protection from heavy falling objects.

Contributory Factors

One of the reasons for the reluctance of heavy-construction-equipment manufacturers to voluntarily develop design criteria for rough-terrain and track-mounted equipment with the operator's cab mounted on a turntable is a fear that adopting operator protection from upset would be costly in terms of retrofit.

Hazard Identification and Prevention Matrix

	Eliminate the Hazard		Guard the Hazard		Provide a Safety Factor		Provide Redundancy		Provide Reliability
	Hazard	Safety	Hazard	Safety	Hazard	Safety	Hazard	Safety	
Natural	Gravity, instability								
Structural/ Mechanical	Structural failure	Steel cab		Steel operator's cab/ ROPS		Anti-tilt meter or indicator		Certified welds and bracing	95%
Electrical									
Chemical									
Radiant Energy									
Biological									
Automated Systems									

Applicable Standards and Literature

OSHA

CFR1926.1000 requires rollover protective structures on industrial equipment manufactured after July 1969 for the following equipment:

(a) All rubber-tired, self-propelled scrapers
(b) Rubber-tired front-end loaders
(c) Rubber-tired dozers
(d) Wheel-type agricultural tractors used in an industrial setting
(e) Crawler tractors
(f) Crawler-type loaders
(g) Motor graders, with or without attachments, used in construction work.

Comment: This requirement does not apply to side-boom pipe-laying tractors. OSHA has not yet adopted the use of ROPS for rollers and compactors, rear- and side-dump wagons, and water trucks, as do the Corps of Engineers and the SAE.

MSHA

CFR56.14130 "Roll-over protective structures (ROPS) and seat belts" and **CFR57.14130** "Roll-over protective structures (ROPS) and seat belts for surface equipment.

(a) *Equipment included.* Rollover protective structures (ROPS) and seat belts shall be installed on:
  (i) Crawler tractors and crawler loaders
  (ii) Graders
  (iii) Wheel loaders and wheel tractors
  (iv) Wheel tractor portion of semi-mounted scrapers, dumpers, water wagons, bottom-dump wagons, rear-dump wagons, and towed fifth wheel attachments
  (v) Skid-steer loaders
  (vi) Agricultural tractors

(b) *ROPS construction.* ROPS shall meet the requirements of the following Society of Automotive Engineers (SAE) publications, as applicable, which are incorporated by reference:
  (i) SAE J1040, "Performance Criteria for Rollover Protective Structures (ROPS) for Construction, Earthmoving, Forestry, and Mining Machines," 1986; or
  (ii) SAE J1194, "Rollover Protective Structures (ROPS) for Wheeled Agricultural Tractors," 1983.

Comment: Because both OSHA and MSHA address only equipment manufactured after July 1, 1969, many people might assume that older models of mobile self-propelled equipment prone to rollover are safe without ROPS if manufactured before this date. For many, this has been a tragic assumption. Steel-wheel road-rollers are known to easily slide sideways and then tip over. Large tractor-like vibratory compactors literally beat the ground from under their own treads, creating a slope that causes upset.

ANSI

ANSI has not addressed operator protection in the event of the hazard of crane upset.

SAE

The **SAE J1040,** "Performance Criteria for Rollover Protective Structures for Construction, Earthmoving, Forestry, and Mining Machines" is similar to a report of the Construction and Industrial Machinery Technical Committee first published in April 1974 and to the International Organization for Standardization's ISO 3471/1-1986.

The SAE does not address requirements for operator-protective cabs that are mounted on the crane's turntable, but has developed design criteria for ROPS (structures) on the following:

(a) Crawler tractors and loaders
(b) Grade
(c) Wheel dozers and loaders, wheel tractors and their modifications used for rolling or compacting, wheel log skidders, skid-steer loaders, and backhoe loaders
(d) Wheel industrial tractors
(e) The tractor portion of semi-mounted scrapers, water wagons, articulated steer dumpers, bottom-dump wagons, side-dump wagons, rear-dump wagons, and towed fifth-wheel attachments
(f) Rollers and compactors
(g) Rigid-frame dumpers with full mounted bodies

U.S. Army Corps of Engineers

EM385-1-1 16.B.12, "Rollover Protective Structures (ROPS)," requires ROPS with seatbelts for the following equipment:

(a) crawler and rubber-tire tractors including dozers, push and pull tractors, winch tractors, and mowers;
(b) off-the-highway self-propelled pneumatic-tire earth movers such as tracks, pans, scrapers, bottom dumps, and end dumps;
(c) motor graders;
(d) water-tank tracks having a tank height less than the cab;
(e) other self-propelled construction equipment such as front-end loaders, backhoes, road-rollers, and compactors.

Other Standards

- **International Organization for Standardization's** ISO 3471/1-1986 for ROPS where there is a risk of rollover (see SAE J1040 APR88); and ISO 3449-1992 for falling object protection (see SAE J231 JAN81).
- **WCB** (Workers' Compensation Board of Canada) Standard: G608 Mobile Equipment Roof Structures- Heavy Duty

  (a) **Scope**

  This Standard pertains to mobile industrial equipment exposed to hazards from falling objects such as bricks, concrete blocks, and hand tools that may fall from relatively low heights encountered in operations such as highway maintenance or landscaping and other services on construction sites.

(b) **Purpose**
This standard is intended to describe the minimum requirement for the design and selection of plates for roof construction.

(c) **Design Criteria**
(i) Option One: Testing Method
The roof shall be designed to the Minimum Performance Criteria for Falling Object Protective Structure (FOPS) for Industrial Equipment—SAE J1043.
(ii) Option Two: Minimum Recommended Size
The roof shall be designed in accordance with the minimum recommended size as outlined in §4.0.

(d) **Design Requirements**
(i) The roof plate shall be designed in accordance with the design criteria as outlined in §2.0.
(ii) All frame members and supports of the Cab Protective Structure shall be designed to resist the applied load. Equipment covered by SAE J1040 shall be designed in accordance with SAE J1040 or its equivalent.

(e) **Weldments**
Weldments shall conform to applicable sections of General Specifications for welding of Steel Structures, CSA W59.1-1970 (or latest revision thereof) and shall be performed by licensed welders only.

(f) **Minimum Recommended Sizes**
Minimum thickness of roof plate shall be 10 Ga., equivalent to 3.4 mm (0.1345 in). The above is designed on the assumption that stiffeners will be used to limit the unstiffened areas not to exceed 7740 cm^2 (1200 in^2).

British Columbia's worker safety agency has developed standards for operator protection for rough-terrain and crawler-track turntable cranes in the logging safety code. These standards are in the embryonic phases. To learn more about them, contact Olaf Knezevic, PE at the Workers Compensation Board, P.O. Box 5350 Stn Terminal, Vancouver, BC V6B 5L5 604.276.3212, fax, 604.279.7407, e-mail oknezevi@wcb.bc.ca.

Other References

- *Automotive Engineering and Litigation, Volume 2*, by George and Barbara Peters; Chapter 1, "Rollover Protective Systems," written by David V. MacCollum, PE, CSP.
- "Lessons from 25 Years of ROPS" *Professional Safety*, January 1984. First Place Technical Paper Award from ASSE/Veterans of Safety, 1983/84 by David V. MacCollum.
- James F. Arndt, Deere & Company, SAE Earthmoving Industry Conference, "Roll-Over Protective Structures for Farm & Construction Tractors—A 50-year Review," April 5–7, 1971.
- "Tractor Canopies—Start of a Trend?" *Western Construction*, February 1959.
- David V. MacCollum, "Tractor Canopies," *Pacific Builder & Engineer*, October 1958.
- U.S. Army Corps of Engineers, Portland District, Safety Branch, "Evaluation of Tractor Canopies in Rollover Accidents," 1958. Study conducted and written by David V. MacCollum.

- E.G. McKibben, "Kinematics and Dynamics of the Wheel Type Farm Tractor," *Research in Agricultural Engineering,* five articles written February–July 1927.
- Edward R. Hewitt, "The Principles of the Wheeled Farm Tractor," SAE *Transactions*, Vol. 14, Part I, 1919, p. 88.

Commentary

The involvement of structural engineers in the development of protective operator canopies for turntable-mounted boomed equipment is consistent with the engineering code of ethics to hold safety as a paramount design requirement.

Recommendations

Structural engineers need to design a crane cab that is consistent with the SAE and the U.S. Army Corps of Engineers W2 requirements in order to provide crush-resistant protection for operators' cabs mounted on a turntable on rough-terrain and crawler-track-mounted cranes. This action would establish a reasonable engineering standard for most upset situations. The presence of long and heavy crane booms should be considered a safety factor, as in most circumstances the boom will absorb some of the energy from an upset. A W3 design criteria should overcome other unknown forces to offer even better protection to equipment operators.

SECTION 6

Powerline Contact

The Hazard Syndrome

Crane powerline contacts with the boom or the hoist line have been a major source of gruesome deaths and injuries from the time that mobile cranes have been in use. The previous illustration shows a crane boom coming in contact with a powerline in an ordinary residential area. The hazard of an overhead powerline is *dormant* when there is no crane in the vicinity of the powerline. The hazard becomes *armed* when any part of a mobile crane has access to intrude into the ten-foot *thin air* danger zone surrounding a powerline. The hazard is *active* when the boom, hoist line, or any other part of the crane comes into contact with the powerline and sends a current of electricity through the crane and everybody in contact with it.

Scope

The ten-foot *thin air* clearances established by OSHA and other agencies, manufacturers, and standard-making bodies have not provided reliable protection from this hazard. *All types of mobile and stationary cranes within reach of powerlines are vulnerable.* Any boomed equipment able to be raised eighteen feet or more can strike an overhead powerline.

Peril from the Failure Mode

Distribution powerlines usually carry from 3000 to 35,000 volts of electricity, with most lines carrying 7200 volts. Under the right conditions, a person can die when exposed to 50 milliamps of current flow. Because of the dangerously high voltages, the injuries and deaths from powerline contacts are indescribably damaging. Damage from powerline contacts occurs when current enters the body, travels through the body, and exits the body through a ground path. Damage from this exposure falls into two categories: exit burns, which occur where the current has entered and left the body, charring flesh and often requiring amputation; and fibrillation of the heart, which occurs when low current flow electrifying the body interferes with the electrical signals that control the heart, causing rapid or erratic heartbeat. This stops the blood flow to the brain and can cause brain damage and death.

Brief Summary of the Occurrence

The bridge-building construction site was very confining on one side of the river, where a large hydraulic telescoping-boom crane was located. The crane operator had raised the long boom to a near-vertical position in order to rotate approximately 180 degrees so as to lift another load. Attached to the hook were two suspended cable straps. As the boom swung past the 7200 volt powerline, one of the straps brushed the powerline, causing the crane to become momentarily electrified. A pneumatic track drill supplied by a compressed-air hose encased in a wire-web sheath that was draped over one of the crane's outriggers also became momentarily electrified. The drill-rig operator was electrocuted and his helper was injured.

Date of Occurrence

June 22, 1998

Source of Information

Circuit Court, Perry County, AL # -99-12

History of the Hazard

For over thirty years, hundreds of crane/powerline contacts have been the basis of liability lawsuits.

1951: The *Accident Prevention Manual for Industrial Operations*, 2nd Edition: National Safety Council (NSC) 12–16: "Commercial warning devices are available which work by induction from a powerline, ringing a warning bell in the cab when the boom or cable approaches too close."

1950s: U.S. Army Corps of Engineers Portland, OR District investigates the cause of powerline contacts and writes regulations for specific projects that included raising powerlines of 90-foot clearance above the ground 10 feet above the limitation boom length of 80 feet for all crane booms on some projects, and incorporating preproject planning to address the issue of powerline contacts before construction began on many others.

1960: Invention of the Electrically Insulated Link.

1961: Invention of the Proximity Alarm.

1967: U.S. Army Corps of Engineers manual EM385-1-1 requires cage guards and insulating links to be used on boomed equipment "where there is a capability of encroachment on specified clearances."

1970s: Industry-sponsored studies were published, many asserting that proximity alarms and insulated links were ineffective when used on construction sites. What these studies did not mention, however, is that the proximity alarm was rendered ineffective only in boom positions (parallel and between phase lines) that are prohibited under OSHA regulations.

1982: Completion of the study "Evaluation of Proximity Warning Devices" by the U.S. Bureau of Mines states that "The data analysis showed that the typical combinations of crane boom elevation and crane boom radius have the best case performance for a single sensor proximity warning device to produce alarms for distances ranging from 20–60′ to the powerline depending on the combination of crane boom elevation and radius."

1983: Robert Cunitz and Lorna Middendorf present the white paper, "Problems in the Perception of Overhead Powerlines" at the Sixth Annual Symposium for the System Safety Seminar. The research determines that under some conditions, humans are not physically capable of accurately spotting powerlines.

1990: "Efficiency of Insulating Link for Protection of Crane Workers," published by the International Society of Electrical and Electronic Engineers (IEEE) states that "The paper concludes that in spite of the shortcomings, the insulator link increases the safety of the crane operation. Furthermore, the developed model and calculation method can be used for the evaluation of new crane insulator systems." This publication both determines the ultimate effectiveness of insulated links, and develops a standardized testing system for them.

1990s: Manufacturers of sailboats began installing insulated plugs in their masts to prevent electrocution.

2004: OSHA-negotiated discussions from industry groups, crane manufacturers, and regulatory bodies continue underway to define new standards for safe use of cranes to avoid powerline contacts.

The 2004 study "Safety Interventions to Control Hazards Related to Powerline Contacts by Mobile Cranes and Other Boomed Equipment," published by the Hazard Information Foundation, Inc. (HIFI), and funded by the Center to Protect Workers' Rights (CPWR), has compiled a detailed historical timeline of how this hazard has always been someone else's responsibility to prevent.

Contributory Factors

The perception that powerlines are open and obvious hazards erroneously assumes that the crane operator and working crew will avoid them. Human-factors specialists have proclaimed for years that two separate visual targets can not be viewed at the same time.[4] Neither the crane operator, signal person, nor those guiding the load into its proper location can also be viewing the boom or hoist line clearance from a powerline. Reliance upon a single signal person is error-provocative. Real lift powerline contacts that have become the subject of litigation have nearly always occurred mid-span, where power poles do not provide good visual clues to a powerline's presence[5].

Hazard Prevention and Identification Matrix

	Eliminate the Hazard		Guard the Hazard		Provide a Safety Factor		Provide Redundancy		Provide Reliability
	Hazard	Safety	Hazard	Safety	Hazard	Safety	Hazard	Safety	
Natural		Beneath the powerline, mark the danger zone on the ground							
Structural/ Mechanical									
Electrical		Remove powerlines		Insulated link		Proximity alarm/ range-limiting		Bury power-lines	100%
Chemical									
Radiant Energy									
Biological									
Automated Systems									

[4]Lorna Middendorf, Judging Clearance Distances Near Overhead Powerlines, October 10, 1978; and Problems in Perceptions of Overhead Powerlines, September 23, 1983.

[5]These circumstances make the proximity alarm an effective safety appliance.

Available Literature

OSHA

1296.550 Cranes and Derricks

a. (15): Except where electrical distribution and transmission lines have been de-energized and visibly grounded at point of work or where insulating barriers, not a part of or an attachment to the equipment or machinery, have been erected to prevent physical contact with the lines, equipment or machines shall be operated proximate to power lines only in accordance with the following:

(i) From lines rated 50 kV or below, minimum clearance between the lines and any part of the crane or load shall be 10 feet;

(ii) For lines rated over 50 kV, minimum clearance between the lines and any part of the crane or load shall be 10 feet plus .4 inch for each 1 kV. over 50 kV, or twice the length of the line insulator, but never less than 10 feet.;

(iii) In transit with no load and boom lowered, the equipment clearance shall be a minimum of 4 feet for voltage less than 50 kV, and 10 feet for voltages over 50 kV up to and including 750 kV.

(iv) A person shall be designated to observe clearance of the equipment and give timely warning for all operations where it is difficult for the operator to maintain the desired clearance by visual means;

(v) Cage-type boom guards, insulating links, or proximity warning devices may be used on cranes, but the use of such devices shall not alter the requirements of any other regulation of this part even if such device is required by law or regulation;

(vi) Any overhead wire shall be considered to be an energized line unless and until the person owning such line or the electrical utility authorities indicate that it is not an energized line and it has been visibly grounded;

The following requirement is specified for live-line electrical work, but a similar requirement should appear when working with cranes and derricks to improve construction safety:

1926.955 Overhead Lines

e.(5) The automatic reclosing feature of circuit interrupting devices shall be made inoperative where practical before working on any energized line or equipment.

U.S. Army Corps of Engineers

EM 385-1-1 (3 November 2003) 11.E Operations Adjacent to Overhead Powerlines

11.E.01—Overhead transmission and distribution lines shall be carried on towers and poles that provide safe clearances over roadways and structures.

(a) Clearances shall be adequate for the movement of vehicles and for the operation of construction equipment.

(b) All electric power or distribution lines shall be placed underground in areas where there is extensive use of equipment having the capability to encroach on the clear distances specified in 11.E.04. (This requirement has been found to be the most effective method of powerline contact prevention.)

(c) Protection of outdoor trolleys and portable cables rated above 600 volts for supplying power to moveable construction equipment such as gantry cranes, mobile cranes, shovels, etc., shall conform to NESC.

11.E.02—Work activity adjacent to overhead lines shall not be initiated until a survey has been made to ascertain the safe clearance from energized lines.

11.E.03—Any overhead wire shall be considered energized unless the person owning such a line or operating officials of the electrical utility supplying the line *certifies* that it is not energized and has been visibly grounded *and tested.*

11.E.04—Operations adjacent to overhead lines are prohibited unless at least one of the following conditions is satisfied:

(a) Power has been shut off and positive means taken to prevent the lines from being energized.

(b) Equipment, or any part, does not have the capability of coming within the minimum clearance from energized overhead lines as specified in Table 11-1, or the equipment has been positioned and blocked to assure that no part, including cables, can come within the minimum clearances as specified in Table 11-1; a notice of the minimum required clearance has been posted at the operator's position (electric line derrick trucks and aerial lifts are not required to comply with this requirement).

(c) In transit with the boom lowered and no load, the equipment clearance *shall comply with Table 7–1.*

11.E.05—Work activity that could affect or be affected by overhead lines shall not be initiated until coordinated with the appropriate utility officials.

11.E.07—Floating plant and associated equipment shall not be cited or placed within 20 feet (6 m) of overhead transmission or distribution lines.

11.E.08—Cage boom guards, insulating links, or proximity warning devices may be used on cranes, but such devices shall not alter the requirements of any other regulation of this part—even if such device is required by law or other regulation. Insulating links shall be capable of withstanding a 1-minute dry low frequency dielectric test of 50,000 volts, alternating current.

Other References

- David MacCollum, "Critical Hazard Analysis of Crane Design," reprinted in *Professional Safety*, January, 1980.

Table 7–1
Minimum Clearance from Energized Overhead Electric Lines

Nominal System Voltage	Minimum Related Clearance
0-50 kV	9.8 ft (3 m)
51-200 kV	14.7 ft (4.5 m)
201-200 kV	19.7 ft (6 m)
301-500 kV	24.6 ft (7.5 m)
501-750 kV	34.4 ft (10.5 m)
751-1000 kV	44.3 ft (13.5 m)

- David MacCollum, "Hunting Down Crane Hazards," *Lift Equipment*, February–March 1992.
- David MacCollum, "Designing Out Electrical Hazards," *CraneWorks*, July 1992.
- David, MacCollum, *Crane Hazards and Their Prevention*, American Society of Safety Engineers, 1993.
- David MacCollum, "Crane Design Hazard Analysis," Chapter 8, George and Barbara Peters, *Automotive Engineering and Litigation*, Volume 1, Garland Law Publishing, 1984.

Commentary

The regulations enforcing no protection system other than *thin air* clearances have been a death warrant to many. The construction-safety planning process must disallow powerline contact by addressing alternate methods so cranes and powerlines are not in the same location. Beyond the removal of all powerlines in the vicinity of a project, providing a verifiable method of powerline de-energization is the most effective line of defense. Regulations state that no redundant safeguards such as links, proximity alarms, boom cages, or range-limiting devices shall be substituted for safe operating procedure, and the minimum clearance distances of the danger zone must be maintained at all times. However, redundant safeguards should be applied even in cases of de-energization to achieve maximum safety. Some occurrences of powerline contact occur when an operator or construction manager incorrectly assumed a powerline was de-energized. An active proximity alarm would prevent such an occurrence from happening.

Recommendations

Construction planning requirements should detail special conditions in the contract for preventing all powerline contacts. Project plans and blueprints should include areas around powerlines marked as a danger zone so equipment use and mobility can be planned accordingly. The proposed requirement that the ten-foot danger zone be mapped and barricaded with warning on the ground, and that the crane shall not in any manner intrude within the danger zone, is explained in the warning sign below. Even using a spray-paint can to paint the limits of the danger zone on the ground is a step toward avoidance of the danger zone.

With planning and the use of safety appliances, the principles of safe design are fulfilled as follows:

- Eliminate the Hazard: Remove powerlines from the work site where a crane will be used.
- Guard the Hazard: Provide an insulated link.
- Provide a Safety Factor: Incorporate a range-limiting device.
- Provide Redundancy: Provide a proximity alarm. With all possible hazard controls in place, the probability of a powerline contact is zero—or 100 % reliability.

DANGER

YOU WILL BE

ELECTROCUTED

FROM BOOM—POWERLINE CONTACT

DANGER ZONE

UNSAFE FOR CRANE OPERATIONS

DO NOT lift or make boom movements inside the barricaded area.

Map and Barricade the 30 foot wide Danger Zone
(15 feet on each side of the powerline poles).

ALWAYS notify the power company before you begin crane operations near powerlines.

Power Company	Telephone

8 Other Equipment Hazards

INTRODUCTION

Different types of equipment abound on a construction project. Many crafts and trades participate in a project, introducing specialized equipment. Often other parties who may be suppliers, service personnel, or even news agencies contribute vehicles or machines that come with their own set of inherent hazards. The following topics are covered in this chapter:

1. Electronic news-gathering vans (to illustrate the variety of powerline contacts)
2. Conveyors
3. Skid-steer and rough terrain telescoping boom conveyors
4. Trucks
5. Aerial lifts
6. Unsafe restraint systems
7. Equipment- leasing practices

Due to the variety of equipment and the myriad hazards they present, the project manager often becomes responsible for their safe use and maintenance. A construction manager must know the hazards inherent in each machine in order to anticipate and prevent conditions that arm the hazards.

SECTION 1

Electronic-News-Gathering Van Powerline Contact

The Hazard Syndrome

An electronic-news-gathering (ENG) van with a pneumatic mast presents a *dormant* hazard. The hazard becomes *armed* when the van is parked underneath powerlines. The *active* hazard is caused by the foreseeable operator error of unintentionally raising the pneumatic mast of the van into an overhead or immediately adjacent powerline.

Scope

ENG van/powerline contact was selected for two reasons:

- Newscasting unions requested the proximity-alarm system to automatically prevent the mast from being raised into a powerline because of a requirement in the state of California, but were denied their request when the California Department of Industrial Relations made the decision not to require proximity-system alarms on ENG vans.
- Proximity alarms have been proven to be an effective warning and interceding device on some ENG vans, and also an effective warning device on cranes in construction.

Since media journalism is such a fast-paced occupation, there are numerous cases of ENG-van drivers or operators unintentionally raising the mast into powerlines. In some circumstances, the ENG van is parked on a slight slope, and the mast is tilted into a powerline.

ENG van powerline contact is a hazard that can be easily prevented by the use of one reliable safety appliance. Electrostatic proximity warning devices on ENG vans sound an alarm when they detect powerlines. Most models inhibit the van's pneumatic mast from being raised when parked underneath or adjacent to a powerline, thus never allowing the hazard of powerlines to become armed. Proximity warning devices are an example of the superior reliability of overhead-powerline detection. Its role as an available operator aid prevents the pneumatic-mast antenna from being raised into powerlines, or warns, the van users of overhead powerlines. It is a device that has saved many lives.

The epidemic of ENG van powerline contact is a blatant record of industry denial of a hazard. The case detailed in the "Occurrence" example was recorded live by a rival news station, but knowledge was not circulated. If news stations had recognized the life-threatening nature of this hazard, they would have taken pains to install the simple electronic device that prevents it.

Peril from the Failure Mode

Massive discharge of *electrical* energy causes electrocution or electrical shock resulting in fibrillation of the heart, loss of limbs, and severe electrical burns. To date there have been ten occurrences of news-gathering vans raising their masts into powerlines, resulting in nine seriously disabling injuries requiring amputations and three deaths.

Brief Summary of the Occurrence

Pending the TV reporting of an important news story, all of the major network news-gathering vans had assembled on the same site. One van pulled into a parking spot immediately behind another station's van, whose mast and antenna were already raised. Due to the slope of the roadway, the van tilted slightly towards the unseen powerline, causing contact. While sitting on the floor of the van with the side door open and with his feet on the ground, the deceased raised the pneumatic mast of his van into the park-way shade trees, unaware of the overhead powerlines. He was immediately electro-cuted, and the terrible occurrence was videotaped by several of the other TV networks while aid was being rushed to the victim.

Date of Occurrence

February 22, 1994 (one of the first litigated incidences of ENG van powerline contact)

Source of Information

Superior Federal Court, District of Columbia, Washington, DC, # 000142 b-95

History of the Hazard

1985: The U.S. Army develops surveillance vehicles for the U.S. Immigration Service and equips some twenty additional vehicles with a SigAlarm electromag-netic-sensor warning system to prevent raising the mast of the vehicle parked under an overhead powerline. For additional information, see the Hazard Information Foundation, Inc (HIFI) research report "Safety Interventions to Control Hazards Related to Powerline Contact by Mobile Cranes and Other Boomed Equipment."

1985, November 6: A pneumatic mast being raised into a powerline led to serious and multiple disabling burns to the arms and hands of a news reporter. (Philadelphia County, Trial Division, # 4502)

1985, November 6: An incident caused the amputation of the injured legs of a news reporter in Denver, Colorado.

1985, December: An operator was electrocuted when attempting to extinguish a fire that resulted from an ENG-van contact in Bakersfield, California.

1994, February 22: An operator was electrocuted while raising the mast into a pow-erline. (U.S. Superior Court, District of Columbia, # 000142b-95) The occurrence was widely publicized at the time of the operator's death; however, the television networks were exceedingly silent on corrective adoption of the proximity alarm when routinely publicizing appropriate safeguards needed to overcome other organizational hazards.

1994, August 9: An operator was able to reach the control to the mast from outside the van, so when the mast contacted the powerline, the current found a ground path through the operator, resulting in the loss of a foot. (District of South Carolina, Charleston Division, # 2:96 2106-1)

1997, June 17: An operator was electrocuted while raising mast. (U.S. District Court for Northern District of Mississippi, Greenville Division, # 4:00 CV 187-M-B)

1997, September 3: Two people sustained serious injuries when a news reporter attempted to rescue a co-worker. One suffered a burned skull and body, including loss of toes; the operator sustained serious arm burns. (U.S. District, Southern District of Iowa, # 4-98-CV-80131)

2000, May 2: A camera operator seriously burned her arms and hands. (U.S. District Court for the District of Columbia, Washington, DC)

2000, May 23: An on-location anchorwoman lost one leg and the opposite foot, one arm and the opposite hand in Los Angeles, California.

2002, May 10: As a public-service action, HIFI prepared a hazard analysis of the unintentional raising of a pneumatic mast of an ENG van into powerlines and forwarded it to the state of California, Department of Industrial Relations, Occupational Safety and Health Standards Board, with these recommendations:

- Provide and install an electric (electrostatic) field detector to prevent the mast from being raised when the ENG van is parked under or immediately adjacent to overhead powerlines. Such detectors should also prevent the mast from being raised until the van is positioned at least thirty feet lateral distance from the powerline. Further, the manufacturer shall certify the installation is calibrated and locked into adjustment for that particular van to ensure that the mast cannot be activated when the ENG van is parked under powerlines or immediately adjacent to them.
- For failsafe redundancy, insulating materials should be incorporated in the design of accessories that are mounted on the top of the pneumatic mast to prevent current flow in excess of 5 milliamperes when contact is made with 7,500 V powerlines.
- Training needs to include (a) a summary of previous injuries due to inadvertent raising of pneumatic masts on ENG vans into powerlines, (b) the propensity for error- provocative circumstances during the use of ENG vans, and (c) the need for proven safety accessories on ENG vans as standard equipment.

2002, July 11: The American Federation of Television and Radio Artists affiliated with the AFL-CIO; NABET-CWS Locals 57 and 53; ISATSE Local 60; and IBEW Local 45 were informed of the following recommendation: EMF (Electromagnetic Field Detectors) must be installed on all ENG vans.

2002, November 18: Cal OSHA adopts a standard requiring the training of operators, a third order of precedence that omits the requirement of the electrostatic proximity detectors, which would prevent the raising of the pneumatic mast when parked under a powerline.

Contributory Factors

The gathering of timely television news is often conducted with exceedingly short travel deadlines to the news site, night or day, regardless of weather conditions. There is a heavy, yet grossly misplaced, reliance upon the news staff to visually avoid parking under powerlines, even though this practice is fraught with error-provocative circumstances, such as ambient light, objects obscuring the view of the powerline, the slope of streets, and the speed necessary to the profession. Operation of the equipment at night with curtailed vision poses an additional hazard. Human-factors engineers have published studies showing the inability of people to see powerlines in many natural conditions.

Hazard Identification and Prevention Matrix

- ENG van/powerline contact is an electrical hazard that can be reliably eliminated with a SigAlarm that automatically prevents the mast from raising when the van is under the powerline.
- An electrical hazard can also be guarded against by installing a nonconductive material around the dish mast attached on the top of the pneumatic mast.
- An electrical hazard can be prevented by eliminating the need for pneumatic masts by using satellite communication to transmit a news story.

	Eliminate the Hazard		**Guard the Hazard**		**Provide a Safety Factor**		**Provide Redundancy**		**Provide Reliability**
	Hazard	Safety	Hazard	Safety	Hazard	Safety	Hazard	Safety	
Natural	Slope, unob-structed view						Dark-ness		*
Structural/ Mechanical						Vanlev-eling device			*
Electrical	Shock	Prox. alarm to prevent raising of mast		Mast/tilt platform insulation				Insulation on each component; provide lights	*
Chemical									
Radiant Energy		Electro-static detection							
Biological									
Automated Systems									

* The combination of the elimination of the hazard, and guarding with a redundant safeguard against the hazard creates 100% reliability, effectively turning an inherently unsafe product into an inherently safe one.

Applicable Standards and Literature

No federal occupational rules currently pertain to news-gathering vans.

The 2004 HIFI research study "Safety Interventions to Control Hazards Related to Powerline Contacts by Mobile Cranes and Other Boomed Equipment" provides a comprehensive, chronological list of available literature on the subject of powerline contact.

Commentary

Though the powerline-detection technology has existed for over fifty years, ENG van powerline contacts provide another example of politics inhibiting safety. To date, no injuries or deaths have occurred by failure to detect energized powerlines with a proximity safety system installed. Two large highway-construction contracting firms have equipped over two-hundred cranes with the SigAlarm, and found that it has been an effective alarm system that allows the crane to be operated outside a measured and marked danger zone as defined in writing above. The news-gathering unions requested the installation of the SigAlarm on all ENG vans, and the state of California Advisory Committee rejected their request. It appears that business bias did not want the expense of retrofitting their news-gathering vans. The State of California Advisory Board appears to be a victim of false information. This is indeed a short-sighted approach, as the liability damages of previous news-gathering van/powerline contact were said to be monumental in nature. The episodes involving ENG vans are clear examples of how looking upstream to provide engineering controls could save lives.

Recommendations

Electrical engineers should create an educational outreach program whereby system-safety concepts and applications are developed for the news-gathering industry. Social awareness and the value of safety appliances should be included in training for upper management of TV news-gathering media.

SECTION 2

Conveyors

The Hazard Syndrome

Belt conveyors are an integral construction machine used to convey bulk materials, such as sand, gravel, and packaged materials, such as shingles, to rooftops. They may be powered or function by gravity to convey construction materials from one location to another.

Conveyors present myriad specific hazards that will be thoroughly covered in the "Peril" section, but the *armed* hazard can be considered to be an activated conveyor that is not equipped with a lockout device and an emergency stop. This hazard is *active* when the worker has no opportunity to shut down the conveyor to remove debris and becomes entangled.

Scope

The most common type of conveyor found in construction is an "endless-belt conveyor." These can be long, short, light, or heavy. All have a continuous strip forming a loop that moves over a track. The illustration below is typical of such a conveyor in a common application of construction.

There are many other types of conveyors, all vulnerable to the hazard of entanglement:

- Tramways (aerial): Conveyances used to transport materials or people over hills, valleys, water, or other inaccessible locations
- Blowers: Used to convey insulation or other light material
- Bucket: Any type of conveyor in which the material is carried in a series of buckets
- Chain: Any type of conveyor in which one or more chains act as the conveying medium
- Mobile: A movable conveyor on wheels
- Pneumatic: Tubes or small pipe ducts that transport carriers for messages, orders, and other paper material from one location to another using air pressure for propulsion
- Pumpcrete: Used to pump concrete short distances
- Roller: Revolving round cylinders that are either powered or nonpowered (For example, powered rollers are used to propel logs and lumber through a sawmill. Nonpowered gravity rollers on bearings allow boxes or other large objects to travel from one place to another.)
- Screw: An enclosed rotary auger used to transport raw materials such as grain (A conveyor screw revolving either in a trough or pipe.)
- Trolley: A series of trolleys suspended from an overhead track

Such a wide variety of conveyors requires special attention to specific types of guards to offer maximum effectiveness.

Peril from the Failure Mode

Conveyors have had a long history of unguarded (dormant) pinch points, and many workmen have been caught by moving parts when retrieving debris, engaged in maintenance, oiling, or wearing loose clothing that catches in a moving belt. A moving conveyor belt with a piece of debris and no lockout is always an invitation for someone to attempt to remove it. This can cause serious injury or death by entangling a body or body parts in the belt. A U.S. Bureau of Mines letter provides "Fatal Grams" to all mining companies, and has listed over fifteen deaths as a result of belt-conveyor systems with an in inadequate or faulty lockout system within recent years.

There are many hazards associated with conveyors, including

- Absent or ineffective lockout systems (*mechanical/electrical*)
- Unsafe control systems and lack of emergency stops (*mechanical/electrical*)
- Lack of guards (*mechanical*)
- Pinch points (*mechanical*)
- Friction—overheating, static and electrical sparks that can cause fires (*natural/electrical*)
- Powerline contacts (being extended or moved into powerlines) (*electrical*)
- Lack of guardrails to prevent the hauled material from falling off (*mechanical*)

Brief Summary of the Occurrence

A gravel-pit worker lost his right arm when caught on a moving gravel conveyor while attempting to retrieve a piece of wood. The employers were using a borrowed gravel conveyor, and a pre-use safety inspection of it revealed that it lacked both a lockout system and an emergency stop. Lockout switches and emergency stops were not installed, and shortly thereafter the worker lost his arm when it became entangled between the moving belt and roller. This case was selected because the employer was found to have "concealed" a hazard, and liability was found to be outside the exclusive remedy of the Nevada workers' compensation law.

Date of Occurrence

July 8, 1985

Source of Information

District Court, Clark County, NV, # A25760

History of the Hazard

1920s: Henry Kaiser gained fame by using a long-belt conveyer system to handle bulk earth material for the construction of an "earth fill," dam instead of costly trucks with drivers, replacing a labor-intensive trucking method.

1980: Broken Hill Mining properties in Oracle, AZ developed a new underground mine with a conveyor system instead of the new, more costly rail system, which shows the widespread use of conveyors by industry.

Contributory Factors

Gravel and crusher plants are portable in nature. They are temporary in arrangement, and designed to be taken down and moved about between many construction sites. Due to their constant relocations, safety components such as lockout systems are not always reinstalled, and might be considered nonessential add-ons before production is commenced. In addition, lockout systems can be inconveniently located and are a source of worker misjudgments of safety, especially if they are not readily visible. In some cases, they lack guarding. In most cases, they rely solely upon voluntary safe work practices.

These operations are usually overseen by a single maintenance person under a supervisor who is involved primarily in directing other activities, including delivery of aggregate rock or sand to the construction site. Often, conveyors are arranged where it might be difficult to see the point where someone is performing maintenance or service. This arrangement, particularly in the absence of a lock system, affords many opportunities for someone to mistakenly start a conveyor when maintenance is being done.

Hazard Identification and Prevention Matrix

The hazard is both mechanical and electrical. The lockout system is electrical. Lockout is a simple electrical safeguard and can be included in design. Safety controls require both lockout and a guarding system to create a reliability rating of 100%.

	Eliminate the Hazard		Guard the Hazard		Provide a Safety Factor		Provide Redundancy		Provide Reliability
	Hazard	Safety	Hazard	Safety	Hazard	Safety	Hazard	Safety	
Natural	Friction								
Structural/ Mechanical	Entanglement/ pinch point						Moving parts	Guards	95%
Electrical	Fire/ shock			Automatic lockout		Closed-circuit TV monitors		Emergency stop	100%
Chemical									
Radiant Energy	Heat from friction					Lubrication			
Biological									
Automated Systems									

Applicable Standards and Literature

OSHA

1926 Subpart N .555 Cranes, Derricks, Hoists, Elevators and Conveyors

1926.550 Cranes and derricks

Department of Labor (DOL)

CFR 1917/1918 Marine terminals/shipbuilding

1917.48 Conveyors

1917.116 Elevators and escalators

1918.63 Chutes, gravity conveyors, and rollers

1918.68 Chutes, gravity conveyors, and rollers

1918.69 Powered conveyors

MSHA

MSHA has numerous standards and references for conveyors in the Code of Federal Regulations (CFR Title 30).

Part 56, Subchapter B: Testing, evaluation, and approval of mining products, 18 - Electric Motor-Driven Mine Equipment and Accessories ensures for explosion-proof electric-powered conveyor systems.

Part 56

56.11002 Handrails and toeboards

56.11009 Walkways along conveyors

56.11013 Conveyor crossovers

56.14108 Overhead drive belts

56.14109 Unguarded conveyors with adjacent travelways

56.14113 Incline conveyors: backstops or brakes

56.14201 Conveyor start-up warning

Part 57

57.10001 Aerial tramways-filling buckets

57.10002 Handrails and toeboards

57.11009 Walkways along conveyors

57.11013 Conveyor crossovers

57.11014 Crossing moving conveyors

57.14108 Overhead drive belts

57.14109 Unguarded conveyors with adjacent travelways

57.14113 Incline conveyors: backstops or brakes

57.14201 Conveyor start-up warning

57.4263 Underground belt conveyors

57.4503 Conveyor belt slippage

ANSI

ANSI B20.1-1996 Safety Standards for Conveyors and Related Equipment

Comment: ANSI/ASME 90.1-1992, which contains safety standards for belt manlifts, should be abolished. Such devices are people traps. Section 1.3 states that this standard applies to manlifts used to carry only *authorized personnel* trained in their use, that manlifts should not be available to the general public, and that they are not recommended for use on construction sites. Belt manlifts are error-provocative for *everyone,* whether they are *authorized personnel* or anyone else.

National Safety Council (NSC)

(a) *Accident Prevention Manual*, 11th Edition.

(b) *Roller Conveyors*, Data Sheet I-528, Rev. 91.

(c) *Belt Conveyors for Bulk Materials-Part I: Equipment*, Data Sheet I-569, Rev. 90.

(d) *Belt Conveyors for Bulk Materials-Part II: Operations*, Data Sheet I-570, Rev. 82.

(e) *Underground Belt Conveyors*, Data Sheet I-725-86.

U.S. Army Corps of Engineers

The following is a summary of pertinent points in the *Safety and Health Requirements Manual*-EM 385-1-1 (3 November 2003) Section 17-Conveyors:

(a) Conveyor systems shall be constructed and installed in accordance with the manufacturer's recommendations.
- (i) No maintenance should be performed when a conveyor is in operation.
- (ii) Safe access should be provided to permit inspection, lubrication, repair, and maintenance activities.

(b) Safety devices
- (i) On all conveyors where reversing or runaway are potential hazards, or the effects of gravity create a potential for hazardous uncontrolled lowering, antirunaway devices, brakes, backstops, or other safeguards should be installed to protect persons from injury and property damage.
- (ii) Conveyor systems shall be equipped with an audible warning signal to be sounded immediately before starting of the conveyor.
- (iii) All conveyors shall be equipped with emergency stopping devices along their full length.
- (iv) Safety devices should be arranged to operate in such a manner that if power failure or a failure of the device occurs, a hazardous condition will not result.
- (v) Take-up mechanism should be guarded as an entity of placing standard railings or fencing.
- (vi) Trolley conveyors should be provided with spill guards, pan guards, or the equivalent if there is a potential for material to fall off the conveyor and endanger personnel or equipment.

(c) All exposed moving machinery parts that present a hazard shall be mechanically or electrically guarded, or guarded by location.

(i) Nip and shear points shall be guarded.
(ii) At all points along the conveyor, except at points where loads are removed from or placed on a conveyor or where a conveyor discharges to or receives material from another conveyor, provisions should be made to eliminate the possibility of loads or material being dislodged from the conveyor

(d) Access
(i) Crossovers or underpasses with safeguards shall be provided for passage over or under all conveyors: crossing over or under conveyors is prohibited except where safe passageways are provided.
(ii) Whenever conveyors pass adjacent to, or over work areas, roadways, highways, railroads, or other public passageways, protective guards should be installed to catch and hold any load or material that may fall off or become dislodged from the system.
(iii) Where conveyors are operated in tunnels, pits, and similar enclosures, ample room shall be provided to allow safe accessway and operating space for all personnel.

(e) Emergency stop devices
(i) Emergency stop devices shall act directly on the control of the conveyor concerned and should not depend on the stopping of any other equipment.
(ii) All emergency stop devices should be easily identifiable and readily accessible.
(iii) Emergency stop devices should be installed so that they cannot be overridden from other locations.

(f) Counterweights
(i) When counterweights are supported by belts, cables, chains, or similar means, the weights should be confined in an enclosure to prevent the entry of personnel beneath the counterweight.

(g) Controls
(i) Conveyor controls should be arranged so that in case of an emergency stop, manual reset or restart is required at the location where the emergency stop was initiated to resume conveyor operations.
(ii) Control stations shall be arranged and located so that the operation of the equipment is visible from them.
(iii) Controls shall be clearly marked or labeled to indicate the function controlled.

(h) Guarding
(i) All openings to the hopper and chutes should be guarded to prevent persons from accidentally stepping into them.
(ii) Dump hoppers having the hopper flush with the floor, which by their use cannot be guarded, should be equipped with grating having a maximum opening (2 in.) and heavy enough to withstand any load which may be imposed on it.

(i) Mobile conveyors
(i) Where mobile conveyors are used, powerlines will be relocated so the conveyors cannot reach, contact or be rolled into them.
(ii) Mobile conveyors should be provided with brakes or other positive locking devices for each degree of motion where movement would present a hazard.
(iii) Mobile conveyors should be designed to be stationary against runaway and stable against overturning under normal conditions of operation.

is driven on a slope or at a tilt, and *active* when it tips over. The hazard is exacerbated by the machine's ability to work close to pedestrian workers who are unaware of its proximity due to lack of warning devices that alert others to the machine's presence.

Scope

The narrow width of loaders and many types of forklifts allow these machines access to many unimproved worksites, such as a backyard with a narrow gate and uneven ground. Common uses include transporting and lifting building materials to the second floor of a housing project. The uneven and sloping nature of the work zone is responsible for most loader upsets, and grading such sites before the loader arrives would significantly reduce the hazard.

Peril From the Failure Mode

When a conventional fork lift or a telescoping-boom forklift upsets, it can strike others working next to the machine. This can result in the hazard of any worker in the proximity being crushed, struck, or caused to fall by the overturning machinery. In addition, an operator who is not restrained within the fall-protective cab of the machine is ejected. This *natural* hazard can be defined as gravity or lack of stability from a slope.

Brief Summary of the Occurrence

A loader was in use to excavate a residential swimming pool, transporting excavated dirt to the street and loading it into a dump truck. The loader upset during the finishing phase of trimming the excavation and incorporating raised landscaping mounds. It was backing up prior to scraping dirt in an attempt to level the patio space next to the swimming pool. The pedestrian worker sustained a crippling back and hip dysfunction when he was hit by the upsetting narrow-width loader as it was backing the left wheels onto a decorative landscaping mound that sloped to the right. The operator looked to his left and made a misjudgment that the mound was not steep enough to cause upset. He did not look to his right and observe the worker who was next to his path of backward travel, and the worker was struck by the overturning loader's cab. The loader operator had previously been traveling another route away from the injured worker, and had changed his task to start working next to the co-worker. A sensing alarm could have warned the operator that the equipment would upset with any further backward movement. Such sensors can be integrated into the design of the machine to limit further movement and prevent upset. Without a back-up alarm signal, the worker was unaware of the presence of the loader, as he was engaged in using a shovel to trim the portion of the pool excavation that would contain the recirculation pump. When struck by the overturning cab of the loader, he was thrown to the bottom of the 10-foot pool excavation. The injured stated that when he knew the loader was nearby, he would move away to work on another task, but without a backup alarm, it is difficult to be aware of the machinery, as the minimal motor noise is never indicative of its change of location.

Date of Occurrence

October 25, 2002

Source of Information

Superior Court of the State of California, County of Ventura, # SC 03794

History of the Hazard

The evolution of small front-end loaders and rough-terrain telescoping-boom forklifts began in the mid 1960s. Today there are a number of manufacturers in this market. The demand for these compact machines continues to grow, and more of them are being produced than ever before. There may be over 100,000 of these machines currently in use.

Contributory Factors

Narrow machines are more vulnerable to upset than their wider counterparts. Narrow-width machines and forklifts with telescoping booms have the capability to raise their center of gravity, decreasing stability and increasing the hazard of upset. When the vehicle starts to ascend a graded surface, it may tip in any downhill direction. Equipment manufacturers have not provided incentive for designs to include antitilt sensors that intercede by preventing travel on dangerous slopes. They assume the purchaser is not interested in a safety appliance that would increase the cost. Manufacturers do not see it as their duty to provide essential safety appliances as standard equipment. Engineers, however, know that narrow machines experience reduced stability.

Applicable Standards and Literature

Currently no design standards exist requiring the use of a tilt alarm on rough-terrain telescoping-boom forklifts. The OSHA standard for backup alarms on this type of vehicle can be included under the 1926.601(b)(4): "No employer shall use any motor vehicle equipment having an obstructed view to the rear unless: (i) The vehicle has a reverse signal alarm audible above the surrounding noise level or: (ii) The vehicle is backed up only when an observer signals that it is safe to do so."

Unfortunately, this standard is ineffective in providing protection against backing occurrences or alerting personnel of the change of position or location of dangerous machines.

Hazard Identification and Prevention Matrix

The hazard of upset is a result of gravity exacerbated by a mechanical condition that affords a tipping or upset leverage. The hazard needs to be eliminated by preventing boom extension/elevation when the forklift is tilted sideways on a slope. Alternatively, the machine could be designed with a lower constant center of gravity to resist the hazard of upset. Additional warning of backing or other travel should be provided to warn workers that the machine is entering their proximity.

	Eliminate the Hazard		Guard the Hazard		Provide a Safety Factor		Provide Redundancy		Provide Reliability
	Hazard	Safety	Hazard	Safety	Hazard	Safety	Hazard	Safety	
Natural	Slope/ instability	Site grading							
Structural/ Mechanical								Provide a simple slope guage	
Electrical				Internal tilt sensor/ boom control				Back-up alarm	90%
Chemical									
Radiant Energy									
Biological									
Automated Systems		Sensors to anticipate upset							

Commentary

The monotonous, low-frequency sound of the loader's engine is a pervasive noise that is an ineffective indicator of the equipment's location. Workers cannot be adequately warned of the machine's location by listening to engine noise; therefore they are not alert to the fact that they may need to relocate. Backing alarms or near-object detection devices on the machines would help alert workers to the presence of a loader. A simple slope indicator mounted on the dashboard of the skid-steer loader alerting the operator to limits for safe travel would help the operator avoid steep grades that lead to upset. A prototype tilt sensor design should be developed to prevent further travel in either direction on high-grade slopes.

Recommendations

As antitilt stability technology becomes commercially available in the automotive sector, it should be promoted as an inherently safe feature that helps to prevent side upset. The marketing of new technology is best accomplished by frequently updating information provided to both news and trade journals. Project managers who provide special conditions on their construction contracts can develop market acceptance of reasonable safety features.

SECTION 4

Trucks

The Hazard Syndrome

A few hazards of construction trucking, without priority of occurrence, are listed below:

- Backing (highest frequency cause of injury and property damage)
- Impaired drivers
- Overturns, rollovers, and cabs that are not crash resistant
- Bad weather (slippery roads due to snow, ice, or rain)
- Reduced visibility due to precipitation, sun glare, fog, dust storms, or low-light conditions (night, dusk, or dawn)
- Use of cellular phones and citizen-band radios while driving causes serious distraction of drivers of both automobiles and trucks, and is as dangerous as drunk driving. Dialing is extremely hazardous as it forces the driver to take his eyes off the road
- Fire (engines, brakes, tires, and cargo). The fire that closed one of the Chunnel tubes under the English Channel to rail traffic for several months was the result of a truckload of styrofoam insulation that caught fire.
- Windshear developed by the truck as it travels
- Load loss
- Unsafe loading or unloading methods and facilities
- Speeding
- Runaways
- Dangerous cargoes
- Roadside repair and maintenance activities
- Roadside obstructions
- Improper and confusing signage for driver information
- Unsafe access on and off trucks and flatbeds
- Carbon monoxide
- Dump beds raised into powerlines
- Approaching driver's vision impaired by glare from headlights of opposing traffic
- Confusing, improper, or lack of traffic control
- Inadequate road shoulders where emergency repairs can be made
- No warning signs of upcoming "low clearance" underpasses

Trucks present too many hazards to describe the *dormant*, *armed* and *active* phases specific to each one.

Scope

There has been an almost complete conversion to subcontracting independent truckers, specialty haulers, and construction suppliers in recent years. Subcontracting trucking tasks relieves the general contractor of a temporary labor force and the costly investment in trucks. Due to enormous omissions in the reporting criteria for self-employed truckers, the hazards associated with trucking are virtually overlooked by the general contractor or construction manager.

The trucking industry involves about 8% of the entire workforce, and construction-associated jobs represent a significant portion of this workforce. Actual deaths in construction trucking could be in the hundreds every year, and perhaps thousands more involve injuries. Truckers who are independent self-employed workers are often outside OSHA reporting oversight. Trucking hazards cannot be remedied until trucking functions are considered under the umbrella of a cohesive construction-safety management system.

Peril from the Failure Mode

National Safety Council (NSC) statistics average nearly 700 fatalities and 1000 truck-related injuries per year. The ultimate hazard of trucking is the industry-wide lack of responsibility and management oversight in construction. Lack of a clearly outlined plan of action or accountable person gives rise to many hazards that could otherwise be identified and prevented with minimal foresight. Both of the cases listed below resulted in death or permanent injury from a falling object that was upset by the imbalance of the materials transported. In one case, a heavy object upset when inadvertently snagged by a large hook. The resulting death and serious injury illustrate a total lack of hazard-prevention measures simply because no one was in charge of load or worker safety.

In the following two scenarios, both cases occurred as a result of the *natural* hazard of gravity and the *mechanical* hazards of a falling load and an unsafe lifting hook.

Brief Summary of the Occurrences

A common trend of trucking-related injuries involves subcontracted haulers. The following two incidents are no exception. The case on load upset involved a contract-trucker fatality. Both occurrences involved an unsafe trucking activity where simple and available design features were not provided.

(A) A man was crushed and instantly killed when he approached a fifth-wheel flatbed trailer carrying two five-foot-wide triangular sections of jib for a tower crane. The first jib section sat firmly on the trailer truckbed. The second jib section was upside down and therefore unstable. The trucking company did not provide supporting stakes to retain the triangular-shaped jib resting against the first section on the truckbed. Instead of the unstable jib sections being supported by stakes to ensure overall stability, they were merely strapped in together, presenting the hazard of rolling off the bed of the fifth-wheel trailer when the straps were undone. The truck driver was verbally instructed not to release the straps until the jib sections were secured to the crane's lifting line, but did

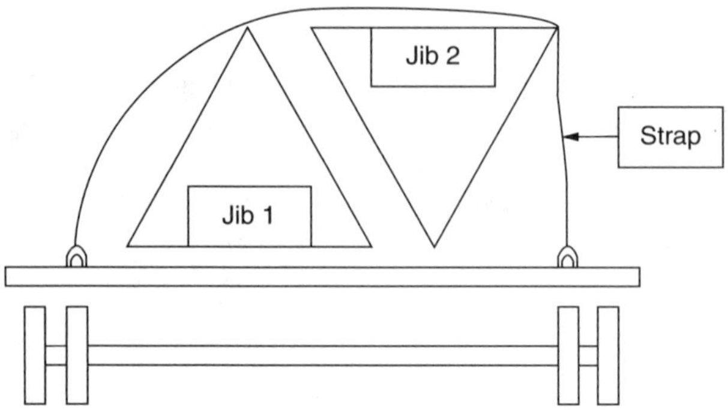

not heed the instructions. Mere hours before, two of his co-workers, also hauling unstable sections of jib boom, had disregarded the same instructions with no dire consequences. The untimely release of the securing straps allowed the unstable jib section to roll off the truck and crush the crane operator, who had approached to check the load, under nearly two tons of steel. In this case, the general contractor's employee was killed as the result of the equipment rental firm's failure to provide supporting stakes on the truck, or other means of stability for transporting triangular boom sections. The image above shows the position of the jib boom on the truck bed.

(B) The contractor was loading large concrete "K-rail" highway barriers weighing approximately two tons onto the fifth-wheel trailer of a contract self-employed truck owner. Rather than use lifting hooks controlled by taglines, the contractors' employees had erroneously requested that the truck driver remove the lifting hooks from the K-rails while standing on the trailer bed. It is not the driver's job to load cargo, and the proper safety precautions were not taken. In the process of lifting the K-rails, one of the hooks caught on the bottom of the K-rail, causing it to upset and crush the driver's knee. The leg was later amputated above the knee. A hook with an eyehole for taglines would have been much safer in this scenario. Such a hook is shown below.

In both of these cases, the failure was the lack of an inherently safe worksite for truckers. The first case was a failure to design a safe cargo space with stakes to support loads being transported. In the second case, lifting hooks and taglines to ensure a load properly controlled from the ground were not provided.

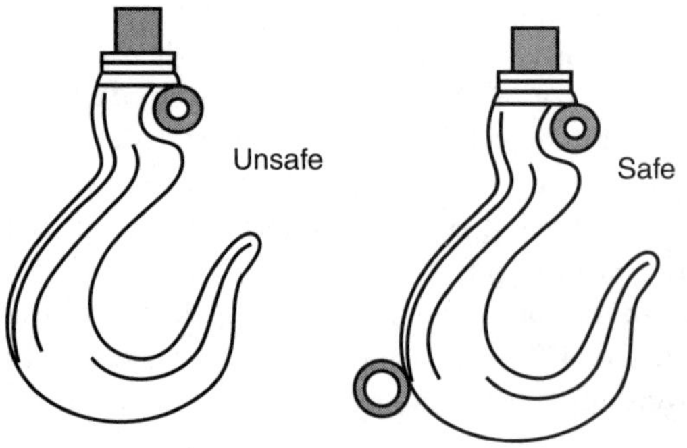

Date of Occurrence

(A) May 16, 2001

(B) November 26, 2001

Sources of Information

(A) Superior Court of the State of California, County of Orange, Central Justice Center, CA, # 01CC10590

(B) Superior Court of the State of California, County of Los Angeles, # TC016169

History of the Hazard

1930s: With the building of Boulder Dam (now known as Hoover Dam), the joining together of six companies started the trend for conduction construction with a group of smaller specialty subcontractors. This trend, whereby the general contractor relies heavily on subs, has continued to grow.

1960s: The term "suitcase contract" was a label given to the incident of an out-of-town general contractor visiting the town of a work project and collecting bids from local subcontractors for specialty work. This term highlights the prevalent use of subcontractors, who are often poorly supervised.

1980s: The trend for companies to reduce heavy capital investments for infrequently needed equipment relies upon rental of costly equipment and the subsequent loss of skilled equipment workers.

1990s: Increasing amounts of trucking for construction involve the small-truck owner and the small-truck companies and specialty hauler, of heavy equipment. The cost per ton-mile of transporting materials has become a first choice of general contractors, as it is an easy "outsource" activity.

Contributory Factors

On most construction projects, no one is "minding the store" when it comes to the safety of trucking activities. The lack of responsibility involved in the subcontracting of trucking services exacerbates the host of other hazards inherent in this industry. This attitude—that responsibility for safety is within the sole control of the trucker—is the cause of many common trucking-related injuries. Improved project safety awareness by the contractor could eliminate or minimize many dangerous circumstances. The cooperation of multiple parties is necessary to achieve improved project method and design, as well as to delineate specific responsibility. To ensure a good relationship with those who hire truckers' services, the drivers are often requested to assist with the loading, unloading, and placement of hauls. In featured case (B), to accommodate the loading of the heavy concrete K-rail barriers, the injured driver was on the fifth-wheel trailer removing the lifting hooks, which should not have been the driver's task. Loading a truck with K-rails requires a trained crew using the lifting hooks with taglines so that the load can be safely guided from the ground in coordination with the crane operator.

Hazard Identification and Prevention Matrix

Gravity causes unstable items to upset, and creates circumstances that allow items to either fall or become vulnerable to being easily tipped over, requiring bracing on a trailer to hold a load in place. The design of the lifting hooks needs taglines to allow disengagement of the hooks from the ground. As a redundancy, the contractor's staff should have included another individual to guide the second hook from the ground. The driver should not be engaged in the loading or unloading activity, and should be required to stay in a designated safe area while loading and unloading are accomplished.

	Eliminate the Hazard		Guard the Hazard		Provide a Safety Factor		Provide Redundancy		Provide Reliability
	Hazard	Safety	Hazard	Safety	Hazard	Safety	Hazard	Safety	
Natural	Gravity					No worker allowed on flatbed for loading or unloading			
Structural/ Mechanical	Load insta-bility	Stability with rigging, stakes		Trailer bed stake, hooks with taglines				Second rigger	100%
Electrical									
Chemical									
Radiant Energy									
Biological									
Automated Systems									

Applicable Standards and Literature

Federal Highway Administration (U.S. Dept. of Transportation)

Federal Motor Carrier Safety Regulations, Parts 383, 390-397, 399.

Note: Regulations Manual (#C0870) is available from American Trucking Associates, 2200 Mill Road, Alexandria, VA 22314-4677, 1-800-ATA-LINE.

International

Brazil, Israel, Switzerland, and two Australian states have passed laws against using hand-held phones while driving.

ANSI

American National Standards Institute Manual on Uniform Traffic Control Devices for Streets and Highways D6.1-1988.

SAE

There are many Society of Automotive Engineers (SAE) design criteria that ensure safe trucks. A select few of those listed in the current SAE Handbook, Volume 4, On Highway Vehicles and Off-Highway Machinery are:

SAE J1967 MAR91 Retroreflective Materials for Vehicle Conspicuity

SAE J1490 JAN87 Measurement and Presentation of Truck Ride Vibrations

SAE J1521 MAR90 Truck Driver Shin-Knee Position for Clutch and Accelerator

SAE J1522 MAR90 Truck Driver Stomach Position

SAE J1948 MAR89 Cab Sleeper Occupant Restraint System Test

SAE J1834 JUN91 Seat Belt Comfort, Fit, and Convenience

SAE J260 JUN90 Rear Underride Guard Test Procedure

SAE J1252 JUL81 SAE Wind Tunnel Test Procedure for Trucks and Buses

SAE J699 NOV85 Average Vehicle Dimensions for use in Designing Docking Facilities for Motor Vehicles

SAE J695 DEC89 Turning Ability and Off Tracking-Motor Vehicles

SAE J848 JAN91 Fifth Wheel Kingpin, Heavy-Duty Commercial Trailers and Semitrailers

SAE J849 NOV85 Connection and Accessory Locations for Towing Multiple Trailers

SAE J1489 JUN97 Heavy Truck and Bus Retarder Downhill Performance Mapping Procedure

SAE J318 SEP 80 Air Brake Gladhand Service (Control) and Emergency (Supply) Line Couplers-Trucks, Truck-Tractors, and Trailers

Other References

- *Transportation and Traffic Engineering Handbook*, Institute of Transportation Engineers, Prentice-Hall, Inc., 1982.
- Joseph E. Badger, *Trailer Underride-Conspicuity, Human Factors, and Rear Bumpers*, Institute of Police Technology and Management, University of North Florida, 1995.
- *Status Report* (monthly newsletter), Insurance Institute for Highway Safety, 1005 North Glebe Road, Arlington, VA 22201, Telephone: (703) 247-1500.
- *Transport Topics* (newsletter), American Trucking Associations, Inc. (ATA), 2200 Mill Road, Alexandria, VA 22314, (703) 838-1800.
- "Large Truck Hazards," *Hazard Information Newsletter,* Volume 1, Issue 12, March 1997.

Commentary

Because the use of trucks in construction is often a stepchild of the use of subcontractors, truck activity is often not included in the experience of the general contractor. To delegate responsibility and improve task coordination, construction scheduling, and planning must be concerned with all the applicable safety-design requirements, in addition to delivery dates and volumes of materials handled. The trucking hazards of construction are numerous and often relegated to a generic classification of "traffic safety," which, in the minds of construction managers, is someone else's problem.

Only continued awareness and greater responsibility by the general contractor will reduce the numerous hazards inherent in current trucking design and practices. The broad observation that responsibility in construction trucking is often ignored becomes the starting place to address the dangerous aspects of inadequate planning. The focus on unsafe loads and unsafe loading hooks without taglines was chosen to show two examples of confusing standards and lack of construction management responsibility that lead to easily avoidable failure modes. As illustrated in case occurrence (B), the OSHA requirements only indirectly refer to the use of taglines when lifting personnel in a basket with a crane in 1926.550 (g)(b)(iii): "Taglines shall be used unless their condition creates an unsafe condition." This regulation makes taglines only a matter of choice, but it is a choice that should be opted for as a redundant safety feature in many situations. Truck drivers should not be engaged in loading and unloading objects, and no one should be positioned on the trailer or truckbed.

Recommendations

Construction safety planning needs to include an examination of the potential trucking hazards encountered by subcontractors and suppliers using trucks on the construction site and list the specific safety requirements to control them. Simple design improvements, such as stakes to support framing for odd-shaped materials, and provision for taglines on loading hooks, should be listed as special provisions in the contract. An acknowledgement of the myriad hazards associated with construction trucking and-methods of elimination should be a subject of general-contractor training. The contracts for trucking services need to clearly delineate how the materials will be secured, and designate that only the truck driver should ensure that the load is secured for safe travel with the use of blocking, chains/straps, and covers.

SECTION 5

Aerial Lifts

The Hazard Syndrome

Aerial lifts are boomed equipment similar to cranes, in that they too present a dormant hazard in their ability to reach overhead powerlines. Aerial lifts can pose many other hazards, including but not limited to upset and collapse. The primary factor that separates cranes from aerial lifts is that many lifts are used to elevate people. Any piece of equipment that raises people presents bodily hazards related to falling in the event of upset or collapse. However, this section's "Case Occurrence" is a good example of a powerline-contact hazard specific to an aerial lift, and shows the lift boom as a *dormant* hazard.

Scope

It is tempting to use a piece of equipment as seemingly versatile as the aerial lift for a variety of applications. This leads to many circumstances wherein the use of an aerial lift is inappropriate. An ongoing misuse of the aerial lift is using an improperly equipped lift in close proximity to a powerline. Whereas powerline contact by cranes can be guarded against by proximity alarms and insulated links, people working on aerial lifts do not have the opportunity for such protection. While aerial lifts are often used specifically for live line and "hot" work, some lifts used for such purposes do not provide proper insulation against the potential contact with current.

Myriad factors that contribute to aerial-lift/powerline contact can be altered to prevent such an incident. When looking upstream to define preventive measures for powerline contact, there are usually a number of parties who could have prevented the injury. The landowner and rental agencies, in addition to the manufacturer, can be contributors in a powerline contact due to lack of planning and precautionary measures regarding foreseeable use near a powerline.

Peril from the Failure Mode

Other inappropriate uses of an aerial lift include overloading or use on a slope, which presents an upset hazard. Yet one of the most common aerial lift hazards is the chance of powerline contact due to use adjacent to powerlines. Inadvertent powerline contact can occur when a worker fails to see the powerline due to glare or distraction by the task at hand. In circumstances wherein the worker on an aerial lift directly contacts the powerlines, the burns can be crippling, disfiguring, or fatal. This hazard is *electrical*. Unless a qualified worker is using a properly equipped lift designed specifically for hot work, removing the powerlines is the best engineering control to prevent contact.

Brief Summary of the Occurrence

The aerial-lift operator sustained serious head, shoulder, and arm burns when he raised the lift into a powerline. He did not see the powerline, as he was blinded when looking into the sun. The operator was remodeling a movie set from the rear service area. He was employed by a movie producer, who leased the movie set and the aerial lift. The operator's burns were so severe and disfiguring that he had to wear a sack over his head for three years while undergoing plastic surgery so as not to frighten his wife and children.

Date of Occurrence

February 1, 1993

Source of Information

State of North Carolina, General Court of Justice, County of New Hanover, # 94-CVS-997

History of the Hazard

1960s: The first aerial lifts with articulating booms were developed for the U.S. Army Missile Command. These lifts serviced rockets before launching.

The design concept was quickly adapted for the electric-utility industry. From that, many applications evolved, including sign maintenance, street-light servicing, and tree trimming operations. Aerial lifts became popular on construction sites for providing mobility and quick access to elevated locations.

Contributory Factors

For cost reasons, the company mentioned in the occurrence did not bury the powerlines in the service area, even though the power company recommended that action. The movie lot was supplied with rental equipment for the movie producers, who knew of the overhead powerlines at the movie lot, yet failed to provide an insulated protective cage on the aerial lift to protect the operator from unintended contact with the powerline. The manufacturer of the aerial lift did not offer an insulated safety cage as a safety appliance, and settled long before the time of trial. The movie-lot owners were found liable for providing a dangerous workplace. This verdict was upheld by the North Carolina Supreme Court.

Applicable Standards

OSHA

OSHA has no distance requirements other than the ten-foot thin-air clearance for aerial devices.

ANSI

92.2 2001

5.1 Electrical Specifications

5.1.1 Insulation: The aerial device manufacturer shall state in the manual and on the instruction plate(s) whether the aerial device is insulating or non-insulating. NOTE: Insulating aerial devices do not protect personnel from phase-to-phase to ground contacts at the platform end.

5.2.1 Insulated Systems: All components crossing the insulated portions of the aerial device shall have electrical insulating values consistent with the design voltage

Hazard Identification and Prevention Matrix

	Eliminate the Hazard		Guard the Hazard		Provide a Safety Factor		Provide Redundancy		Provide Reliability
	Hazard	Safety	Hazard	Safety	Hazard	Safety	Hazard	Safety	
Natural									
Structural/ Mechanical									
Electrical	Power-line contact	Remove/ bury powerlines		Insula-ted canopy/ cage				Proximity alarm	
Chemical									
Radiant Energy									
Biological									
Automated Systems									

rating of the boom, and when provided, of the chassis insulating system. The insulating system shall maintain the electrical insulating values in all working boom positions as defined by the manufacturer.

Commentary

This case was selected for review because it highlights several upstream actions that could have prevented the injury. Unfortunately, this is not a standout case of aerial-lift powerline contact. A more recent case involved a building restoration project that used an uninsulated, unguarded aerial lift in an unsafe location wherein the powerlines could have been removed or buried before the use of the lift (see Lorain County Court of Common Pleas, OH, # 02CV113622).

Use of a lift on rental property should stipulate in the rental agreement that the power-lines will be buried or relocated if the user is to use boomed equipment.

Recommendations

The landowner should be party to upstream safety planning and bury or relocate powerlines where boomed equipment will be used. An aerial lift should never have been used for such an application.

SECTION 6

Unsafe Restraint Systems

The Hazard Syndrome

The hazard of restraint systems is attributed to uncomfortable or ineffective design that leads to misuse. A restraint system is a safety accessory that has the function of preventing the dislocation of a seated person so as to avoid injury. The conventional seat belt relies upon voluntary use, so the most effective design concept is a system that automatically provides restraint. A mobile vehicle without a restraint system is a *dormant* hazard. A vehicle with defective restraint system is also a dormant hazard. A vehicle in use without a restraint system or employing a defective restraint system becomes an *armed* hazard. Someone who is injured or experiences ejection from a vehicle due to lack of or defective restraints is the victim of an *active* hazard.

Scope

A vehicle coming to a sudden stop, suddenly accelerated, or subject to overturning, creates kinetic energy that is transferred to the operator or passenger. Restraint systems for vehicles in ordinary traffic decrease the likelihood and severity of an injury so effectively that their use is a law in many U.S. states. Though vehicles on a construction site often move more slowly than secular traffic, an effective restraint system is still important for operator protection. Any seat belt that allows the user to slip out or under it is a defective system. Construction vehicles are the likely subject of upset or overturn, conditions that could displace or eject the operator if protection is not offered. An operator who is ejected from a tractor with ROPS or a forklift with FOPS may be easily struck down or caught by the overturning vehicle.

Peril from the Failure Mode

Lack of an effective restraint system is a *structural/mechanical* defect that is exacerbated by the *natural* condition of kinetic energy, which causes a body in motion to remain in motion, even when the surrounding object (the vehicle) comes to a sudden stop.

Brief Summary of the Occurrences

(A) The operator of a crawler tractor was crushed by the ROPS structure of his vehicle when it overturned. The operator was unaware of the presence of a seat belt, as it had fallen underneath the seat and was not visible. There was no warning to alert the operator to a seat belt, so the operator did not look for one. The operator sustained a paralyzed left side when ejected.

(B) An operator of a scraper with a low-backed bucket seat experienced extreme whiplash. The operator was in the lead, or front, scraper and was in tandem with a scraper pushing from behind. The front scraper was stuck in the mud and jolted free by the rear scraper. The impact caused the operator to be propelled out from under the seat belt, up and over the seat's low back, resulting in an injury to his lower spine.

Date of Occurrences

(A) January 7, 1977

(B) June 25, 1984

Both these early cases established legal precedents for seat belts on industrial equipment.

Source of Information

(A) Kansas Supreme Court, 2/21/86, # 57,864

(B) 1st Judicial Circuit, Williamson County, IL, # 85-L-180

History of the Hazard

There have been numerous injuries as a result of the absence of or ineffective seat-belt restraint systems. The first case listed here was a landmark case that made notification of the presence of seat belts necessary on all subsequent equipment of that type. It was affirmed on behalf of the injured. The second case is illustrative of a construction-equipment seat-belt failure.

1914–1918: Seat belts were first introduced in WWI on aircraft.

1930s: Adoption of seat belts in race cars.

1950s: Seat belts become a feature in secular automobiles. At the same time, J.P. Stapp of the U.S. Air Force began tests involving restraint systems. There is ample technology on restraint systems suitable for all vehicles.

Contributory Factors

Since seat belt use has been voluntary for so long, it has taken years to gain significant instances of usage, as the equipment operator may have numerous reasons for nonuse. Provision of seat belts is a start of the reduction of ejection and motion-related injuries, but is not totally effective without a guarantee of 100% application.

Hazard Identification and Prevention Matrix

	Eliminate the Hazard		Guard the Hazard		Provide a Safety Factor		Provide Redundancy		Provide Reliability
	Hazard	Safety	Hazard	Safety	Hazard	Safety	Hazard	Safety	
Natural	Kinetic energy								
Structural/ Mechanical	Disengage-ment			Automatic restraint				Seat belt warning for use	
Electrical								Sensors to assure total vehicle stoppage	
Chemical									
Radiant Energy									
Biological									
Automated Systems									

Applicable Standards

U.S. Army Corps of Engineers

EM385-1-1, November 3, 2003, 16.B.08: Seat belts and anchorages meeting the requirements of 49CFR 571 shall be installed and worn in all motor vehicles (installation and usage on busses is optional). Two piece seat belts and anchorages for construction equipment shall comply with applicable federal specifications or Society of Automotive Engineers (SAE) standard B56.1.

Commentary

In view of the number of automobile, pick-up truck, and other vehicle injuries in construction, the design of seat belts is critical. Some small front-end loaders, such as are discussed in Section 3, are equipped with a restraining bar. To enter these machines, the operator must step over the bucket and climb into the front of the machine. To start the machine, the operator must sit down in the seat and pull the restraining lap bar down over the lap. Before exiting, the operator must raise the restraining bar to the top of the cab to turn the machine off. This type of system is automatic, as it requires the operator to engage it in order to start the machine. This system also requires full stoppage of the machine before an operator can exit. Restraint of this nature guarantees protection of an operator. Such automatic restraints can be adapted to fit many other construction vehicles.

Recommendations

Design engineers need to develop seat belts and automatic restraint options that are suitable for mobile construction equipment.

SECTION 7

Equipment Leasing

The Hazard Syndrome

The hazard of leasing equipment when unsafe use is foreseeable requires due diligence on the part of the agency that provides rental equipment. Equipment rental firms hold a unique position in construction safety. They are not directly involved with the construction, and are not subject to conventional methodological requirements. In short, their main legal responsibility is to ensure that their fleet of equipment is maintained to the degree specified in manufacturers' and users' manuals. However, these firms must involve themselves in the process of contractor and operator competence, because to lease equipment to a construction firm or contractor that does not employ a qualified operator is a violation of legal and business specifications to which the rental firm can be held accountable.

The hazard most commonly encountered by cranes that are leased out is powerline contact, followed by upset. These incidents are almost always caused by incompetent operators or contractors who do not specify a list of qualified operators. A crane always has the *dormant* hazard of being able to reach powerlines. When a crane is leased for a location where it is known that it will be used adjacent to powerlines, the hazard becomes *armed.* To prevent an *active* contact with a powerline, the crane rental agreements should include specific limitations of use.

Scope

The issues of where and how the crane will be used must be addressed in the rental agreement. These agreements will alert *all parties* to any *dormant* or *armed* hazards. Unsafe working conditions involving rental equipment can be prevented by the actions of any one of a number of parties. Optimum safety must be a concern of the rental firm, the architect/engineer, landowner, controlling contractor, and the construction manager, who would all be liable in the event of an *active* hazard.

Peril from a Failure Mode

A powerline contact by a rental crane involving injury or death can result in negligence claims against the rental agency, equipment dealer, manufacturer, general contractor, construction manager, or other third party. When specific safety tasks are not delineated to responsible parties, they often go overlooked. When a rental agency exercises no limitation over the hazardous use of a piece of equipment, responsibility is often forced upon said company in the form of liability litigation.

Brief Summary of the Occurrence

A worker who was injured was employed by a concrete subcontractor. He was standing on a partially completed elevated highway access ramp underneath another access ramp, guiding the hoist line that held a suspended concrete placement bucket to a ground level. The crane rental firm was monitoring the use of the crane, as they were responsible for daily maintenance. The rental firm was aware that it was to be used in the danger zone, violating the bounds of the lawful ten-foot clearance, yet made no effort to intercede. The worker lost both arms when the hoist line touched the powerline. The state highway department and the general contractor neglected to bury the powerlines, even though reimbursement funds for this activity were available.

Date of Occurrence

August 21, 1992

Source of Information

Montgomery County, MD, # 117274-V

Contributory Factors

The attitude that the crane user/operator has total responsibility is erroneous and does not protect the rental company or construction manager from vicarious liability. This case was settled, charging the rental firm as a contributor to the damages.

History of the Hazard

Since the 1950s, hundreds of crane/powerline contacts have become a matter of litigation. Many of these injuries ended in complex third-party liability with numerous defendants, all of whom were responsible for contributing to the injury or fatality that occurred on the worksite. Rental companies were found liable in many of these cases. These cases set a precedent that extends safety responsibility beyond the equipment user.

1970s: Equipment-rental firms started serving the construction industry. This industry has steadily grown, as it reduces the capital investment for contracting firms.

Hazard Identification and Prevention Matrix

	Eliminate the Hazard		Guard the Hazard		Provide a Safety Factor		Provide Redundancy		Provide Reliability
	Hazard	Safety	Hazard	Safety	Hazard	Safety	Hazard	Safety	
Natural	Incom-petent operator (human error)	Specify limitations							100%
Structural/ Mechanical		Specify limitations of use						Require danger zone be marked on ground and crane will not be used in danger zone as part of rental contract	100%
Electrical	Shock/ power-line contact	Specify limitations of use		Range-limiting device		Proximity alarms		Insulating link	100%
Chemical									
Radiant Energy									
Biological									
Artificial Intelligence									

Applicable Standards

OSHA

OSHA has only a ten-foot *thin air* clearance rule and does not address the responsibility of equipment-rental companies.

ANSI

ANSI does not identify responsible parties.

Commentary

Construction managers involved in planning often lack training in how to anticipate foreseeable hazards, such as crane/powerline contacts. Their primary relationship with crane-rental companies is the establishment of time and cost. Crane- and equipment-rental-firms need to be aware of where their equipment will be used, whether it is leased without an operator, and the conditions pertaining to who is authorized to operate the crane must be delineated. The rental contract should include a mutually-agreed upon list of persons qualified to use rental equipment.

Recommendations

An equipment rental firm's involvement in construction planning needs to start upstream at the time of material allocation. Topics addressed should include inherently safe use, in addition to the requisite time and cost analysis. The rental company should require:

- operator certification,
- a contract that agrees to prevent use in dangerous locations, and
- identification and a statement of purpose regarding the user's proposed use of the equipment.

9 Universal Hazards

INTRODUCTION

The following twelve sections are examples of hazards that plague all construction sites. These hazards constitute small parts or inherent functions of commonly used equipment, making them both hard to detect and hard to avoid. The "perils" presented by the hazards in the following sections may at first seem unavoidable, but can be easily controlled by careful attention to detail. When these hazards have been isolated and targeted, the inherent safety of a construction site will be fundamentally improved.

These hazards can be addressed by the facility designer, equipment designer, and the construction manager. They can be overcome with the inclusion of design instructions as special conditions in the project specifications using the critical-path planning schedule. Each example will show design and managerial personnel how to recognize and control the hazard on the construction site so the project workers are free from potential danger. Hazards include:

1. Killer hooks
2. Blind zones
3. Dangerous nuts, bolts, pins, and connectors
4. Unsafe control systems
5. Guarding moving parts
6. Noisy equipment and machines
7. Mold
8. Dust
9. Ventilation
10. Pinch points
11. Compressed gases
12. Secondary voltages

SECTION 1

Killer Hooks

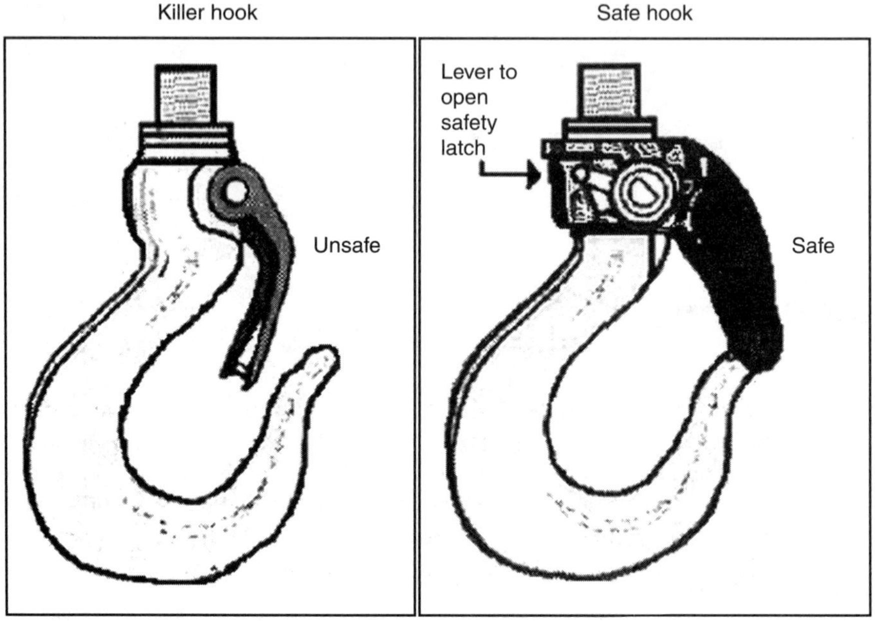

The Hazard Syndrome

Hooks that cannot be locked shut are a *dormant* hazard. An example of such a hook, along with its safe counterpart, is depicted above. These unsafe hooks are dangerous and should not be used. The line movement inherent to a load lift causes these hooks to lose their loads; or the load can snag on an item and cause the hoist line to slack, disengaging the strap from the hook. The hazard becomes *armed* when the load is raised from the ground by an open hook or one with a sheet-metal latch of inadequate strength. The hazard becomes *active* when the load disengages from the hook.

Scope

Killer hooks are classified by the ease with which the load can be lost. The loose load is able to fall on someone in close proximity and cause him or her harm. The solution to this hazard is a strong and sturdy steel latch made out of the same material as the hook itself. As this mechanical advancement is easily achieved and commonly used on construction sites, there is no excuse to use an unsound hook. A strong and safe construction hook must be equally structurally sound in all places. Hooks like the one shown below have a sturdy clasp to hold the load with integrity.

Peril from the Failure Mode

A falling load jeopardizes the entire construction site. The worker guiding the load is in immediate peril from a falling object. If a load falls from an elevation, any worker performing a task in the vicinity can be struck. Loads or parts of loads left on the ground can cause tripping hazards or other barriers that impede work. The project can also be affected by the value of the materials in the dropped load. This hazard is caused by the *natural* factor of gravity and the *mechanical* hazard of structural failure caused by incongruity in the strength of the latch.

Brief Summary of the Occurrence[1]

The injured was struck by a falling tote basket. He sustained brain damage and severe crippling injuries when the eye of one of the two straps fell out of an open hook, causing the basket to fall. A dynamometer (weight scale) and hook were above the headache

[1]This case was test recorded on videotape and shows the actual failure mode of an open hook.

ball on the main hoist line so the content in the tote basket could be weighed. The upper hook had a swivel latch that required a button to be depressed before it could be opened. Such a safeguard reliably protects the load from disengaging. However, the lower hook remained open and the danger from a falling load remained high. In this occurrence, the upper hook with the swivel latch was inherently safe, while the lower hook that held the straps from which the tote basket was suspended was inherently unsafe.

Date of Occurrence

May 28, 1997

Source of Information

U.S. District Court, Western District of Washington at Tacoma.

History of the Hazard

1867: The first of over 180 patents was issued for various types of safety hooks in hopes of overcoming the hazard of open hooks.

1940s: A maritime practice up through World War II was to "mouse" open hooks with wire to prevent straps from falling out of them.

Contributory Factors

There are a number of reasons hooks will disengage. The most common occurrences take place on open hooks or hooks with thin sheet-metal latches. The following illustration shows how an open hook on a boom can fail.

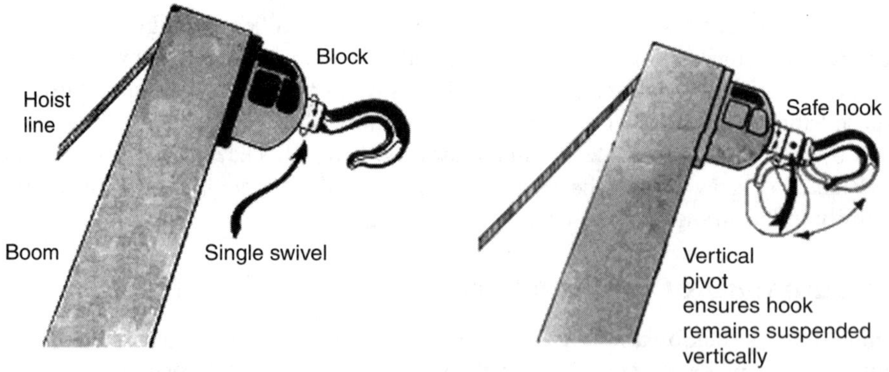

The next illustration shows how a hook on a backhoe bucket can fail. The knuckled-under open hook lets the load fall.

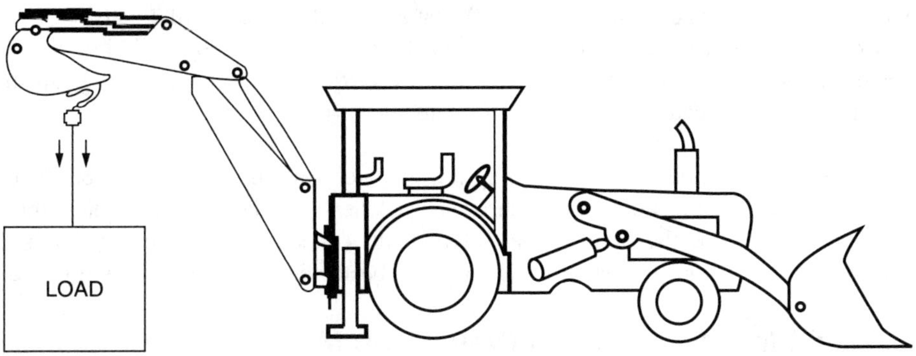

A better method would be to weld a ring instead of a hook on the bucket. This would require the use of a shackle to connect the strap, as shown below.

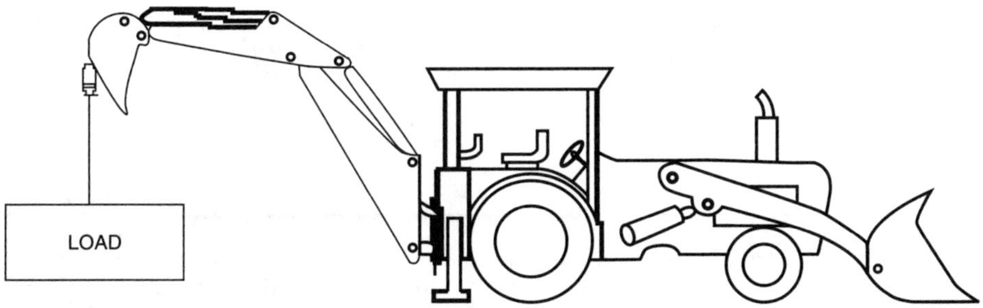

A two-blocked headache ball and open hook can lose the load. The next picture shows how a load can fail in a two-blocking situation.

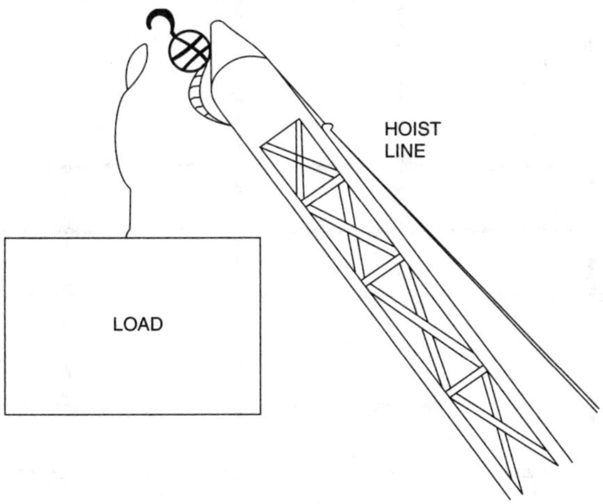

Some cranes have load hoist blocks that fit flat against the boom. With this type of hoist block, the rigging rests deep in the throat of the hook when the boom is in a lowered position; but when the boom is raised to a nearly vertical position, the rigging can slide out of the throat because the swivel can rotate and turn the hook downward.

On a safety belt, a hook without a safety release latch roll-out is also a source of hook failure. On this type of hook, the latch is easily depressed when pushed against the D-ring, which attaches the hook to the safety belt. A hook with a latch that must be depressed to disengage the load is useless if it can be accidentally depressed while in use.

Hazard Identification and Prevention Matrix

Gravity is a natural hazard that causes the load to fall from an open hook. Few endeavors still use an open hook, but the wide use of easily compromised, thin sheet-metal latches does not eradicate the danger of a falling load or an open hook. In addition, an open hook always presents the hazard of snagging an object, as discussed in Section 4 of Chapter 8. The hazard can be eliminated by a hook design that has a heavy steel latch with a safety release button.

	Eliminate the Hazard		Guard the Hazard		Provide a Safety Factor		Provide Redundancy		Provide Reliability
	Hazard	Safety	Hazard	Safety	Hazard	Safety	Hazard	Safety	
Natural	Gravity								
Structural/ Mechanical	Structural failure of sheet-metal flipper	Latch of sufficient and uniform strength		Locking latch		Strength of latch exceeds weight of load		Release button to be depressed before latch opens	100%
Electrical									
Chemical									
Radiant Energy									
Biological									
Automated Systems									

Applicable Standards and Literature

OSHA

1910.179 Overhead and Gantry Cranes (h)(viii)(4); (j)(2)(iii); (l)(3)(viii)(a)

1910.180 Crawler Locomotive and Truck Cranes (d)(3)(v); (h)(3)(xiii)(c)

1910.181 Derricks (a)(29); (d)(2)(vi)

1910.184 Slings (e)(9)(ii); (f)(5)(vi); (e)(2)(i)

1910.263 Bakery Equipment (i)(12)(iii)(iv)

1910.265 Sawmills (c)(24)(ii); (c)(24)(viii)

1926.251 Rigging Material for Material Handling (b)(2); (b)(3); (f)(1)(2)

1926.452 Scaffolds (o)(3)

1926.550 Cranes and Derricks (g)(4)(iv)(B)

1926.800 Underground Construction (r)(15); (t)(1)(vii)

Comment: The listing above provides most of the key references to hooks in the OSHA standards. It would be far better and simpler to define the safe use of hooks in one basic standard as to what constitutes a safe hook. A hook can be a killer wherever it is used if it does not have a safety latch or is damaged.

29 CFR PART 1915 Occupational Safety and Health Standards for Shipyard Employment

1915.71(b)(10)
1915.114(b)
1915.113(b)(3)

29 CFR PART 1917 Marine Terminals

1917.42(a)(1)
1917.42(a)(3)
1917.45(e)(2)

MSHA

CFR Title 30, Parts 56 and 57:

56.9301/57.9301 Dump Site Restraints
56.19075/57.19075 Use of Open Hooks

CFR Title 30, Part 77:

77.1605 Loading and Haulage Equipment; Installations
77.1907 Hoist Construction; General

Shipbuilding and MSHA, which are also under the U.S. Department of Labor, should also have one standard for hooks to eliminate the need to search-and-find.

ANSI

Requirements for safety hooks were referenced in ANSI's early standards. Current requirements are listed in the following page.

B30.2-1996 Overhead and Gantry Cranes (Top Running Bridge, Single Girder, Underhung Hoist)

B30.3-1990 Hammerhead Tower Cranes

B30.4-1990 Portal, Tower and Pillar Cranes

B30.5a-1995 Mobile and Locomotive Cranes

B30.6-1995 Derricks, American Society of Mechanical Engineers

B30.7-1994 Base Mounted Drum Hoists

B30.8-1993 Floating Cranes and Floating Derricks

B30.10-1993 Hooks

B30.11-1993 Monorail Systems & Underhung Cranes

B30.12-1992 Handling Loads Suspended for Rotorcraft

B30.14-1996 Side Boom Tractors

B30.15-1973 Mobile Hydraulic Cranes

B30.16-1993 Overhead Hoists (Underhung), American Society of Mechanical Engineers

Comment: These crane standards generally state that "hooks with safety latches should be used where possible," which is a very weak guideline.

C2-1997 *National Electrical Safety Code*, Section 42, General Rules for Employers, Part K, "Fall Protection," Institute of Electrical and Electronic Engineers (IEEE). Note on page 202 goes into great detail on the hazard of rollout of personnel snap hooks from the D-Rings.

U.S. Army Corps of Engineers

EM 385-1-1 *Safety and Health Requirements Manual*, November 3, 2003.

15.A.05 Rigging, Inspection, and use.

"When hoisting loads, a positive latching device shall be used to secure the load and rigging."

15.F.07 Hooks

a. "The manufacturer's recommendations shall be followed in determining the safe working loads of the various sizes and types of specific and identifiable hooks. Any hook for which the manufacturer's recommendations are not available shall be tested for twice the intended safe working load before it is put into use; the employer shall maintain a record of the dates and results of such tests."

b. "Open hooks are prohibited in rigging used to hoist loads."

National Safety Council (NSC)

Data Sheet I-478-Rev. 87, (first revised in 1969) "Alloy Chains for Overhead Lifting," "Safety Procedures . . . O. Replace broken safety latches."

Other References

- W.E. Rossnagel, *Handbook of Rigging for Construction and Industrial Operations*, 3rd edition, McGraw-Hill, 1964.

"To avoid the possibility of a sling accidentally slipping off the load hook if the load being lowered should foul an obstruction, it is good practice to install on the hook a rugged safety latch. If such is not readily obtainable, make a mousing of several wraps of manila rope, yarn or twine on the hook."

- *International Organization for Standardization*, "Single Lifting Hooks with Shank-Capacity up to 25 Tonnes-Grades M, P. S (T.T)-Hammer and Drop Forged Hooks," December 1, 1973.
  "4.1 Boss. The term refers to a shoulder intended to accommodate a safety catch, if this is fitted . . . Whether or not a boss is provided depends on the wishes of the purchaser and on the regulations of the countries in which the hook is to be used."
- D.E. Dickie, *Rigging Manual,* 1st Edition. Construction Safety Association of Ontario, 1975.
  "Hooks: Make sure that all hoisting hooks, except grab and sorting hooks are equipped with safety catches."
- David V. MacCollum, *Professional Safety,* "Critical Hazard Analysis of Crane Design," *January* 1980.
- *PSCA Standard No. 4*, "Mobile Power Crane and Excavator and Hydraulic Crane Standards," Power Shovel and Crane Association, 1983.
- David V. MacCollum, *Automotive Engineering and Litigation*, Volume 1, "Crane Design Hazard Analysis," edited by George A. Peters, Garland Law Publishing, 1984.
- "British Standard Code of Practice for Safe Use of Cranes, Part 1. General," *BS 7121: Part 1: 1989.* British Standards Institution, effective May 31, 1989.
  "16.3 Hooks and hook blocks . . . To prevent displacement of the sling or load the hook should be provided with a safety catch or other efficient device."
- E.E. Halmos, *Lifting and Transportation International,* "Crane Attachments," May 1991.
  "And 'hooks' vary considerably . . . to hooks with spring-loaded metal pieces that snap shut once the load has been gripped, to prevent slippage."
- *Reclamation Safety and Health Standards*, "Shackles and Hooks," Bureau of Reclamation, 1993.
  "17.7.2 Hooks (a) Manufacturing, testing, inspecting and use of hooks will conform to the more stringent requirements contained in ANSI B30.10, "Hooks," manufacturer specifications and recommendations, or these standards.
  "Hoisting hooks should preferably be used with swivels and headache balls and shall be equipped with safety keepers."
  "4.01.1.9. Load Hooks . . . All load block hooks and ball assemblies shall be equipped with latches and permanently labeled with their rated capacity and weight. Latches are intended only to retain loose rigging. The latch is not a load supporting device."
- David V. MacCollum, *Crane Hazards and Their Prevention*, "Killer Hooks," American Society of Safety Engineers, 1993.
- David V. MacCollum, *Construction Safety Planning*, "Safety Latches," Van Nostrand Reinhold, 1995.
- "Killer Hooks" *Hazard Information Newsletter,* Volume 2, Issue 8, November 1997.

Commentary

Open hooks and hooks with a thin sheet-metal latch are clearly dangerous and should be called killer hooks. The best type of safe latches should be those with a swing hook, shown in the first illustration in this section. The safety latch that must be depressed in order to open the hook is the *only feature* that ensures a truly secure load. Safety requirements for hooks should be heavy steel latches with a separate release button.

Recommendations

Engineers who are designers or construction managers are prudent to specify the use of hooks with a swing latch, which requires depression of a button to open. The use of open-throated hooks and thin sheet-metal latches should be outlawed. Simple, special conditions in the contract outlawing open hooks and thin sheet-metal flipper latches, combined with the requirement for a heavy-duty hook with a safety latch and a separate release button, will make lifting operations safer.

SECTION 2

Blind Zones

The Hazard Syndrome

Blind zones are the most common cause of injuries and deaths from backing equipment on a construction site. All large trucks and vehicles with an obstructed view directly behind them create a *dormant* hazard. The hazard of blind zones becomes *armed* when the truck is engaged in a backing motion. Any person or object that can be struck by a vehicle whose operator cannot see to the rear creates an *active* hazard. There are three overriding considerations concerning backing:

- Partial vision and mirrors require the operator to voluntarily look to the rear or use mirrors to see behind.
- Absolutely blind areas can be overcome by a variety of electronic devices to extend vision or warn the operator of an object in the blind zone.
- The individual in the blind zone needs warning, and conventional backup alarms have been successful for that. Today there are "smart" backup alarms that sound only when someone is in the blind zone, alerting both parties—those in the blind zone and the operator.

Many pieces of mobile equipment have blind zones to the rear, and many pieces of stationary equipment used in construction need to overcome blind zones. The same technology used to overcome blind zones in backing can be modified and applied to overcome most other blind zone situations.

Scope

Large vehicles with obstructed views to the rear are found on virtually every construction site. Though many types of heavy equipment are equipped with backing alarms, many other mobile pieces of equipment without backup alarms remain silent and therefore unsafe. The dynamics of a construction job require careful attention to the task at hand. It is inevitable, then, that a worker's attention is focused on a given task, leading to diminished awareness of vehicles in the vicinity. The absence of a smart alarm makes it impossible for both the worker and the vehicle operator to be adequately warned of a dangerous backing condition.

These incidents most commonly involve trucks. Related deaths are normally identified as "backing accidents," but the term "accidents" is a misnomer, as injuries and deaths are predictable due to the driver's obstructed view to the rear of the vehicle, and the speed at which the vehicles can travel. Blind zones also include a space on heavy earth-hauling trucks *in the front*, as the driver cannot see the ground immediately in front of the vehicle.

Peril from the Failure Mode

Workers can be backed over or crushed between the vehicle and a stationary or heavy object, resulting in severe injury or death. This hazard is a combination of the *natural* hazard of impact caused by lack of vision and the *mechanical* hazard of defective design that creates the blind zone.

A segment from the June 2005 edition of *The Jere Beasley Report* addresses the matter of blind zones with the following information:

> "The National Highway Traffic Safety Administration (NHTSA), which compiles statistics about traffic fatalities, has no concrete numbers regarding the scope of the back-over problem because it lacks a central repository of information about the incidents.
>
> A recent NHTSA study of 1998 death certificates from selected cases estimated that 120 people die annually from accidental back-overs, mostly young children or the very elderly. NHTSA is preparing a regulation that would set back-up safety standards for large, commercial-sized trucks. However, NHTSA isn't prepared to require changes from the auto industry. Too many lives are at stake for further delay. If NHTSA fails to act, Congress should legislate in this area of concern."

Brief Summary of the Occurrences

(A) A pothole repair job by a state highway department of transportation required the use of a large pickup truck. The truck was a major vehicle-manufacturer's factory model of a long bed pickup. It featured a crew cab and dual rear wheels. It did not include a reverse signal alarm as either optional or standard equipment. The local highway-repair administrators, who had little professional safety knowledge, assumed that a highway repairman near the truck would also serve as an observer to guide the backing vehicle. The repairman was walking toward the oncoming traffic on the freeway, flagging the cars to move to the inner passing lane. The long-bed pickup truck was using the rear dual wheels to compact patching asphalt in potholes on the edge of the outer lane of the freeway. The driver did not know that his co-worker was behind the truck's left rear wheel, and the worker was unaware that the truck was backing towards him. The repairman's hips were crushed under the vehicle's left rear tire.

(B) A 19-year-old worker was crushed to death on a forest homebuilding site. This part-time employee for an insulation installer was walking well off the edge of the mountain driveway that was located between two dead trees. The driver of the backing truck strayed from the driveway towards one of the trees and crushed the young man between the rear of the truck and one of the trees. The large van used by the subcontractor to install insulation was driven by a driver who tested positive for marijuana consumption immediately after the fatal occurrence. The driver did not have a commercial driver license for hauling hazardous material. Previously, the driver had been apprehended for transporting approximately 200 lbs. of marijuana, but his private driver's license had not been lifted by the county for this felony. This occurrence involved substance abuse, which is a known problem that should be adequately remedied on every job site. Technology may not be able to alleviate the danger of a worker under the influence, but an audible back up alarm could have provided warning to the victim.

Date of Occurrences

(A) January 22, 1996

(B) December 19, 2003

Source of Information

(A) Johnson Superior Court, IN, # 41001-9705-CT-00171

(B) This case fell outside the range of liability. No complaint was filed, as the general contractor and employer were not cited by the state OSHA investigator. State law requires that liability only exists when "willful misconduct" occurs, an intention that is much harder to prove and a great deal more severe an offense than "gross negligence."

History of the Hazard

1915: A patent for a backup light for cars was issued.

1916: A patent for a vehicle audible signal was issued.

WWII: West Coast shipyards installed travel alarms on the large track-mounted "whirly" crane able to rotate 360 degrees. These tower cranes had travel alarms with different sounds for moving forward or backward to warn people on the ground that the large crane was moving, as the operator in the tower could not see the ground beneath the crane. These alarms effectively stopped injuries caused by running into pedestrians. In the early 1950s the U.S. Army Corps of Engineers installed reverse signal alarms on all large trucks, tractors, graders, and road-rollers. This action immediately and dramatically reduced backing injuries and deaths from such equipment.

1970s: Closed-circuit TV rear video viewers were installed on the mobile waiting rooms at the Dulles Airport in Washington DC, allowing the driver a rear view from both the front and the rear operators' stations. Soon RVs (recreational vehicles) and some large vans of a few express delivery companies were installing rear-view closed-circuit TVs with a monitor at the drivers' stations.

1980s: Ultra-high frequency (UHF) near-object detection and infrared were also developed as systems to assist drivers in overcoming blind zones when backing. In 1990, some delivery companies installed near-object detection systems to aid the driver in blind-zone detection and injury prevention. In 2001 GM and Ford began to provide UHF near-object detection for their luxury cars, but not long-bed pickup trucks, as standard equipment.

1986: Some fifty-five backup alarm patents have been issued by the U.S. Patent Office.

2004: MICO developed a near-object detection and braking-interlock system. This system detects objects within the radar detection zone and automatically applies brakes at a predetermined distance from the object while alerting the driver to the danger.

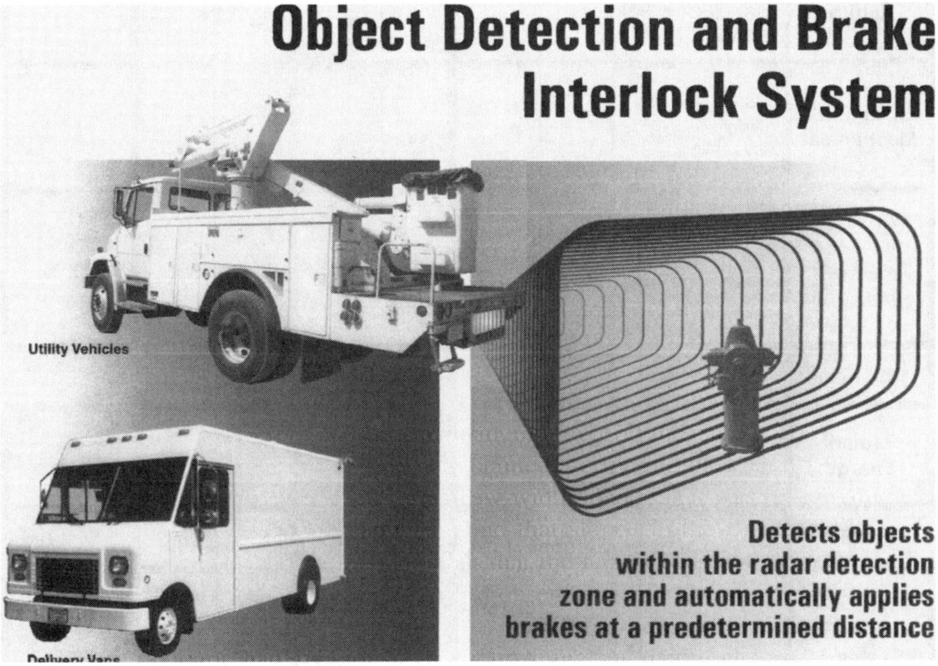

Contributory Factors

Most people assume that they can always see a moving vehicle, but fail to consider that they may be distracted and unaware of one. Beeps and other alarms are fairly common on a construction site and the noise from a conventional backup alarm may be seen as an irritant for those associated with construction. Even though other forms of overcoming blind zones are available with near-object detection systems that automatically stop the vehicle, or with closed-circuit TV, the public is unaware of their options unless they own a luxury car or an RV. Manufacturers have generally considered that appliances such as backup alarms are the purchaser's responsibility. Most of the time manufacturers are reluctant to use safety appliances, as they feel threatened by liability in the event of failure for the lack of user maintenance. In case **(A)** above, after a long trial in 2003 on the verge of being decided by a jury, the automobile manufacturer paid damages to the victim for failure to provide backup alarms on their large pickups. However, the settlement is cloaked in protective orders and cannot be openly discussed.

Hazard Identification and Prevention Matrix

	Eliminate the Hazard		Guard the Hazard		Provide a Safety Factor		Provide Redundancy		Provide Reliability
	Hazard	Safety	Hazard	Safety	Hazard	Safety	Hazard	Safety	
Natural	Impact / no rear view								
Structural/ Mechanical	Defective design: blind zone								
Electrical		Alarm that automatically stops the vehicle		Audible UHF alarm/ backup alarm				Closed-circuit TV	100%
Chemical									
Radiant Energy									
Biological									
Automated Systems		Automatic motion and near-object sensors						Smart alarm	100%

Applicable Standards and Literature

OSHA

OSHA 1910.179(H)(4)(I) Warning Device, states: "Except for floor-operated cranes, a gong or other effective warning signal shall be provided for each crane equipped with a power traveling mechanism."

OSHA 1926.601 Motor Vehicles, Paragraphs 3 and 4 read,

(a) All vehicles shall be equipped with an adequate audible warning device at the operator's stations and in an operable condition."

(b) No employer shall use any motor vehicle equipment having an obstructed view to the rear *unless*:

- The vehicle has a reverse signal alarm audible above the surrounding noise level *or*
- The vehicle is backed up only when an observer signals that it is safe to do so."

Comment: This requirement eliminates the need for a backup alarm, as anyone can be an observer.

MSHA

MSHA, Subpart M, Machinery and Equipment, 56.14132 Horns and backup alarms states:

(a) Manually-operated horns or other audible warning devices provided on self-propelled mobile equipment as a safety feature shall be maintained in functional condition.

(b) When the operator has an obstructed view to the rear, self-propelled mobile equipment shall have:

(i) An automatic reverse-activated signal alarm;

(ii) A wheel-mounted bell alarm which sounds at least once for each three feet of reverse movement;

(iii) A discriminating back alarm that covers the area of obstructed view; or

(iv) An observer to signal when it is safe to back up.

(c) Alarms shall be audible above the surrounding noise level.

(d) An automatic reverse-activated strobe light may be used at night in lieu of an audible reverse alarm.

This standard does not apply to rail equipment.

ANSI

Cranes, Derricks, and Hoists, **ASA B30.2-1943**, reaffirmed 1952, Section 1143(a), Truck Fenders, states: "Bridge Trucks shall be equipped with fenders which extend below the top of the rail and project in front of the truck wheels." Section 1143(b) states: "Trolley trucks should be equipped with similar fenders."

Overhead and Gantry Cranes, **ANSI B30.2.0-1976**, Warning Devices, states: "On cab-operated cranes, a gong or other warning devices shall be provided for each crane equipped with a power traveling mechanism. On cranes other than cab-operated, a gong or other warning device should be provided for each crane equipped with a power traveling mechanism."

ASME/ANSI B56.1-1993 Industrial Trucks, 4.15 Warning Device; 5.3 Traveling and "Part III - For the Manufacturer," 7.32 Warning Device:

- **4.15.1** Every truck shall be equipped with an operator-controlled horn, whistle, gong, or other sound-producing device(s).

  Comment: There is no mention of the need for reverse signal alarms to alert workers and those around backing equipment that they are in a danger zone.

- **4.15.2** The user shall determine if operating conditions require the truck to be equipped with additional sound-producing or visual (such as lights or blinkers) devices, and be responsible for providing and maintaining such devices.
  Comment: There is no mention of the need for mirrors to assist operators in improving their vision while backing machines.
- **5.3.3** Do not pass another truck traveling in the same direction at intersections, blind spots, or at other dangerous locations.
- **7.32.1** Every truck shall be equipped with an operator-controlled horn, whistle, gong, or other sound-producing device(s).
  Comment: Horns require voluntary action being taken by the driver and are not as reliable as automatic travel alarms.
- **7.32.2** In addition, other devices (visible or audible) suitable for the intended area of use may be installed when requested by the user.

SAE

Improved Visibility System for Large Haulage Trucks, Guy A. Johnson, SAE Technical Paper Series 800934, 1980.

SAE J1050a *Describing the Driver's Field of View*, September 1973.

SAE J985 *Vision Factors Consideration in Rear View Mirror Design,* October 1988.

SAE J834a *Passenger Car Rear Vision*, first published in 1962.

SAE J964 *Test Procedure for Determining Reflectivity of Rear View Mirrors,* June 1992.

U.S. Army Corps of Engineers

Safety and Health Requirements Manual, EM 385-1-1 (U.S. Army Corps of Engineers, November 3, 2003)

Section 8.B.04: "A signal person shall be provided when the point of operation is not in full view of the vehicle, machine, or equipment operator; when vehicles are backed more than 100 feet; when terrain is hazardous; or when two or more vehicles are backing in the same area."

Section 16.B.01 Reverse Signal (Back-up) alarm

(a) All self-propelled construction and industrial equipment, whether moving alone or in combination, shall be equipped with a reverse signal alarm. *Equipment designed and operated so that the operator is always facing the direction of motion does not require a reverse signal alarm.*

(b) Reverse signal alarms shall be audible and sufficiently distinct to be heard under prevailing conditions.

(c) Alarms shall operate automatically upon commencement of backward motion. Alarms may be continuous or intermittent (not to exceed 3-second intervals) and shall operate during the entire backward movement.

(d) Reverse signal alarms shall be in addition to requirements for signal persons.

Safety and Health Requirements Manual, EM 385-1-1, (October, 1992)

Section 16.B.02 : "A warning device or signal person shall be provided where there is danger to persons from moving equipment, swinging loads, buckets, booms, etc."

Comment: The U.S. Army Corps of Engineers has had requirements for Reverse Signal Alarms since March, 1958.

Other References

- *Manual of Accident Prevention in Construction*, The Associated General Contractors of America, Inc., 18-7 General Safety Rules states: "Operators should always check to make certain that other workers are in the clear before starting the machine. . . . Heavy earthmoving equipment should be equipped with a reverse signal alarm which will operate automatically with backward movement. The alarm should give an audible signal suitable for the conditions. The signal may be continuous or intermittent, but should provide suitable alarm during the entire time of backward movement. Exceptions may be made for light-service trucks, power shovels, draglines, and double front end loaders."
- MESA Mine Safety, *Truck Blind Spots Can Kill.* January 1976.
- *Proceedings* ~ 17th Annual Human Factors Conference, October 1973:

  "e. An Exploratory Study of Driver Mirror Usage, David Hoffmeister."

  "f. Visual Search by Drivers in Freeway Merging, Vivek D. Bhise."
- *Proceedings* ~ 16th Annual Human Factors Conference, October 1972: Evaluation of Visual Field Requirements of Vehicles in Freeway Merging Situations, Bhise & Rockwell.
- National Safety Council's *Data Sheet 558*, Revision 1991; Paragraph 7 reads, "Equip pendant-operated trucks with traveling warning alarms."
- *Surface Mine Truck Safety,* IC 8827, Bureau of Mines, Information Circular, 1980.
- *Handbook of Mirror Technology for Large Haulage Trucks*, U.S. Bureau of Mines, 1980.
- *Improved Visibility Systems for Large Haulage Vehicles*, MB Associates, San Ramon, CA, prepared for U.S. Bureau of Mines, PB-286065, April 30, 1978.
- *Human Engineering Guide for Equipment Designers*, 2nd Edition, Woodson/Conover, 1964.
- *Industrial Accident Prevention,* by David Stewart Beyer, PhD, (Houghton Miffiln Company, 1916). Page 182 states: "All gantry cranes shall be equipped with automatic warning bells."
- "Blind Zones on Backing Equipment," *Hazard Information Newsletter*, Volume 1, Issue 7, September 1996.

Commentary

Manufacturers should be involved in safety leadership by making backup systems standard equipment. Ultra-high frequency smart backup alarms sound only when someone is in the blind zone, calling attention to an immediate hazard and eliminating unnecessary noise. The OSHA requirement has a loophole, which states anyone can act in the role of observer. However, not all people are qualified or alert observers, leaving the situation open to human error. Mechanical alarms would reduce but not eliminate the opportunity for blind-zone dangers, as mechanical alarms do not completely overcome human limitation. With today's technologically advanced electronics, such as near-object detection systems, smart alarms, and closed-circuit TVs, blind zones can be totally eliminated.

MICO, Incorporated of North Mankato, MN currently supplies an "Object Detection and Brake Interlock System," as shown below. For more information, contact Mico/Preco at (507) 525-6426, or visit www.mico.com.

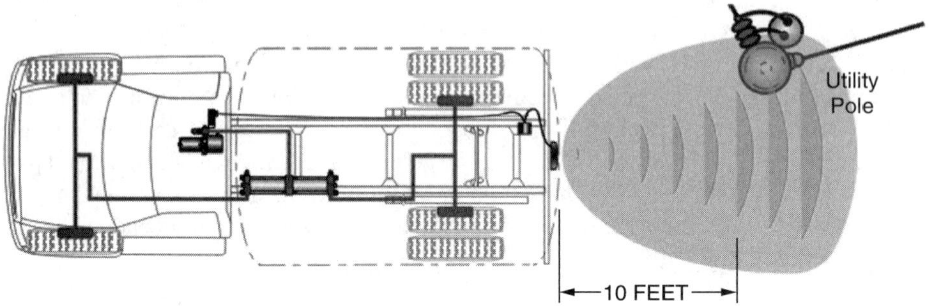

Recommendations

Backup alarms of some type should be standard for all mobile equipment. News of new technology should be spread throughout the industry, with updates at regular intervals. The act of providing data on types of appliances that overcome blind zones (such as ultra-high frequency near-object detection) is crucial to advancement of industrial safety, and facilitates adaptation of such devices to different types of trucks and other mobile equipment for more complete coverage.

Project contract specifications should outline specific requirements for vision in the blind zone or compel the use of UHF near-object detection systems.

SECTION 3

Dangerous Nuts, Bolts, Pins, and Other Connectors[2]

The Hazard Syndrome

Connection systems are perhaps the most important component process existing in construction. Loose nuts, bolts, pins, and other connectors come apart and result in critical failures that cause costly damage, serious injury, and fatalities. A loose connector is a *dormant* hazard when the equipment is not in use. The hazard is *armed* when the equipment is in use and becomes *active* when the connector fails.

[2]Many thanks to Ben Lehman for his assistance in providing information on The Design of Joints with Screw Threads. For more information on connectors, contact Mech Elex Tex, Inc., PO Box 3480, Stateline, NV. Phone (702) 588-7765.

Scope

Wherever major structural components are bolted together or connected by fasteners, failure can occur due to the inadequacy of the bolts to withstand foreseeable vibration or repetitive tension stress. Not all designers use the design phase to calculate the total stress on a bolt in terms of both weight and tension. A bolt's strength in tension can be exceeded even below maximum strength from repetitive vibration, causing the bolt to fail. A bolt's integrity can also be lost because it can be easily sheared or fail from hydronization erosion.

In many cases, design flaws and miscalculation directly contribute to bolt or connector failure. Failure to properly assess the initial strength of connectors, or failure to incorporate connector strength into design plans leads to structural instability and disaster. Nowhere are the effects of the failure to plan and incorporate the principles of inherently safe design more obvious than in instances of connector failures. Though not mentioned in the sample matrix, reliability calculations in this process are indispensable for establishing the probability of connection failure.

Peril from the Failure Mode

Major structural components can become separated when a bolt or nut breaks and lets the component fall. Personnel are vulnerable to being crushed by falling objects or falling from an elevation. Damage is due to the *natural* hazard of gravity, but arises from the *mechanical* hazards of structural failure due to tension, vibration, and shearing, and the *chemical* hazard of hydronization.

Brief Summary of the Occurrences

The examples are presented to provide an overview of the widespread and potentially devastating implications of bolt failure.

(A) An experienced miner was struck by one of the chain-feed zoom-booms with a rock-drill assembly that fell from a mobile two-boom jumbo mobile drill rig. The miner was crushed against the mine drift wall by 2000 pounds of steel machinery as he was walking alongside the machine to perform a task at the mine face while the equipment was idle. The equipment motor was running, causing vibration. Some months earlier the other zoom-boom had fallen off without warning when four bolts that connected two steel plates had loosened and failed. A maintenance mechanic welded the two plates together after that particular incident, but did not do the same on the other boom. These same machines are used on tunnel construction, for both drilling shaft blasting holes, and for roof bolting. Examination of a new machine that had just arrived revealed that the plates had been welded together. The dealer was aware that the plate was poorly designed, with a steel "hockey puck" to absorb the shear stress. The plate relied on four bolts found in each corner of the square connecting plates, which proved to be inadequate to sustain a sound connection. Competitors who manufacture similar drill rigs use as many as twelve bolts to connect this type of plate.

(B) A worker using a hundred-foot aerial lift fell to his death when all the supporting bolts on the turntable broke because of repeated tension stress. The turntable design of the hundred-foot lift was the same as it was for a fifty-foot aerial lift. The failure mode was created by vibration the boom sustained in the travel mode, as the folded boom would vibrate in the carrier frame when traveling down the highway. The repeated vibration caused the turntable bolts to sustain continued tension stressing. The continual stressing led to a small crack in the bolts that slowly grew and eventually led to bolt failure and collapse of the boom from the turntable.

Date of Occurrences

(A) October 3, 1992

(B) February 11, 1997

Source of Information

(A) Federal District Court, Reno, NV, # VC-N-94-783-DWH

(B) N/A: Case settled under protective order.

History of the Hazard

Failures or misuse of bolts, nuts, or pins is generally not the most newsworthy subject. However, these failures can result in enormous catastrophes. Some early bridge failures were the result of shearing or tension failures of either bolts or rivets that were insufficient for the forces they had to sustain. The collapse of the suspended lobby walkway at the Hyatt Hotel in Kansas City in 1981 was a classic example of an inadequate review of a design change that allowed the substitution of a bolting assembly that was not strong enough for the total stress placed on it. A comprehensive safety program must ensure that the initial design and *change orders* have an independent review by a licensed engineer qualified to evaluate design of structural substitutions.

The space shuttle *Columbia's* recent mission was cut short when astronauts were prevented from accomplishing two space walks as a result of a jammed hatch. NASA reported that engineers discovered that a half-inch screw had fallen from its hole and become embedded in the gears of the hatch. This incident reveals the costly consequences when fasteners loosen. One wonders what the aftermath would have been had the astronauts made their space walk and, upon their return to the shuttle, were unable to open the hatch from the outside.

A great variety of methods have been developed and used commercially to increase the reliability of joints dependent upon threaded fasteners. One of the simplest of these is redundancy—the use of more than the minimum number of fasteners. Larger bolts are also an important and easy method of achieving a more secure connective system. Oversize bolts are especially desirable for joints such as pipe flanges and structural connections, provided there is adequate space. Experience shows that not all of the fasteners loosen at the same rate: if one or two are observed loose in a joint where redundant fasteners have been used, often no damage will have been done before the loosening is observed and corrected.

Contributory Factors

Designers generally have little training or experience in the field of system safety. When designing a component of a machine, the designer usually only examines the static stresses of normal loading encountered during operation, and remains unaware of shear and tension stresses that arise from vibration and may result in failure.

In most designs, fasteners holding down the cover of a vessel containing pressure, or supporting a cantilevered fixture up against the ceiling are loaded in tension. The studs that connect the outer wheel to the driving axle of a powered vehicle are loaded in shear. These two situations create different requirements for the joint designer. If any loosening will permit a potentially dangerous situation, it is of primary importance to maintain the tension in the fasteners to produce compression of the parts fastened. This compressive force on the joint will compress gaskets or tightly lock the metal-to-metal sealing adhesives that bind surfaces together and keep slack from occurring. Failure occurs when too much slack is created.

In critical applications, the design of the joint should be analyzed to determine if loosening of the joint will cause an opportunity for unsafe loading. Bending of the joined parts, shock loading, and manufacturing imperfections may create loads greater than anticipated. Where inspection and maintenance in service is difficult, a design review should be requested of the manufacturer and a reassessment made on total connection capacity and maximum weight.

Designing and maintaining the correct tension in the fasteners is important:

- To ensure for a high level of friction force that will keep the fastener from loosening.
- To prevent failure by "fatigue." The design of the fastener must be of sufficient strength so that any repeated variation of the stress imposed is only a small fraction of the total strength of the fastener.
- To avoid stressing the fastener beyond its inherent strength and causing either immediate cracking or a tendency to crack in service (i.e., using a wrench with a four-foot handle when all that is needed is an eight-inch handle).
- To know the chemical content of the steel connector and understand its vulnerability to hydronization.

Another method of improving joint reliability is to design the joint so that deformation of some of the metal occurs when the joint is tightened. A nut that has been slightly pinched out of round provides a high level of friction as it is threaded onto a bolt and is less likely to loosen.

Detailed descriptions of many of the devices available to make reliable threaded joints can be found in mechanical engineering handbooks, in technical design literature, and in federal specifications (see SAE Standard J7 below).

Hazard Identification and Prevention Matrix

	Eliminate the Hazard		Guard the Hazard		Provide a Safety Factor		Provide Redundancy		Provide Reliability
	Hazard	Safety	Hazard	Safety	Hazard	Safety	Hazard	Safety	
Natural	Gravity								
Structural/ Mechanical	Vibration		Shear and tension forces	Larger, sturdier bolts		Regular mainte- nance	Shear or tension	Many/more connectors	98%
Electrical									
Chemical	Hydroni- zation					Anneal the bolt thread		Bolt insulation	95%
Radiant Energy									
Biological									
Automated Systems									

Applicable Standards and Literature

OSHA/MSHA

OSHA/MSHA have no standards for the design of connecting systems.

Comment: Government safety regulations should include a provision that equipment used in the workplace be certified by the manufacturer that it has been analyzed, tested, and evaluated as safe for use.

SAE

SAE Standard J2270 Surface Vehicle Standard, Ships Systems & Equipment, "Threaded Fasteners-Inspection-Test, and Installation Requirements," January, 1996.

SAE Standard J429 Mechanical and Material Requirements for Externally Threaded Fasteners, January 1980.

SAE Standard J1199 Mechanical and Material Requirements for Metric Externally Threaded Steel Fasteners.

SAE Standard J121 Test Methods for Metric Threaded Fasteners.

SAE Standard J82 Mechanical and Quality Requirements for Machine Screws June 1979.

SAE Standard J995 Mechanical and Material Requirements for Steel Nuts, June 1979.

SAE Standard J110 Mechanical and Material Requirements for Wheel Bolts.

SAE Recommended Practice J43 Mechanical and Chemical Requirements for Non-threaded Fasteners (Rivets).

SAE Recommended Practice J933 Mechanical and Quality Requirements for Tapping Screws, June 1979.

SAE Standard J78 Steel Self-Drilling Tapping Screws, June 1979.

SAE Standard J81 Thread Rolling Screws, June 1979.

SAE Standard J1237 Metric Thread Rolling Screws.

SAE Recommended Practice J1061a Surface Discontinuities on General Application Bolts, Screws, and Studs.

SAE Recommended Practice J123c Surface Discontinuities on Bolts, Screws, and Studs.

SAE Recommended Practice J122a Surface Discontinuities on Nuts.

SAE Recommended Practice J121a Decarburization in Hardened and Tempered Threaded Fasteners.

Military Specifications and Standards

MIL-HDBK-60 Threaded Fasteners-Tightening to Proper Tension, 1990.

MIL-N-25027G Nut, Self Locking (Service Temperatures 250 to 800 degrees F).

MIL-F-18240E Fastener Element, Self-Locking, Threaded Fastener, (250 degrees F Maximum).

Note: The Department of Defense Index of Specifications and Standards (DODISS) contains more than 50 Military Specifications covering all types of fasteners.

ISO 9000 Quality Management

A key issue is the development of a customer feedback system of pertinent hazard information on a continuous basis. This would include chronic loosening of bolts that lead to a failure mode. With such traceability at hand, manufacturers could then alert other users of the problem *before* further failures occur.

Other References

- Harold Rothbart, Editor-in-Chief, *The Mechanical Design & System Handbook,* McGraw-Hill, New York.
- Howard Beyer, *Selection of Materials for Component Design*, American Society for Metals.

- T. Baumeister, E.A. Abalone, and T. Baumeister, III, Eds., *Marks Mechanical Engineers Handbook,* McGraw-Hill, New York.
- R.O. Parolee, *Standard Handbook of Fastening and Joining*, McGraw-Hill, New York.
- John H. Bickford and Marcel Dekker, *An Introduction to the Design and Behavior of Bolted Joints,* New York.
- "Fastening and Joining Reference Issue," *Machine Design Magazine*, Penton/IPC Cleveland, OH, Nov. 17, 1983.
- "Bolts, Nuts, Screws, Washers, Nails & Spikes," *Machinery's Handbook*, 21st Revised Edition (see p. 1131), Oberg, Jones, & Horton, Industrial Press, Inc., New York, New York, 1979.
- "Nuts, Bolts, Pins, and Other Connectors," *Hazard Information Newsletter*, Volume 1, Issue 10, January 1997.

Commentary

Design of fasteners is critical to the success of the construction project. The best method of control is to perform a system safety analysis at the time of design to ensure that each component does not create inherent hazards with foreseeable use. Further, operators' manuals should include specific instructions on how to examine and test critical bolts at regular intervals of time. Unfortunately, the communication of failure modes of component parts between users and manufacturers of equipment is usually inadequate. Often, the first time the manufacturer hears about product defects is when they are revealed in personal-injury litigation.

Recommendations

No matter how slight a nut or bolt connection failure may appear, the user should always immediately notify the manufacturer, as they must be made aware so they can initiate appropriate notification to other users and provide appropriate retrofits.

SECTION 4

Unsafe Control Systems

The Hazard Syndrome

Construction workers are regularly exposed to overtly dangerous equipment, machines, and processes. Yet additional dangers can be found in places where the average worker does not think to look. The very controls that are used to operate machines are prone to a number of failure modes that create an ever-present *dormant* hazard. The following are some common failure modes.

- The control is positioned so it can be inadvertently activated by an operator or someone else during another operation.
- The size, shape, or configuration of the control does not immediately, upon touch, tell the operator the purpose of the control.
- The movement of the controls mirrors the desired movement of the equipment in operation.
- The control does not have a detent that requires two distinct hand motions before it can be activated. (This is a mechanism for temporarily keeping one part in a certain position relative to that of another part that can be released by application of force to one of the parts.)
- The control does not automatically return to neutral when pressure is released.
- The control is not labeled as to its function.
- Guards have not been installed to protect exposed controls located where they can be unintentionally struck and activated by outside objects.
- Sequential motions are not required to activate a control or perform a specific function, such as manually changing gears.
- The hydraulic system allows controls not properly set in neutral to inadvertently slip out and activate.
- An intermittent "sneak circuit" causes an unexpected action.
- Improperly programmed or malfunctioning computers manage control systems.
- Robotic controls are positioned inside the envelope of the robot's reach so the robot is able to strike the operator if the controls are inadvertently activated.
- There is no reset button to prevent automatic starting after a power failure to avoid jamming of materials being processed.
- There is no emergency stop.

This hazard is *armed* whenever the equipment is used. When the defective control fails, the hazard is in the *active* mode and may cause injury.

Scope

During the early years of United States' participation in World War II, the training of Navy fighter pilots to take off and land on the small decks of aircraft carriers resulted in the loss of many lives and aircraft simply because of a confusing throttle control. The full throttle on land-based training planes was a pull back towards the pilot; full throttle on the tactical aircraft was a push all the way forward. These reversed controls were error-provocative with fatal consequences. Ross McFarland, of the Massachusetts Institute of Technology, was an early researcher of this syndrome for the U.S. Navy.

In today's world, this type of failure mode may seem almost archaic. Unfortunately, it isn't, as it still appears on some industrial equipment. Human factors, or ergonomics, play a crucial role in making the controls more user-friendly by eliminating confusing or hard-to-maneuver arrangements of controls. Special caution must be taken when designing specific control systems.

Peril from the Failure Mode

Unguarded controls are always vulnerable to unintentional activation. A control on an aerial lift that will raise the operator's basket when depressed is a frequent cause of injury. The operator can inadvertently depress the control on the basket, causing it to suddenly move upward and trap the operator against ceiling trusses, electrical cable trays, or other objects, crushing the individual. These tragic occurrences continue to be the result of the hazard of unguarded or incorrectly guarded control designs. The *mechanical* hazard of ineffective design and ineffective guards can cause unintentional activation and can be compounded by the *natural* hazard of human error that results from inept ergonomic design.

Brief Summary of the Occurrence

An aerial lift operator leaned forward in the basket and his stomach accidentally depressed the toggle switch and suddenly raised the boom. The operator died of suffocation when he was crushed between the basket control panel and a cable tray. It is believed he had been approaching the ceiling to install electrical conductors when he became entrapped by the unintended upward movement of the aerial-lift basket.

Date of Occurrence

March 22, 2004

Source of Information

Pending case in CA

History of the Hazard

1963: Analysis of one of the human-factors aspects of control systems led to the development of system safety concepts that were incorporated into many military specifications.

1995–2006: Many U.S.-made cars use rocker toggle switches to raise and lower car windows. These switches resemble the toggle switches that inadvertently raised the boom of the aerial lift cited in the incident above. Several small children have been killed by an upward guillotine movement of the window. European and Japanese cars have push/pull window switches, down to the open window or pulled up close to the window. This precaution prevents child deaths and eliminates the cause for lawsuits for such design defects.

Contributory Factors

The designer/manufacturer often selects the cheapest control available to install on a unit, as they have difficulty perceiving the hazard of unintentional control activation. Since many designers of controls never use the controls in a field situation, they have little idea of what useful or lifesaving features to include.

Hazard Identification and Prevention Matrix

Elimination of the hazard can be achieved by simple redesign that requires manual activation.

	Eliminate the Hazard		Guard the Hazard		Provide a Safety Factor		Provide Redundancy		Provide Reliability
	Hazard	Safety	Hazard	Safety	Hazard	Safety	Hazard	Safety	
Natural	Unintended activation (human error)								
Structural/ Mechanical	Defective control design	Push/pull				Control guard			99%
Electrical								Electronic circuit-check system	*99%
Chemical									
Radiant Energy									
Biological									
Automated Systems		Automatic shutoff							*99%

*When every safeguard mentioned is in place, the control systems are heavily foolproof.

Applicable Standards and Literature

OSHA

When writing the Occupational Safety and Health Act, Congress addressed only the employer's responsibilities for a safe workplace, and made no attempt to ensure for the safe design of tools, machines, or equipment used by industry.

U.S. Bureau of Mines

The U.S. Bureau of Mines is publishing a free *Mining Health and Safety Update.* It includes this new control-design criteria listed next.

- Design controls to comply with anthropometric data on human operators. For example, push buttons should be large enough to activate easily with a gloved hand. The force required to engage a panic bar should not exceed the capabilities of the operator in his or her normal operating position.
- Ensure that the operator can identify the proper controls quickly and accurately. For example, critical controls should be larger than noncritical controls.
- Where feasible, the speed of a vehicle or component should be proportional to the displacement of the control from its rest position and in the same direction.
- Controls should have sufficient resistance to reduce the possibility of inadvertent activation by the weight of a hand or foot.
- Design controls to withstand or guard against abuse, such as from falling roofs and ribs, or from the forces imposed during a panic response in an emergency stop. Also, surfaces should be designed to prevent hands or feet from slipping.

ANSI

Various standards for different types for equipment include uniform placement of controls and give guidelines for what they must achieve.

Comment: It appears that ANSI has not developed a generic standard for "user-friendly" control systems consistent with established behavioral patterns of the general population when using other controls (for example, the controls of an automobile), and not subject to malfunction from other external or internal failure modes.

ISO 9000 Quality Management

The Standards Compendium of the International Standards Organization has a number of requirements that apply to control design for the European Economic Community (EEC), which places new marketplace pressure to design safer controls on all producers worldwide who wish to trade with European countries.

Consistent with the objectives of this organization, equipment manufacturers need to develop an ongoing system for user feedback so they can find out about irregularities and malfunctions of their machines, including feedback on control systems. Additionally, these international standards require traceability of owners and the users of machines so that in the event a product hazard is identified, the manufacturer can inform each owner/user of the hazard and the necessary preventive measures or retrofit.

In summary, ISO requirements as they apply to any product, but particularly in this case to control systems, require the following general steps:

- Design review
- Customer feedback
- Traceability

When the manufacturer becomes aware of a hazardous condition on a machine or piece of equipment, it needs to notify immediately and directly all owners or operators, and not rely upon dealers, thus avoiding a possible communications gap.

U.S. Department of Defense

Military Standard 992B, System Safety Program Requirements, U.S. Department of Defense, March 30, 1984.

Military Standard, Human Engineering Design Criteria for Military Systems, Equipment, and Facilities, MIL-STD 1472, pp. 54 & 67, February 9, 1968.

"Controls," *Human Engineering Guide for Equipment Designers,* sponsored by joint Army-Navy-Air Force Steering Committee, pp. 2–29, Revised Edition, 1964.

SAE

SAE Recommended Practices, J983, of the Society of Automotive Engineers, "General Notes," paragraph 2 states that all controls for all other functions shall be positioned to avoid operator confusion and physical interference, October 1980.

Other References

- The U.S. Department of Energy (DOE) has written guidelines for the design of robotics. See (DOE) *OSH Technical Reference (OTH)* Chapter 1: "Industrial Robots," which is also available by e-mail at support@tis.eh.doe.gov.
- George Peters and Barbara Peters, *Human Error: Causes and Control*, Taylor and Francis, 2006.
- David V. MacCollum, *Crane Hazards and their Prevention*, Chapter 10, "Controls," American Society of Safety Engineers, 1993.
- *Health and Safety Executive (HSE)*, "Guidance on Regulations (Provision and use of Work Equipment Regulations)," HSE Books, Sudbury, Suffolk, England, 1992.
- William English, "Human Factors Engineering: Going Beyond Regulation," *Occupational Health and Safety*, September 1981.
- Wesley E. Woodson, "Controls, Amplifiers/Transmitters/Receivers," *Human Factors Design Guidebook,* McGraw-Hill Professional 1981.
- William English, "Designing Equipment for Human Use—Human Factors Engineering, a Neglected Art," *Professional Safety*, March 1978.
- Alphonse Chapanis, *Human Engineering Guide to Equipment Design,* Chapter 8, "Design of Controls," McGraw-Hill 1972.
- D.W. Conover and W.E. Woodson, "The Design Implications of Product User Behavior," study conducted for the National Commission on Product Safety, Letter Contract # 70-168, pp. 125, December 30, 1969.
- *Hazard Information Newsletter,* "Unsafe Equipment Control Systems," Volume 1, Issue 9, December 1996.

Commentary

The best way to eliminate unsafe controls is to perform a thorough review and analysis of the system. This should be performed by system safety engineers and human-factors psychologists at the time of design and development to identify possible hazardous failure modes, so that appropriate design improvements can be made.

Because construction supervisors and operators must finish their work on schedule, sometimes there is neither time nor resources to make control modifications on machines that have faulty controls. They can take three important steps:

- Have the company immediately advise the manufacturer that the controls are unsafe and request assistance (such communication should be sent as a registered letter).
- Advise all operators or anyone else who must work with the equipment of the deficiencies or shortcomings of the control system. Place warning labels directly on the problem areas of the equipment.
- Research and specify the type of control desired for the equipment.

Those responsible for operator safety need to ask operators to inform them regarding malfunctioning or error-prone controls. In addition to the failure modes listed in the definition, the following laundry list for safe controls should be considered when designing, purchasing, or evaluating equipment or safety modifications, or when instructing and warning operators of its hazardous limitations:

- Controls should have enough resistance to reduce the possibility of inadvertent activation by the weight of a hand or a foot.
- Control movements that seem "natural" for the operator and require normal, typical motions are more efficient and less fatiguing than those that seem awkward or difficult.
- Control movements should be as short as possible, consistent with the requirements for accuracy and "feel."
- The number of controls should be kept to a minimum, and control movements should be as simple and easy to perform as possible.
- Controls for electronic equipment should include a test circuit to show system malfunctions.
- The operator should know instantly that the desired system response has been triggered.
- Controls should be designed to stand abuse.
- When an operator cannot apply enough force to operate the controls unaided, power-boosted or fully powered control systems are necessary, and artificial resistance should be provided.
- The maximum force, speed, accuracy, or range of body movement required to operate a control should not exceed the limits of the least capable operator. Normal requirements for control operation should be considerably less than the maximum force that can be exerted by most operators.
- Remote controls that are electrically controlled with conductive tethers are inherently hazardous if crane booms, boomed conveyors, pumpcrete booms, or other components can reach powerlines.
- Controls used to perform the same function on different but closely associated equipment should be consistent in size and shape.
- Controls that duplicate the same or similar functions should be arranged consistently from panel to panel, and from operating position to operating position on all machines.

- Controls should have labels (on the panel or control) that
  - Identify what they control
  - Show how to operate the control (if it is not obvious).
- Controls should be differentiated from each other by color, size, shape, or location so they can be easily identified by sight or touch or both.
- Push button controls should have Braille-type surfaces so the operator knows when he or she is pushing the right button.
- Critical controls should be larger and more accessible than noncritical controls.
- Emergency stop buttons or panic bars must always be placed where they are easily accessible to those who may be endangered by moving parts when maintenance or service is being performed.
- Some sort of backup power supply is needed for the control system in case of power failure.
- The most frequently used operating controls, meters, or other devices should be placed where they are most accessible and visible. However, the most important controls may not be the ones that are used most frequently; therefore the control's criticality must also be considered.

Use of one control should not interfere with use of another control, unless they are deliberately interlocked in sequence. Here are some guidelines for placement of hand controls:

- Separate infrequently used controls from those used in a normal sequence.
- Recess or place physical barriers around controls to prevent hands from being pinned between controls and other objects.
- Install a detent on controls so the operator must make an extra movement or must operate another control first to prevent unintended action.
- Operate only by definite or sustained effort. To continue movement, the lever or button must be held in constant deflection.
- Design the surface to prevent the hand, finger, or foot from slipping off of the control.
- Designers and manufacturers of controls should ensure that a user information feedback system be provided.

Recommendations

Manufacturers must address design of control systems by concentrating on ergonomics to incorporate as many of the aforementioned design improvements as necessary into a specific product. Control systems should conform to principles of ergonomics, feasible dimensions, and the capabilities of the average human worker. The image below shows a control system that is error-provocative. On this system, the operator can activate the toggle switches and raise the aerial-lift boom just by leaning into the control panel.

SECTION 5

Guarding of Moving Parts

The Hazard Syndrome

Moving parts of machinery have always been a hazard to people, animals, or objects that can reach them. A table saw hazard is *dormant* when the saw is not in use. It becomes *armed* when the saw is running without an effective guard. This hazard is *active* when the saw is inadvertently touched and does not stop, causing the user harm.

Unguarded machinery presents many opportunities to catch objects or extremities in moving parts. The ability to recognize hazards presented by any machine with reachable moving parts varies from machine to machine. The blade of a table saw is an obvious hazard to be wary of, but moving gears accessible for repair or maintenance are often overlooked. Any bystander is exposed to the unguarded auger shown below.

Machine guarding is an effective means to avoid contact with open gears, rollers, and reciprocating parts. Construction machines present many opportunities to come into contact with moving parts of powered equipment, and in most instances they can be guarded to prevent entanglement. This section introduces a revolutionary product that stops the hazard of a moving table saw from becoming active by stopping the saw when it comes into contact with flesh. (Section 2 of Chapter 8 discusses similar situations in further detail.)

Scope

Many different types of saws are used in many types of construction, particularly for wooden homes, apartments, or mall-located businesses. Conventional guarding of power table saws has always been a contentious safety subject, as the guards seem to present a hindrance to precision cuts for many carpenters. Current design of some saws is not ergonomic or practical for their intended use. Workers are often forced to make a decision between working safely and working effectively.

Peril from the Failure Mode

The U.S. Consumer Product Safety Commission reports approximately 33,000 injuries from table and cut-top saws in 1998 that were sufficiently serious to require a hospital visit. Of these injuries, approximately 95% were hand and finger injuries, with approximately 3,000 amputations of one or more fingers. In some cases the slip of a portable power saw has resulted in the severing of a main artery, and the victim has bled to death before the bleeding could be stopped. Moving parts are a *mechanical* hazard.

Brief Summary of the Occurrence

The homebuilding carpenter was making a special cabinet using a table saw. It was a "shop" table saw with a conventional metal guard that covered the moving saw blade. He removed the saw guard because it created a visual obstruction to precision cuts. His hand slipped and he lost his forefinger, most of his second finger, and the tip of his ring finger. This common example of a foreseeable finger amputation provides an illustration of how the concept of guarding can be expanded to utilize simple electronics to afford effective guarding in a variety of ways. Design technology is now surpassing physical guarding and developing automatic shutdown.

Date of Occurrence

N/A

Source of Information

One solution to this hazard is an electronic sensor that detects an electrical change when a human appendage is touching the saw. These injuries can now be minimized or completely avoided with a guard such as the SawStop safety system. This innovative technology is an example of the triumph of the quest to find an engineering solution for a common hazard. The author of this book visited manufacturers of the product and witnessed SawStop's safer operations. The illustration below shows the unique SawStop braking mechanism that uses the electrical impulse from a human finger to stop the saw in milliseconds. To date no complaints have been filed for the failure of a table saw manufacturer to provide a new concept in saw guarding.

Currently, SawStop is marketing its own saws equipped with the electrical safety device. Thousands of their saws have been sold and their market is growing. The SawStop example is used to emphasize that with creative thinking, innovative and lucrative concepts can be developed to ensure inherently safe design. The image below displays the SawStop table saw.

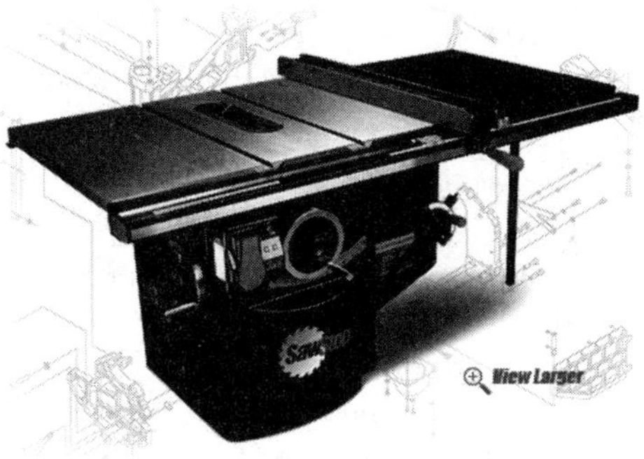

The SawStop company is aware of fifty-two occurrences wherein someone inadvertently touched the saw and was not injured, as their system worked flawlessly. With the order for the replacement brake, each saw owner stated in their letter how impressed they were by the SawStop's ability to prevent injury.

History of the Hazard

Since the invention of electric power, table saws, cut-off saws, and planers have been commercially available. Conventional guards are often considered hampering to precision woodwork. It is still common to see a lack of saw guards in many cabinet shops that cater to specialty work.

1844: The English Factory Act of 1844 was enacted to protect workers, especially those who worked in mills and factories. The first laws did not specifically define what parts of a machine were dangerous; therefore, machinery was used in an unguarded state until enough injuries occurred to draw attention to the dangerous parts.

1868: The first U.S. patent granted for a machine guard was an interlocking guard for a beverage bottling machine. Verne L. Roberts, in his book *Machine Guarding*, published by the Institute for Product Safety, 1980, notes that "the concept of guarding did not begin in 1868, for the patent acknowledges prior art in terms of the use of a guard . . . [however] the acceptance by designers of the need to guard has lagged behind the state-of-the-art from the beginning, and unfortunately in many instances it continues to this day." The machine designer "should consider the guard an integral part of his design, rather than an add-on . . . required by standards, law or litigation."

1878: The first English law on machine guards, the Factory and Workshop Act, 1878, 41 & 42 VICT. CH 16, May 27, 1878, required fencing (i.e., guarding) for hoists, flywheels, wheel races, and mill gearing. It defined mill gearing as "every shaft, whether upright, oblique or horizontal, and every wheel, drum or pulley by which the motion of the first moving power is communicated to any machine appertaining to a manufacturing process."

1895: Belt guards became mandatory.

1901: The English Factory and Workshop Act, 1901.1 EDW.7 CH22, required guarding for machines that traverse and might pin someone against a wall, and placed restrictions on cleaning a machine while it was in motion.

1921: ANSI published the Safety Standard for Mechanical Power Transmission.

1935: Walter S. Gundeck, a Studebaker department manager, in an article "Let the Builder Guard the Machine," *National Safety News*, Vol. 40, No.1 (Jan.1935): 35–36, 2pp. PSN Doc. No. 0002468, makes a firm case for the manufacturer's responsibility to design guards into the machinery product before sale, rather than leaving it to the buyer to add an improvised guard. If the machine manufacturer includes its own guards as part of the machinery design, then "the safety devices are more likely to function correctly and without interference to the productive capacity of the machine."

Post-WWII: System safety has become a well-recognized engineering discipline that eliminates hazards at the time of design, preventing hazards before they can cause injury.

1971: OSHA was enacted, which oversaw industry standards for machine guarding.

1990s: An ingenious inventor developed a "Saw Stop" that works like a touch lamp switch. If someone touches the moving saw, it stops in one-quarter of a turn and causes no injury.

Contributory Factors

It is foreseeable that a cabinetmaker, experienced or inexperienced, will remove a table-saw guard to ensure precision cuts. Human-factors research indicates that it is impossible to concentrate on two tasks at once. Performing the task of cutting material with a table saw requires the attention to be on the cut, not the constant location of the worker's hands or fingers. Novice workers sometimes become so engrossed with the shaping of the sawed material that they do not remember the proper placement of their hands or fingers. Experienced workers can sustain injuries when they become complacent about their usage of the saw.

Hazard Identification and Prevention Matrix

	Eliminate the Hazard		Guard the Hazard		Provide a Safety Factor		Provide Redundancy		Provide Reliability
	Hazard	Safety	Hazard	Safety	Hazard	Safety	Hazard	Safety	
Natural									
Structural/ Mechanical	Moving parts			Guards				Saw blade drops away	*1
Electrical				Capacitance sensor stop switch				Interlock stop	*2
Chemical									
Radiant Energy									
Biological									
Automated Systems									

*1 and 2 = 100%

Applicable Standards and Literature

There are no standards addressing the new concept of electronic guarding; however, the existing requirements for conventional guarding are lengthy.

OSHA

OSHA 1910, Subpart O, (.147, .212, .217, .219) **OSHA 1910,**

Subpart P, H, Portable Powered Tools and other Hand Held Equipment.

(a) OSHA Fact Sheet #91-36; New OSHA Civil Penalties Policy; U.S. Labor Dept.

(b) OSHA Bulletin #3067; Concepts & Techniques of Machine Guarding, 1981.

(c) *The Principles and Techniques of Mechanical Guarding*, OSHA Booklet 2057.

(d) *Concepts and Techniques of Machine Guarding*, OSHA Booklet No. 3067, 1980.

OSHA Subpart N, 1926.550(a)(9) requires "Accessible areas within the swing radius of the rear of the rotating superstructure of the crane, either permanently or temporarily mounted, shall be barricaded in such a manner as to prevent an employee from being struck by the crane."

OSHA 1910.180(i)(6) requires that locomotive cranes should not be swung when the possibility exists that they might strike other railroad cars on adjacent tracks and defines the need to barricade the swing area effectively to avoid striking people with the crane cab and counterweight.

MSHA

MSHA 57.14112 Construction and Maintenance of Guards.

MSHA's Guide to Equipment Guarding, published by U.S. Department of Labor Mine Safety and Health Administration.

NIOSH

Safe Grain and Silage Handling, 95-109.

Health and Safety Guide for Sawmills and Planing Mills, 78-102.

Health and Safety Guide for Construction and Related Machinery Manufacturers, 78-103.

Health and Safety Guide for Textile Machinery Manufacturers, 78-146.

Health and Safety Guide for Manufacturers of Industrial Trucks and Tractors, 78-147.

Machine Guarding-Assessment of Need, U.S. Department of Health, Education and Welfare, June 1975.

U.S. Department of Labor

Bureau of Labor Standards; No. 463: *Safety Code for Mechanical Power Transmission Apparatus*

ANSI

ANSI B11 Series

Since 1927, ANSI's, Safety Standard for Mechanical Power Transmission Apparatus B15.1-1996, is the primary guide for protecting people from motion hazards. The intent of this standard is to prevent bodily injury due to contact with moving parts by anyone walking into, reaching over, falling, slipping, tripping, or being struck by flying objects set in motion by moving parts or the result of breakage of any component. The standard is the basic guideline applicable as necessary to all other ANSI safety standards and machines not specifically covered by ANSI standards.

Comment: This absence of references to guarding in other ANSI Standards should not be misinterpreted to infer that guarding is not necessary.

U.S. Army Corps of Engineers

Safety and Health Requirements Manual, EM 385-1-1 (U.S. Army Corps of Engineers, November 2003) Section 16.B.03 outlines requirements for guarding machine parts, hot surfaces, equipment having charging, charging skip, overhead protection and guard rails, handholds, etc. . . Section 16.C.08.b. and c. require the following clearances:

(a) Adequate clearance shall be maintained between moving and rotating structures of the crane and fixed objects to allow the passage of employees without harm: the minimum adequate clearance is 16 inches.

(b) Accessible areas within the swing radius of the rear of the rotating superstructure of a crane, either permanently or temporarily mounted, shall be barricaded to prevent an employee from being struck or crushed by the crane.

National Safety Council

Accident Prevention Manual for Industrial Operation, 9th Edition, Engineering and Technology, Chapter 8, pp. 197–215, 1991.

Data Sheets #211, 237, 534, 603, 624, & 637, 1983–1986.

"Guarding and Shielding Farm Equipment," *Rural Accident Prevention Bulletin*, published by the Farm Department, which outlines safe practices, types of accidents, and safeguarding procedures for farm equipment.

Power Press Safety Manual, 1974, 3rd Edition.

Other References

- Allan St. John Holt and Hugh Andrews, *Principles of Health and Safety at Work*, IOSH Publishing, Ltd., pp. 2–5.
- Marine Terminal Standard 1917. 45(i)(2) and Longshoring Standard 1918.74(a)(10) and 1918.74(c) address guarding of the swing radius.
- *Machine guarding, A Historical Perspective*, by Verne L. Roberts, published by the Institute for Product Safety (1980). This gives an overview of many varied patents to assure for the guarding of a variety of machines with moving parts.
- *Agricultural Machinery Safety*, Fundamentals of Machine Operation, Chapter 3 "Recognizing Common Machine Hazards," published by John Deere Service Publications, Dept. F., John Deere Road, Moline, IL 61265, 1974.

Commentary

Innovation in inherently safer design requires thinking outside the box, rather than simply following the rules. The SawStop safety brake system detects unintentional contact of the hand and the saw blade. The device applies the same electrical concept that is used for "touch" electric lamp switches. The system detects the electrical capacitance charge of the human body and stops the saw in 5 milliseconds, 25 times faster than people can react. This piece of electronics is an excellent example of the marriage of technology and innovation. It is a product that effectively saves fingers by replacing outdated guards with the incorporation of available technology.

Our current culture of behavior-based safety, which forces the use of antiquated guarding and relies upon voluntary use, leaves little incentive to develop a more user-friendly design feature to prevent injury. Administrators of effective safety programs are the individuals who can foster the use of new technological answers to prevent injuries by adopting their use whenever applicable and, where successful, recommend their use to create a demand for the product. Major saw manufacturers had the opportunity to be licensed for manufacturing the SawStop, but were unable to grasp the new concept of using electronics to stop the saw before it causes injury. The typical response of "safety doesn't sell" is being challenged by the rapidly accelerating sales of the SawStop.

Recommendations

SawStop is an excellent example of how inherently safe design can be achieved by application of available technology. This concept of using a capacitance switch to guard against moving parts can be applied to other circumstances as well. To overcome manufacturers' knee-jerk response that safety is the user's personal responsibility, and that crudely constructed guards and warning labels are sufficient, the safety culture must recognize the value of an innovative, inherently safer appliance and how it increases the worth of a product. For further information regarding the SawStop contact.

Steven F. Gass, President
22409 SW Newland Rd.
Wilsonville, OR 97070
(503) 638-6201
fax: (503) 638-8601
e-mail: steve@sawstop.com
www.sawstop.com

SECTION 6

Noisy Equipment and Machines

The Hazard Syndrome

Low-level noise of less than 20 decibels is a *dormant* hazard. It cannot cause distraction or injury. As the sound level increases it becomes *armed.* It also becomes an annoyance, or if unusual, can cause distraction. Under these circumstances it can be considered armed. Excessive noise is often part of the construction site. Noise levels from activities such as rock drilling or machinery such as compressors, trucks, and many others can cause noise in excess of 80 decibels. A noisy machine presents an *active* hazard when people are nearby. In other words, a noisy machine is always in the active mode when a person must endure excessive noise levels.

"According to one expert, the sound level of society in general has been increasing at the rate of one decibel a year. Dr. Vern Knudsen, a pioneer in acoustics, recently stated that 'the loudest noises to which we are exposed have increased some 20 decibels in the past 20 years, and if this rate of increase continues for another 20 years, they will become lethal.' "[4]

---

[4]Rita Braver, "*Noise = Threshold of Danger,*" *Noise, The Environmental Problem: A Guide to OSHA Standards,* a U.S. Dept. of Labor OSHA Publication, July-August 1972.

Scope

In practice, a noise seldom comprises a single frequency; it is usually the combination of a number of frequencies commonly classified as wide-band (incorporating a wide range of frequencies) or narrow-band (incorporating only a few frequencies). The frequency composition of the noise is known as the spectrum. Noise may also be described by its occurrence in time as steady-state (i.e., continuous and unvarying); impact (single impulses of short duration at regular or irregular intervals); or intermittent.

Environmental noise is noise of measurable intensity, such as is generally encountered in workplaces. Although technology and available soundproofing techniques have reduced noise levels in many of our confrontations with everyday sounds, such as automobile, truck, and aircraft engines, excessive noise, nevertheless, still presents itself at the workplace through the use of older machinery in construction and mining (drilling equipment, rock crushers, power saws, air compressors, air hammers), manufacturing (power presses, conveyor systems, forklifts, etc.), and agriculture (tractors, chainsaws, pumps, etc.). All these noises can be prevented or guarded against with the right protection and technology.

Backup alarms on vehicles and machines engaged in reversing produce a useful noise employed to provide hazard awareness. These devices are tuned to specific frequencies to pierce ambient sound levels, and to be distinctively heard at harmless levels behind the backing vehicle or machine. Some people find them an annoyance, but even the continuous sounding reverse-signal backup alarms can be overcome with technological innovations such as the infrared sensor, which can sense persons in or approaching the "danger zone," and cause the alarm to sound only when someone is detected.

Roughly ten million American workers are exposed to what OSHA considers excessive noise levels on the job (see Table 9-1). Unwanted noise often masks speech, unless appropriate design and operational procedures are adopted. Studies conducted by NIOSH indicate that 25% of persons 55 or older who have been exposed to occupational noise with an average level of about 85 dBA during their careers have developed some hearing impairment. Manufacturing and transportation employees constitute the largest at-risk groups, followed by military personnel and employees in construction, mining, and agriculture.

The Farm Credit Services issued a Health Fact (#11) stating that "Farmers and ranchers have increased levels of hearing loss when compared to other occupations. Hearing loss is prevalent among farmers who are exposed to loud and continuous noise from machinery and equipment which have not been properly maintained or when hearing protection is ignored." According to otologist Dr. J.B. Gregg, it takes about five to ten years of exposure to farm tractor noise to produce evidence of inner ear damage. Often boys begin operating farm equipment at 9-years-old or younger.

Table 9-1 identifies various industrial noise levels, while Table 9-2 lists the typical sound level that causes hearing damage.

Table 9–1
Noise Levels in Industrial Environments.[5]

dB(A) Level	Environment
120–150	Aircraft jet engines
115–130	Rock drilling equipment
105–115	Heavy crawler tractors
100–115	Concrete mixers, blast furnaces, pile drivers
100–110	Diesel locomotives, large highway trucks, riveting, ship engine rooms
95–105	Outboard motors, car radios, and boom boxes
95–100	Printing press
90–100	Sheet metal shops, workshops
85–100	Power equipment
85–95	Small farm tractors
75–80	Computer room
Below 75	Library reading room

Table 9–2
Typical Sound Levels Hearing Damage.[6]

dBA	Decibel Sound Source	Hearing Damage
140	Shotgun blast, nearby jet taking off, 50hp siren at 100 feet	Nearly immediate damage
120	Jackhammer, chainsaw, thunder clap, crowd yelling in a stadium, rock concert music	Nearly immediate damage
115		Damage after 15 minutes
110		Damage after 30 minutes
105	Axle fan	Damage after 1 hour
102		Damage after 1½ hours
100	Farm machines, tractor w/o soundproof cab, ATV, snowmobile, motorcycle, subway car	Damage after 2 hours
97		Damage after 3 hours
95		Damage after 4 hours
92	Lawnmower	Damage after 6 hours
90	OSHA limit	Exposure to noise above 90 dBA hearing damage occurs

[5] Some of these sound level values appeared in Brian C. Myers, *"Sound Strategies," ASSE Professional Safety,* October 1994.

[6] " 'Sound off' down on the farm," *Ohio Monitor,* March/April 1997.

Peril from the Failure Mode

Noise is a *radiant* hazard that causes damage through waves. A common source of noise at levels that can cause damage is the typically *mechanical* hazard of vibration.

Continuous exposure to noise can result in severe ear damage, a condition that is irrevocable. Further, noise can mask communication and audible warning signals. Noise itself can be annoying and downright irritating, which is error provocative. Scientific studies have been conducted in an attempt to show that noise causes stress, high blood pressure, migraine headaches, constriction of the blood vessels, fatigue, sleeplessness, and stomach and intestinal ailments.

On many occasions, heavy equipment emitting considerable engine noise has caused serious injury and fatalities by backing or running over workers or nearby pedestrians who were oblivious to the moving equipment or could not hear the warning shouts. The pounding of pile-driving equipment, a street worker's jackhammer, a tree trimmer's chainsaw, a high-speed roller-conveyor system operating in a construction warehouse, loaders, and other equipment can far exceed OSHA noise limits. It is also interesting to note that loud music has conditioned many workers to loud noises that leads to the disregard of personal protective devices intended to protect them from hearing loss. On the other hand, one must also be aware that personal noise protection may protect the ears from damage, but may also suppress conversation, instructions, verbal shouts or warnings for one's safety.

Another recent danger has surfaced with stereo music headsets worn by a growing number of workers in boring or monotonous jobs. According to an article in NSC's *Safety & Health*, some workers wearing headphones are oblivious to sounds that are clues to impending danger and have walked into the paths of forklifts and stayed at workstations after fire alarms have sounded.

Noise has a profound effect on a worker's productivity. Hearing loss is generally a slow, degenerative process that diminishes the worker's ability to hear. By the time the condition is noticeable in most people, the damage cannot be stopped and will continue. Continual exposure to noise degrades the quality of the workplace and is a source of worker stress.

Brief Summary of the Occurrence

Excessive noise has always been a psychological irritant. The lighter side of a serious subject presents the hazard information in a more meaningful manner. The following account concerning events caused by excessive noise took place on a seagoing hopper dredge that was removing silt from a harbor channel. It is an account of an anonymous crewman who reported the occurrence in a local newspaper in the city where the dredge was working. Shipboard stress is a common part of a sailor's everyday life. Part of the problem can be attributed to the constant loud hum of 90 dB from propulsion motors vibrating through the vessel's hull 24 hours a day or a noisy auxiliary engine while underway.

Jeff was a crew member aboard a harbor and river dredge that steadily worked its way from the mouth of the harbor entrance up river. The ceaseless drone of engine noise from the dredge pump and onboard generators made the living conditions intolerable at times. Jeff developed a continual headache and was always complaining about the meals served in the galley, which always seemed to include beans in the main dish. The pressure mounted, and one night after a dinner heavily laden with bean sauce, Jeff went over the edge. Picking up a fire axe, he lunged at the captain, who was temporarily engrossed in stuffing a bean burrito into his mouth. Shouting at the top of his lungs, "Beans, beans, they're good for your heart, the more you eat, the more you . . ." he swung the axe, just missing the captain's head by inches. By this time, the rest of the crew joined in the melee and started throwing five-bean salad around the cabin. The captain, sensing something was amiss, quickly ran to the wheelhouse and turned off the power to the engines and generators. A strange, unfamiliar quiet descended on the dredge—a quiet that was unusual to all those aboard. Control and a sense of calm slowly came back to the crew. In time, these hardened sailors returned to the mess table and finished eating their final course, which consisted of the cook's favorite dessert . . . bean mousse! Jeff was finally at peace with himself. This story could have had tragic results, but quick thinking on the part of the captain prevented a rebellious mutiny and a messy ship, all because of the unnerving noise the engines made—or was it the beans?

Fortunately, the unwanted publicity reached the office of the agency overseeing the dredging process, and the safety branch, which had been requesting improvements for several years, had the funds to retain an engineering consultant to look into how to reduce vibratory noise. This noise often continually exceeds 90 to 100 dB on most parts of the ship. Design improvement of engine and pump mountings on the bulkheads and ceilings reduced the noise to an ambient level of 65 to 75 dB.

The bad PR did some real good, as funds were created for research and a sound-reduction design improvement. Priority for safety often occurs in strange ways. Needless to say, no more bean outbreaks occurred aboard this dredge.

Date of Occurrence

Mid-1950s

Source of Information

A local newspaper caused unwanted publicity and afforded funds to an engineering sound-reduction study and ship renovation to eliminate hazardous noise levels.

History of the Hazard

> "Noise is a common hazard in almost all industries. The fact that sound can be destructive was recognized centuries ago, and the fight against it is not new. An old story states that due to the noise it caused, Julius Caesar tried to have daytime chariot racing banned from the streets of ancient Rome. As early as 1700, a physician named Ramazzini described

> how persons hammering copper 'have their ears so injured by that perpetual din . . . that workers of this class become hard of hearing and, if they grow old at this work, completely deaf.' Blacksmiths, too, were identified in the last century as another occupational group likely to become deaf from their work."[7]

Ever since the invention of explosives, the military has been confronted with dangerous overpressures that our ears cannot withstand. Not only are our ears in jeopardy, but during the Vietnam conflict, it was found that aircraft pilots' radio microphones were subject to failure due to overpressures from enemy shelling or mortar fire. Pilots often left their headgear with attached earphones and mouthpiece microphones in the aircraft ready rooms next to the aircraft bunker shelters. Incoming enemy artillery created overpressures in these microphones that caused them to fail a few minutes after takeoff, leaving pilots without any means of communication. Subsequently, replacement microphones that could resist 11 lbs. of overpressure were issued after a safety engineering review. Unfortunately, medical science has not been able to overcome all types of hearing loss as easily as microphones can be redesigned.

1971: Occupational Safety and Health Act resulted in federal regulations for noise at employment locations. The May 29, 1971, issue of the Federal Register contains OSHA standards on noise.

1972: John Deere Co. introduced its modular, isolated "Sound Gard" cabs on its "Generation II" tractors.

1974: White Farm Equipment Co.'s 4-wheel-drive, articulated tractor (4-150 Field Boss) had a noise level of 78 dBA under full load with a cab blower at normal speed, and 81 dBA with the cab blower on high. The company made a quiet rollover protection structure (ROPS) standard equipment on the 4-150, eliminating the problem of design compromise it had when it offered both cab and noncab versions. The new cab had a tightly fitted and sealed instrument/control panel mounted on the tractor chassis, which lowered the inside noise level and prevented dust from entering the cab. It also provided the opportunity to install air conditioning. By making the equipment user-friendly through the reduction of these hazards, they sold more tractors. Also, farmers found they were more productive when enclosed in these new cabs.

1979: The Environmental Protection Agency (EPA) proposed a method of labeling hearing protectors that would enable users to determine the approximate noise reduction capability (attenuation). The resultant regulation required that all manufacturers of hearing protectors place a Noise Reduction Rating (NRR) label on their products.

Contributory Factors

Often, noise is not considered a work condition that can be rectified with engineering improvement. Noise itself is not often diminished in industrial processes; instead,

---

[7]Rita Braver, "Noise = Threshold of Danger," Noise, The Environmental Problem, A Guide to OSHA Standards, a U.S. Dept. of Labor OSHA Publication, July-August 1972).

it is dimmed with the use of personal protective equipment. This equipment is not always adequate for the external decibels it must protect against, and hearing loss can occur despite the careful use of hearing protection equipment. In addition to the damage caused by hearing loss, noise diminishes the ease of communication between workers, sometimes resulting in a lack of warning of a condition that is fatal. Noise can also cause stress that leads a worker to be more prone to error than that the individual would be in a quieter environment.

The issue of noise as a source of error-provocative actions is one of the least investigated human-factors aspects of hazardous noise conditions. Workplace injury and death rarely address the subject of noise as the cause of communication failure. Workplace studies on the effects of noise in terms of stress and other psychological effects in addition to hearing loss should be conducted.

Hazard Identification and Prevention Matrix

	Eliminate the Hazard		Guard the Hazard		Provide a Safety Factor		Provide Redundancy		Provide Reliability
	Hazard	Safety	Hazard	Safety	Hazard	Safety	Hazard	Safety	
Natural									
Structural/ Mechanical	Vibration			Insulation/ isolation muffler				Ear protection	50%
Electrical									
Chemical									
Radiant Energy	Noise/ hearing loss							Eliminate source of noise	100%
Biological									
Automated Systems									

Applicable Standards and Literature

OSHA

OSHA 29 CFR 1910.95 requires employers to protect workers from noise levels above 90 dBA averaged over eight hours, and to institute a training program for all employees exposed to noise at or above an 8-hour weighted average of 85 dBA.

OSHA 29 CFR 1910.95

Appendix A: Noise Exposure Computation

Appendix B: Methods for Estimating the Adequacy of Hearing Protector Attenuation

Appendix C: Audiometric Measuring Instruments

Appendix E: Acoustic Calibration of Audiometers

Appendix F: Calculations and Application of Age Corrections to Audiograms

Appendix G: Monitoring Noise Levels Non-Mandatory Informational Appendix

OSHA 29 CFR 1926.52 Occupational Noise Exposure

OSHA 29 CFR 1918.95 Longshoring Operations in the Vicinity of Repair and Maintenance Work. Subparagraph (a) states: " Longshoring operations shall not be carried on when chipping or scaling of decks, bulkheads or sides of vessels creates excessive noise which interferes with communication of warnings or instructions."[8]

National Institute of Safety and Health (NIOSH)

New criteria released August 1996 revises NIOSH's occupational-noise-exposure standard, keeping with its 85 dBA recommended 8 hours exposure limit, but parting with OSHA over how much to automatically lower a given hearing-protection device's noise-reduction rating to reflect actual workplace performance.

MSHA

30 CFR, **Subchapter N** Metal and Nonmetal Mine Safety and Health, Part 56, Surface Mines, 56.5050 Exposure limits for noise.

30 CFR, **Part 57** Safety and Health Standards, Underground Metal and Nonmetal Mines, 57.5050, Exposure limits for noise.

Department of Transportation

Federal Motor Carrier Safety Regulations, Title 49, Parts 393.94 Vehicle Interior Noise Levels. This standard allows up to 90 dBA for the inside of the truck cab at the driver's position.

Comment: This is an excessive truck-generated noise level for continuous exposure, and also isolates the driver from other warning sounds.

U.S. Army Corps of Engineers

EM 385-1-1 (November 3, 2003) Section 32, Noise Control, states that protection against the effects of noise exposure must be provided whenever the sound level exceeds 85 dB (A) continuous or 140 dB impulse, regardless of the duration of exposure.

NSC

The NSC recommends the noise level should be less than 85-dBA for eight hours of exposure.

---

[8]Rita Braver and Marcia J. Hovey, *Noise-The Environmental Problem - A Guide to OSHA Standards* (U.S. Department of Labor, OSHA 2067, reprint of two articles by from the July-August 1972 issue of *Safety Standards.*

Accident Prevention Manual for Industrial Operations, 9th edition, Chapter 17, National Safety Council.

Industrial Noise and Hearing Conservation, National Safety Council, 1975.

ANSI

ANSI Z24.22-1957 Method of Measurement of Real-Ear Attenuation of Ear Protectors

ANSI S12.3 (Rev. 1/1/90) Statistical Methods for Determining and Verifying Stated Noise—Emission Values of Machinery and Equipment

ANSI S12.16 (Rev. 1/1/92) Guidelines for the Specification of Noise of New Machinery

SAE

SAE J2102 (Rev. 6/26/91) Acoustics—Measurement of Airborne Noise Emitted by Earthmoving Machinery—Method for Determining Compliance with Limits for Exterior Noise-Stationary Test Condition.

International Organization for Standardization (ISO)

ISO 11690-1 (Rev. 11/1/96) Acoustics—Recommended Practice for the Design of Low-Noise Workplaces containing Machinery-Part 2: Noise Control Strategies

ISO 11690-2 (Rev. 11/1/96) Acoustics—Recommended Practice for the Design of Low-Noise Workplaces containing Machinery-Part 2: Noise Control Measures

ISO 11202, 11204 & 11201 (Rev. 12/15/95) Acoustics—Noise Emitted by Machinery and Equipment—Measurement of Emission Sound Pressure Levels at a Work Station and at other Specified Positions

ISO 11200 (Rev. 12/15/95) Acoustics—Noise Emitted by Machinery and Equipment-Guidelines for the use of Basic Standards for the Determination of Emission Sound Levels at a Work Station and other Specified Positions

ISO TR 11688-1 (Rev. 3/15/95) Acoustics—Recommended Practice for the Design of Low-Noise Machinery and Equipment-Part 1: Planning

ISO 6396 (Rev. 10/1/92) Acoustics—Measurement of Operator's Position of Noise Emitted by Earth-Moving Machinery—Dynamic Test Conditions

ISO 6395 (Rev. 9/1/88) Acoustics—Measurement of Exterior Noise Emitted by Earth-Moving Machinery—Dynamic Test Conditions

ISO 6081 (Rev. 12/15/86) Acoustics—Noise Emitted by Machinery and Equipment

ISO 5131 (Rev. 8/1/96) Acoustics—Tractors and Machinery for Agriculture and Forestry-Measurement of Noise at the Operator's Position-Survey Method

ISO 4871 (Rev. 8/1/84) Acoustics—Noise Labeling of Machinery & Equipment

ISO 7029 (Rev. 12/15/84) Acoustics—Threshold of Hearing by Air Condition

Other References

- Lynda A. Bennett and Edward Tessler, "Insurance Coverage for Noise-Induced Hearing Loss," *Occupational Health & Safety*, March 1997.
- Laura Avers, " Sound off' down on the Farm" *Ohio Monitor*, March/April 1997.

- Charles Agnoff, "New Challenges for Live Roller Conveyor Systems" *Occupational Health & Safety*, May 1996.
- Greg Smith, "Noise? What Noise?" *Occupational Health & Safety*, March 1996.
- Graydon Megan, "Sound Advice for a Quieter Workplace" *Safety & Health*, February 1996.
- Brian C. Myers, "Sound Strategies" *Professional Safety*, October 1994.
- Rob Crackel, "Noise Control and Hearing Protection" *Ohio Monitor*, January 1993.
- Rick Burson, "Nowhere to Hide" *Ohio Monitor*, September/October 1993.
- "Headphone Hazards," *Ohio Monitor*, May 1991.
- *Encyclopedia of Occupational Health and Safety,* International Labour Office, Geneva, 1983.
- Rob Crackel, "Danger: Excessive Noise May Be Hazardous" *Ohio Monitor*, October 1982.
- Harry C. Smith, "Hear today, gone tomorrow" *Ohio Monitor*, January 1981.
- Wesley E. Woodson, *Human Factors Design Handbook,* Chapter 3, Component and Product Design, "Auditory Displays" and Chapter 4, Human Factors Data, "Auditory Response," McGraw-Hill Company 1981.
- "You can lose your HEARING before you know it," *Ohio Monitor*, March 1979.
- *Human Engineering Guide to Equipment Design,* Chapter 4, "Auditory and Other Sensory Forms of Information Presentation," American Institutes for Research, Washington, DC, 1972.
- "Noise" *Hazard Information Newsletter*, Volume 2, Issue 5, August 1997.

Commentary

Management generally assumes that ear protection is sufficient and never really addresses alternate strategies; however, some trade associations have been exceedingly progressive. In the 1950s Dan Adair, safety director for a paper and pulp industry association, organized a very effective intercompany program that was years ahead of its time. He arranged for older employees who had already lost their hearing and who were assigned to newer or renovated plants designed with noise suppression equipment to have the opportunity to bid on promotion to any of the plants with high noise levels. This system gave younger paper mill workers intercompany opportunities for advancement into newer mills without harmful noise levels. Dan Adair's innovative program resulted in over four decades of paper mill workers in the Northwest able to avoid hearing loss, while enjoying opportunities to advance in their trade.

Noise control is not an exact science with a single answer. One noise consultant has devised the following series of ten steps for attacking a noise control problem:

1. Define the problem
2. Establish a noise-reduction target
3. Design noise control strategies
4. Determine how material handling and operating procedures are improved by the strategy

5. Repeat step 3 to minimize step 4
6. Estimate new reduction target
7. Assess technical feasibility
8. If necessary, construct a model or mock-up of the control
9. Assess economic feasibility
10. Implement design

Workplace injury and death investigations should always list whether noise could have been a contributing factor in order to document the severity of this hazard. The owners of the equipment can reduce noise levels at the time of design by insulating parts or providing an external muffler.

Recommendations

The best way to reduce noise and hearing loss is to design the noise away by engineering controls. Worksafe Western Australia (the worker safety bureau) has devised a very cost-effective way to reduce noise in the workplace. The following tips have been taken from their Worksafe and Health Solutions program "Buying Quiet:"

Engineering controls:

- Conduct engineering studies to determine the sources of excessive sound.
- Define the design feature that will reduce the sound levels:
  - Substitute a quiet process for a noisy one
  - Install mufflers or silencers on noisy parts
  - Install sound-absorbing materials on the walls and ceilings
  - Reduce engine vibration or machine noise-producing vibration with proper mountings
- Confine any noisy equipment within an acoustical enclosure.
- Confine the operator in an acoustical enclosure.

When ordering equipment, include specifications for acceptable interior or exterior noise levels. Most people would choose the quieter machine, but does the quieter machine cost more? Specify manufacturing tolerances that are tighter, gears that are made to mesh better, quieter cooling fans and a soundproof operator's cab with electronic communications. The question is, "Does the decibel reduction justify the extra cost?" If a machine that is 5 decibels quieter costs $100 more, the extra cost is negligible at $20 per dBA. If the equipment has a life span of 20 years, a dollar a decibel cost per year, or $5.00 per year for a quiet machine, can be considered cheap noise control. There are also some hidden benefits in a quieter machine:

- It is easier to hear warning sounds.
- It is less stressful and tiring to use.
- You can be less reliant on personal hearing protectors.

Before buying equipment, ask, "Is there a quieter way of doing the job?" For example, before buying a pneumatic impact wrench, consider the various hydraulic and torque-controlled units now available. While these units may cost more, they last longer and cause less damage to the nut, as well as lowering noise and hand/arm vibration levels. If you think the machine you are buying may be noisy, ask for noise level information from the supplier.

To make sense of noise level information from a supplier, the contractor must ask several questions:

- Is this information complete?
- Should I compare brands?
- Is there a quieter version of the same machine?
- What will this noise level mean in my workplace?

The answer to the last question will determine how much of a priority noise control should be. As specific tasks require greater concentration than others, noisy processes should be kept to a minimum on days when intricate tasks are being performed. Likewise, some tasks require more communication than others. Plan the noise level accordingly.

Although all of the above steps could prevent ear damage from excessive noise, as previously stated, one must realize that some hearing protection equipment may also prevent effective voice communication. To the crane or heavy-equipment operator, effective communication between workers and operators is mandatory for safe and efficient operation. The use of hand signals, radio communications, and video images must be considered when working in a noisy environment that prevents normal voice communication.

SECTION 7

Mold

The Hazard Syndrome

Black-colored toxic mold, illustrated in the next figure, has been in the news lately as a cause of litigation. Mold is the common name for fungi which appears on decaying materials. It is a large group of one-celled or multicelled organisms that can be partially compared to algae. The spores of fungi are carried long distances in the atmosphere, then collect and multiply in wet, dark locations, often where there is little air circulation. Types of mold can be found inside the walls of almost every building over ten-years-old, especially in geographic areas with a high annual rainfall or a high relative humidity.

Exposure to excessive fungi, especially in an enclosed space, can lead to myriad health problems with a wide range of symptoms. A broad term used for toxic molds is *stachybotrys*. Locations that can produce molds present a *dormant* hazard. The presence of mold can mean an *armed* hazard that is *active* when it assaults a person's health with an infestation in the lungs.

Scope

In the past ten years, over 9,000 toxic-mold lawsuits were filed in the United States and Canada, affecting building owners, property managers, general contractors, architects, and insurance companies. According to a September 2002 article in *Professional Roofing,*[5] "Liability for mold growth in buildings is currently the most serious environmental concern facing the construction industry." A number of building tenants have been affected by indoor collections of mold that manifest in allergies, coughs, lung infections, neurological conditions, and tissue damage.

Peril Presented by the Hazard

A building with harmful mold levels becomes a devastating experience to both the injured and the builder. The failure mode is both *structural/mechanical* and *biological.*

The health risks caused by mold can be placed into three categories: allergy, toxicity, and infection. Though the Center for Disease Control (CDC) has identified over 3,000 molds, only about 25 of them are toxic. The toxicity is caused mainly by a group of chemicals

[5] Stephen M. Philips, "Uncovering Mold", *Professional Roofing,* September 2002.

called *mycotoxins,* known for both their carcinogenic properties and their ability to cause damage to internal organs such as the liver or kidneys.

Molds can cause irrevocable ill-health conditions in humans, including fibromyalgia, respiratory problems, and damage to renal function. Other symptoms caused by mold infestation are fatigue, skin rashes, chest tightness, tremors, heart palpitations, and gastrointestinal problems. Mold has also been linked to the tripling of the national asthma rate over the past twenty years.

Infants and children exposed to mold can contract chronic respiratory ailments, including lung hemorrhages. In some cases, as in the incident listed in this section, the exposure to mold causes damage or detriment to neurological development, leading to lifelong brain damage. Exposure to mold decreases immune function, making a person more susceptible to conditions that would not otherwise be a threat.

Toxic accumulation of mold is most commonly a result of poor construction and persistent water leaks. Old or faulty pipe systems within the walls of a building, and air-conditioning vents that seep fluid are two of many ways that structural components can be exposed to constant moisture. Porous materials and improper insulation also contribute to the initiation and accumulation of moisture. The most well-known toxic mold, *stachybotrys chartarum* (often called "black mold"), needs an ongoing water source or very high humidity to exist. Many buildings, through leaks, improper design, or faulty building materials, supply enough water to allow this danger to flourish. The image below shows mold growing on a damp pipe.

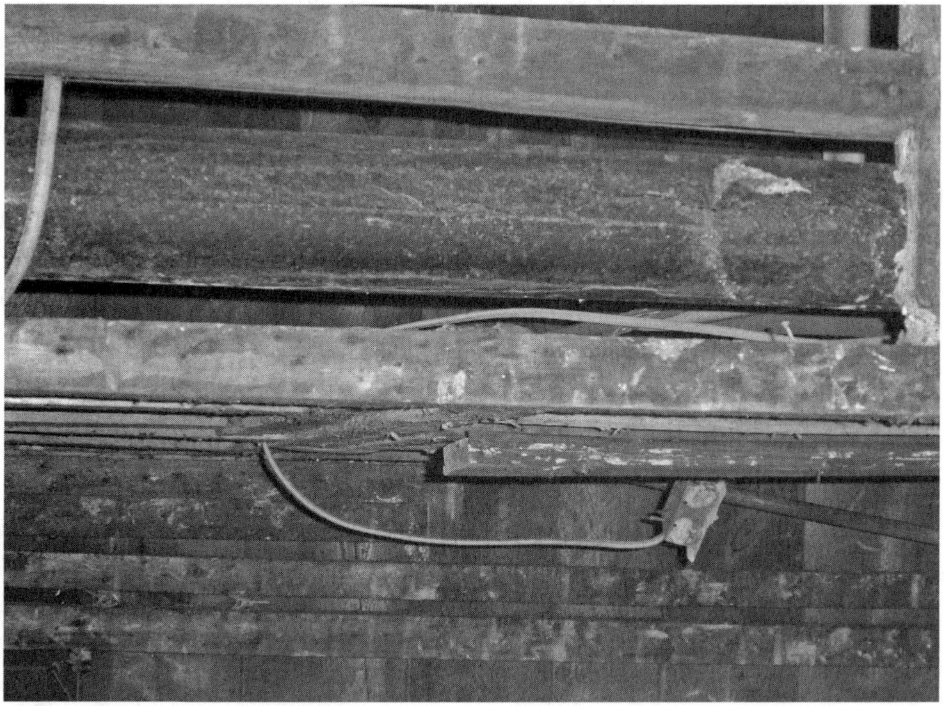

Brief Summary of the Occurrence

A family claimed that toxic mold in their home was the cause of brain damage to their infant. Moldy lumber was installed in the home, and dampness allowed the mold to multiply and spread its spores into the living quarters. This case resulted in a $22.6 million verdict against seventeen defendants. The child will need 24-hour care for the rest of its life.

Date of Occurrence

1999

Source of Information

Arizona Daily Star, Saturday, November 5, 2005

History of the Hazard

Fungus is one of the earliest forms of life; it is neither animal nor vegetable. The hazard of mold and mildew dates back through over 4,000 years of human history. There is evidence that the former USSR used aflatoxins, a type of mycotoxin, in their biological weaponry arsenal. As the demand for energy efficiency increases, there is a greater demand for airtight buildings. The combination of reduced airflow and certain building materials has increased the level of mold exposure in the past twenty-five years.

Contributory Factors

The temperature for ideal mold growth ranges between 40°F and 100°F. Extremes of temperature outside this range are rarely found in insulated buildings, allowing mold an easy living space. Modern efficiency desires have caused some buildings to be built airtight, allowing little to no ventilation inside and between walls. These factors create a climate hospitable to mold. In addition, high-cellulose materials commonly hold water when they are exposed to moisture through leaks or high humidity. Drywall or gypsum board, fiberboard, ceiling tiles, and most wooden building materials are at risk for containing mold colonies when they are exposed to moisture.

Once a mold spore has found a suitable environment, it can reproduce with astonishing speed, often creating a firm and potentially toxic toehold within forty-eight hours.

Applicable Standards

OSHA and the U.S. Army Corps of Engineers have not mentioned mold.

Other References

- Nicholas P. Money, *Carpet Monsters and Killer Spores: A Natural History of Toxic Mold*, Oxford University Press, 2004. (**Note:** This book goes into great detail on the history and the effects of mold.)
- Mold Across America: http://www.moldacrossamerica.org/toc.htm
- Stephen M. Philips, "Uncovering Mold," *Professional Roofing*, September 2002.

Hazard Identification and Prevention Matrix

	Eliminate the Hazard		Guard the Hazard		Provide a Safety Factor		Provide Redundancy		Provide Reliability
	Hazard	Safety	Hazard	Safety	Hazard	Safety	Hazard	Safety	
Natural									
Structural/ Mechanical	Source of moisture in an unvented area	Use moisture-proof/ resistant building materials		Provide adequate ventilation				Fulfill proper mainten-ance schedule	99%
Electrical									
Chemical									
Radiant Energy									
Biological	Mold spores	Check for mold on construc-tion materials before building							99%
Automated Systems									

Commentary

Toxic mold is both a health threat and an excess cost to builders. Building owners, architects, and design engineers are increasingly responsible for designing systems and buildings that can stay relatively mold-free. Leaks in water systems, installation of poor or inadequate ventilation, or use of building materials corrupted by moisture have all been sources of mold-based litigation. Since mold does not always immediately appear in a building or enclosed space, this hazard is retroactive. It is important to keep careful records and perhaps perform mold tests at the time of construction. If mold later develops in a building, it is important to be able to show that the construction was carried out in a way as to ensure the absence of mold or a mold-hospitable climate.

Recommendations

Industry standards should be developed for the treatment of lumber and other building materials. Materials should be stored in a clean, dry environment with minimal exposure to outside elements.

Ventilation systems should be checked both for potential leaks and adequate air flow throughout the construction process. A maintenance and inspection schedule should be developed and enforced throughout the life of the building.

SECTION 8

Dust

The Hazard Syndrome

Dust can be described as suspended particles of solid matter in such a fine state of subdivision that they may be inhaled, swallowed, or absorbed into the body. Dangerous airborne dusts are considered to be an *active* hazard, as health dangers are cumulative and long term. Dusts do not diffuse in the air but settle under the influence of gravity. A surface where dust has collected is a *dormant* hazard. Activity that makes the dust airborne *arms* the hazard. The major hazard to workers involved in industries that produce such dust particles is the exposure to specific illnesses caused by dust. Such ailments include silicosis in the case of quartz dust, asbestosis in the case of asbestos dust, black lung in the case of coal dust, and allergies in the case of allergenic dusts or pollen. Carcinogenic dusts arising from either chemical or biological origins can provide a serious long-term threat to health. Dust can also be a tremendous irritant to the eyes. Certain dusts like coal, grain, or other carbonate substances can be very explosive by nature, and accumulation of these dusts in confined spaces can be very dangerous. Additionally, when it becomes wet, dust accumulated on surfaces can become slippery and contribute to fall hazards.

Scope

Dust is a hazard prevalent to many construction activities. Underlying the potential dust from cutting, sawing, boring, and other industrial processes that can take place on-site, the site itself can present a dust hazard. Dirt and dust that form as a result of earth being moved do not sound particularly life threatening, but they have the potential to be. Homebuilders in northern California recently came under legal restrictions when their building sites were discovered to be releasing dormant asbestos dust back into the air and polluting neighborhoods and schools. Depending on what was there before construction began, the soil on a site can range from innocuous to deadly.

Peril from the Failure Mode

One of the dangers of dust is its constant presence. Constant exposure to airborne particulates is an irritant at best, a killer at worst. Literally thousands have died of pulmonary

and respiratory diseases as a result of the effects of silicosis, asbestosis, black lung, pollen, or other dusts. Because of their irritating qualities, some dusts can also cause either an allergic reaction or inflammation when they come in contact with the skin. The particular dust from sanding laminated wood used in molding development was found to cause irreversible stomach cancer for persons engaged in this craft for approximately twenty years. Likewise, explosions from combustible dust have been the source of numerous, well-publicized catastrophes in grain elevators, coal mines, and many other places in which dust is present. Sometimes subtle warnings go unnoticed and allow these particles to enter and damage workers' bodies.

Silica ranks as the second most common mineral in the earth's crust. The National Industrial Sand Association (NISA) reports that its silica-producing members employ approximately 3000 workers. This makes silicosis a risk in mining, sandblasting, rock drilling, masonry, foundry work, ceramics, glass manufacturing, stone cutting, shipbuilding, farming, and myriad other activities. NIOSH estimates that more than a million workers encounter silica in the workplace; other government estimates run as high as three million. Silicosis kills an estimated 200 to 250 people a year.

Many workers and employers do not realize that dust containing silica is dangerous to health on a continuous basis, as it can cause permanent lung damage and even death. Silicosis occurs when microscopic silica particles lodge deep in the lungs, forming nodules. The diseased lungs have trouble fighting infections such as bronchitis and tuberculosis. Chronic silicosis, the most common form of the disease, usually occurs ten or more years after exposure. Workers may have no symptoms at first, except a nagging cough. Heavier exposure can result in accelerating the silicosis during the first five to ten years. Where silica exposure is exceptionally heavy, acute silicosis can occur in mere weeks or months. The disease is progressive, disabling, and often fatal. Pneumoconiosis and silicosis continue to be very serious threats to hard-rock miners and tunnelers. Today great progress has been achieved in controlling exposure to asbestos that causes asbestosis, and coal dust that causes black lung disease. In the past, these two irritants have had similar death rates.

The woodworking industry also has a serious dust inhalation problem, especially in small specialized activities. Sawdust and sanding in a woodworking facility increase the chance of serious lung problems for workers where there is no effective dust-removal system, or where they do not wear personal protective clothing and equipment. Some wood dusts, such as walnut dust, have been linked to leukemia. Toxic dusts should always be announced by prominently displayed warning signs, like the one on the opposite page.

Workers in the agriculture and shipping industries are exposed to the hazard of grain dust, especially in silos, bins, barges, ship storage facilities, and other confined spaces. Grain dust is explosive. One-eighth of an inch of dust accumulation is more than enough to cause an explosion.

Nontoxic dust is a *natural* hazard that becomes harmful in large quantities. Toxic dust is a *chemical* hazard that causes severe harm when ingested, especially over a period of years.

Brief Summary of the Occurrence

A pattern-and-die worker ingested toxic dust in the workplace over a period of twenty years. He was exposed to laminated-wood particles and dust at an auto manufacturer. The manufacturer did not adequately warn him of the dangers of dust or of preventive measures. The toxins caused irreversible colon cancer that led to his death. Exposure to this hazard was quite innocent, as workers would eat their lunches in the shop where they worked. Dust in the air and on their hands was transferred to their food and ingested. Unfortunately, when the first case was diagnosed, it was determined that all the car manufacturers were using the same laminated wood for a mold, as it would not shrink or swell. Both the glue manufacturer and the laminated-wood manufacturer failed to warn of the toxic nature of this dried glue dust.

Date of Occurrence

1981

Source of Information

Claim filed in Detroit, MI

History of the Hazard

Early 1930s: Francis Perkins, President Franklin Roosevelt's labor secretary, launched a war against silicosis after about 700 workers died of the disease during and after excavating a hydroelectric-plant tunnel in West Virginia.

1960s and 1970s: According to NIOSH, silicosis deaths dropped, although the disease still claimed nearly 1200 lives in 1968.

1972: OSHA adopted its first standard on respirators.

1977: NISA developed its first occupational health program on silica.

1980s: Fatalities leveled off, but because there was no requirement to report silicosis cases, NIOSH believes the disease was underreported.

1994: MSHA issued a new standard cracking down on drill dust.

1995: NIOSH established new regulations for testing and certifying nonpowered air-purifying particulate-filter respirators.

1997: OSHA's Special Emphasis Program on silicosis completed 640 inspections nationwide, and inspectors found more than 1600 violations involving silica exposure.

1998: Under NIOSH's new regulations, manufacturers must produce and distribute only Part 84-certified respirator products.

Contributory Factors

Although some health professionals may think that crippling and sometimes fatal respiratory diseases are a thing of the past, the continued persistence of silicosis in the workplace is forcing them to think otherwise. Many workers and employers also continue to think that dusty conditions in the workplace are merely a part of the job, do not recognize the potential danger of continuous inhalation or ingestion of such particles, and do nothing to alleviate dusty situations.

The extent to which any type of dust represents a health risk depends on the exposure, the nature of the dust, its concentration, the duration of exposure, and the state of the worker's health. All these factors will play a part in the severity and onset of disease.

Hazard Identification and Prevention Matrix

	Eliminate the Hazard		Guard the Hazard		Provide a Safety Factor		Provide Redundancy		Provide Reliability
	Hazard	Safety	Hazard	Safety	Hazard	Safety	Hazard	Safety	
Natural	Particle irritants					Warning signs/ PPE			
Structural/ Mechanical				Wet guards/ "dust tree"					
Electrical									
Chemical	Toxic particles	Change formula				Air tests			100%
Radiant Energy									
Biological									
Artificial Intelligence									

Applicable Standards and Literature

OSHA

1910.19 Special provisions for air contaminants
Subpart B, Adoption and Extension of Established Federal Standards

1910.94 Ventilation

Subpart G, Occupational Health and Environmental Control

1910.134 Respiratory protection

Subpart I, Personal Protective Equipment

1910 Subpart I

App B, Non-mandatory Compliance Guidelines for Hazard Assessment and Personal Protective Equipment Selection

1910.146 Permit-required confined spaces

Subpart J, General Environmental Controls

1910.261 Pulp, paper, and paperboard mills
Subpart R, Special Industries

1910.265 Sawmills
Subpart R, Special Industries

1910.272 Grain handling facilities
Subpart R, Special Industries

1910.272 Grain handling facilities
Subpart R, Special Industries

1910.399 Definitions applicable to this subpart
Subpart S, Electrical-Definitions

Publication 3079 Respiratory Protection

NIOSH

42 CFR Part 84 Testing and Certification of Respiratory Protective Devices

Publication No. 96-101 NIOSH *Guide to the Selection and Use of Particulate Respirators Certified Under 42 CFR 84*

NIOSH 0500 Nuisance Dust, Total

NIOSH 0600 Nuisance Dust, Respirable

MSHA

CFR Title 30

ANSI

ANSI Z12.16 Prevention of Dust Explosions

ANSI Z12.3 Dust Explosions in Feed Mills

ANSI Z12.4 Dust Explosions in Grain Elevators

ANSI Z33.1 Blower & Exhaust Systems for Dust, Stock & Vapor Removal or Conveying

ANSI Z88.2 Respiratory Protection

ASTM

D 4532 Respirable Dust in Workplace Atmospheres

D 4930 Dust Control Material on Calcined Petroleum Coke

D 520 Zinc Dust Pigment

E 1132 Health Requirements Relating to Occupational Exposure to Quartz Dust

Note: Numerous other ASTM standards and directives are available concerning test methods, sampling, and evaluation of dust

U.S. Army Corps of Engineers

EM 385-1-1 (3 November 2003) *Safety & Health Requirements Manual*

Section 16 Machinery and Mechanized Equipment

16.M Drilling Equipment

16.N.09 Equipment Operation

Section 21 Safe Access and Fall Protection

21.I Access and Haul Roads

National Fire Prevention Association (NFPA)

NFPA 61 Prevention of Fires and Dust Explosions in Agricultural and Food Products Facilities

NFPA 61B Prevention of Fire/Dust Explosions in Grain Elevators and Facilities Handling Bulk Raw Agricultural Commodities

NFPA 61C Prevention of Fire/Dust Explosions in Feed Mills

NFPA 653 Prevention of Dust Explosions, Coal Preparation

NFPA 654 Prevention of Fire and Dust Explosions in the Chemical, Dye, Pharmaceutical, and Plastics Industries

Other References

- Andrew Schneider, and David, McComber, *An Air that Kills*, GP, Putnam's Sons, 2004.
- *Encyclopaedia of Occupational Health & Safety,-* Volume 1, International Labour Office, Geneva, Switzerland, 1983.
- *The Dictionary of Terms Used in the Safety Profession,* Third Edition, American Society of Safety Engineers, 1988.
- "Choose the Right Respirator," *Safety & Health*, February 1995.
- "Dust," *Occupational Health & Safety*, August 1995.
- "Dust/Mist Masks," *Occupational Health & Safety*, November 1995.
- Philip Mann, "Dry Particulate Protection: New technologies Mean More Choices," *Occupational Health & Safety*, December 1995.
- William Newcomb, "Fit Testing Under the New 42 CFR Part 84," *Occupational Health & Safety*, February 1997.
- Craig Colton, "Workplace Performance Studies of Half-Mask Respirators," *Occupational Health & Safety*, February 1997.
- Robert J. Sheppard, "It's Time to Change to 42 CFR Part 84 . . . And it's Easier Than You Think," *Occupational Health & Safety*, May 1997.
- Alan Johnston, "Guiding the Transition to Part 84," *Occupational Health & Safety*, May 1997.
- Zane N. Frund Jr. and John W. Cobes III, "Protecting Against Multiple Hazards," *Occupational Health & Safety*, May 1997.
- Janet Willen, "Respirator Standard Spurs Safer Products, Lower Prices," *Safety & Health*, May 1997.
- "Say Hello to Your New Respirator," *Safety & Health*, May 1997.
- Jeffery Birkner, "Bringing Your Program Into 42CFR84 Compliance," Occupational Health & Safety, August 1997.
- Sandy Graham, "Dust War," *Safety & Health*, March 1998.
- Robert A. Gadon, "Interpreting 1910.134," *Occupational Health & Safety*, May 1998.
- Craig E. Colton, "The ABC's of 1910.134," *Occupational Health & Safety*, August 1998.

- "Anger Over Asbestos Contamination in a Wealthy Sacramento Suburb," Brian Melley, July 1, 2005.
- "Dust," *Hazard Information Newsletter*, Volume 3, Issue 9, December 1998.

Commentary

Dust control is a safety-design issue, in that dust can be reformulated (if toxic) or controlled with guards. For complete protection, a redesign of dust-producing machinery would eliminate the possibility of many toxic particulates from entering the air. Although mining, manufacturing, and construction will continue to expose workers to the hazard of dust inhalation, the exposure can be controlled, or at the least, those exposed can receive the maximum protection available. NIOSH continues to screen miners and construction workers, participates in outreach efforts, and helps disseminate educational materials on silicosis to workers and employers. MSHA also inspects mines on a regular basis to ascertain which have silica overexposure, and which are complying with drill-dust standards.

Recommendations

Long ago it was found that walnut dust in cabinetmaking facilities was carcinogenic. Many cabinetmakers suffered from cancer that led to leukemia. Likewise, there is a perception that many dusts are harmless, a mere nuisance, when in reality they could prove fatal. It is very important to know the quality and characteristics of any and all dust the worker may be exposed to throughout the life of the construction cycle. The best approach to protect worker health is to substitute a safer material. Develop a listing of alternative substances that are either harmless or nearly so. Next, elimination of the dust can be performed with a device such as an exclusive dust-free system that captures dust and exhales it into collection points, preventing it from becoming airborne. The next-best solution would be the development of dust-free-machines systems that rely on advanced air-filtration systems to catch the dust before it can be dispelled into the atmosphere.

Common controls to minimize dust include:

- Wetting rock while drilling
- Substituting other substances for sand in sandblasting
- Dampening a dusty area or using a vacuum to suck up dust
- Enclosing cabs on work vehicles to shut out dust
- Ventilating a dust-generating work area
- Changing dirty clothes and showering at the end of a shift

SECTION 9

Ventilation

The Hazard Syndrome

Dead air spaces can collect dust, fumes, and mold that can be explosive or toxic. Such accumulation of explosive dust or vapor presents an *armed* hazard that, when air is ignited or inhaled, becomes *active*. Construction of tunnels and underground facilities presents an ongoing challenge to remove toxic fumes. Ventilation is especially important in confined spaces, even if the fumes the worker is exposed to are not toxic. Dead air spaces create a hazard for the life of a facility that starts at the time of construction.

Scope

The need for ventilation goes far beyond dead air spaces in false ceilings. All enclosed areas in buildings, tanks, or underground spaces created by tunnel construction or mines can contain dangerous air if not properly ventilated.

Peril from the Failure Mode

All too often, exhaust fumes from outside engineering processes are easily sucked into the excavation. This type of air contamination represents a *mechanical* hazard, where dangerous air is not properly removed. However, the toxicity of the air is often the result of an underlying *chemical* hazard**.** The U.S. Chemical Study Investigation Board, an independent federal agency, estimates that nearly 200 dust fires and explosions resulting in approximately 100 fatalities and 600 injuries have occurred in the last twenty-five years.

Brief Summary of the Occurrence

In Kinston, SC a pharmaceutical plant exploded, killing six workers and injuring thirty-eight others, including two firefighters. Lack of ventilation had allowed several inches of dust to settle in the ceilings, creating a source of explosive fuel. The blast was felt twenty-five miles away and rained charred materials over a two-mile radius while it burned for two days.

Date of Occurrence

January 29, 2003

Source of Information

National Safety Council, *Safety and Health*, August, 2005

History of the Hazard

Tunnel and mining disasters continue to occur from the same lack of ventilation that led miners to carry canaries to test for methane, carbon dioxide, and other toxic gases deep inside mines. Additionally, in the 1970s a series of grain-elevator disasters were linked to the accumulation of thick layers of wheat that turned into explosive dust in the elevators and attributed to lack of ventilation.

Contributory Factors

The U.S. Chemical Safety and Hazard Investigation Board concluded that the two major contributors to many chemical explosions are the lack of engineering assessment to identify combustible dusts and fumes, and the failure to design adequate ventilation systems.

Hazard Identification and Prevention Matrix

	Eliminate the Hazard		Guard the Hazard		Provide a Safety Factor		Provide Redundancy		Provide Reliability
	Hazard	Safety	Hazard	Safety	Hazard	Safety	Hazard	Safety	
Natural	Dead air, dust (nontoxic)								
Structural/ Mechanical	No artificial circulation	Ventilation systems		No leakage of gas/dust					1*
Electrical						Monitoring system			2*
Chemical	Toxic fumes/ gases/ dusts	Substitute noncombus-tible materials							3*
Radiant Energy									
Biological									
Automated Systems									

*1, 2, & 3 combined can provide 95% reliability.

Applicable Standards and Literature

OSHA

1926.800 addresses air monitoring for underground work and section k(2) only requires a minimum of 200 cubic feet of fresh air for each employee underground.

U.S. Army Corps of Engineers

EM 385-1-1 (November 3, 2003)

0.6G Ventilation and Exhaust Systems:

0.6.G.01 Design

(a) Ventilation systems shall be designed to prevent dispersion into the air, or drawing through the work area, of dusts, fumes, mists, vapors, and gases in concentrations causing harmful exposure

(b) The design of proposed engineering controls shall be approved by the GDA before acquisition or installation of the equipment; design data and drawings shall accompany the request for approval.

(c) Airborne contaminants created by portable equipment (such as drills, saws, and grinding machines) in concentrations exceeding acceptable safe limits shall be effectively controlled at the source.

06.G.03 Duration of Operation

(a) Ventilation systems shall be operated continuously during operations when persons are exposed to airborne contaminants or explosive gases at or above acceptable safe limits as defined in 06.A.01 or as otherwise specified by this manual, referenced standards, or regulations.

(b) Ventilation systems shall remain in operation for a time after the work process or equipment has ceased to ensure the removal of any contaminants in suspension in or vaporizing into the air.

Commentary

The design of ventilation systems requires the analysis by an engineer who has specialized in the science of moving air in different circumstances. Explosion-proof air movers and the placement of detectors are vital considerations in the design of ventilating systems.

Recommendations

The design engineer and the construction manager should view clean air as a vital part of safe construction, which includes the need for ventilation for removing toxic and volatile fumes, even when painting the interior of a new or remodeled building. Cleaning air by mechanically removing dust, fumes, or vapors is a function wherein the engineer can, by design, ensure that safety is paramount to professional practice.

SECTION 10

Pinch Points

The Hazard Syndrome

A pinch point is any point at which it is possible to be caught between the moving parts of a machine, between moving and stationary parts of a machine, or between the material and the moving parts of a machine (*ASSE Dictionary of Terms, 1971).* Many machines have a pinch point from an articulating boom or steering mechanism. Sometimes called a "shear point," the danger presented by a rotating crane cab and stationary object is very real. The danger sign depicted below shows that shear points and pinch points should never be taken lightly.

Without a guard, these locations are accessible and can catch someone, presenting an *armed* hazard that becomes *active* when someone is entrapped.

Scope

Pinch points are a very broad hazard. They range from doors, where the individual might catch a hand on the frame and sustain an injury when the door closes, to being crushed between the counterweight of a crane cab and a wall. These hazards are generally easily identified and can be measured to determine the clearance necessary to allow the entry of a finger, hand, or arm. In most cases this hazard falls into two categories: unpowered movement of a machine against a stationary object, and powered machine parts. Powered moving parts must be guarded, and mobile rotating machines should be barricaded, because the victim can be trapped.

Peril from the Failure Mode

Pinch point hazards range from a squeeze on a finger to the loss of life from fatal crushing. Pinch points are *mechanical/structural* hazards that are often triggered by the *natural* hazard of human error.

Brief Summary of the Occurrence

The deceased was crushed in a monstrous pinch point by a swinging backhoe boom when attempting to investigate why the boom would not swing to the left. Investigation revealed that the deceased was the victim of a crushing blow to his whole body in a pinch point between the raised upright backhoe boom and the rear of the backhoe. The drive chain was attached to the hydraulic cylinders on both sides of the boom: one to pull the boom to the left, and the other to pull it to the right. The full swing of the boom was 180 degrees. At the time the victim went to investigate the failure, the boom was pulled almost all the way to the left side. When he dismounted the backhoe and was standing on the ground, he could easily reach the control to swing the boom to the right. At that moment the boom trapped the victim and smashed him against the rear of the backhoe. It was found that that the chain controlling the rotation of the boom movement from left to right was broken, thus allowing unrestricted movement to the right. This chain was unguarded and susceptible to clogging with small rocks and debris.

Date of Occurrence

1977

Source of Information

Claim was settled before trial in Oklahoma City, OK

History of the Hazard

Unguarded belts, pulleys, meshing gears, rotating augers, and even doors and doorjambs have been known to provide pinch points since the beginning of the industrial age.

1844: The English Factory Act was enacted to protect factory and mill workers.

1868: First U.S. patent was granted for an interlocking guard. This guard would stop the machine when opened.

1878: Passing of the first English law that defined what machine parts had to be guarded.

1895: Belt and pulley systems had to be guarded.

1921: ANSI published Standard for Mechanical Power Transmission to require that moving parts be either inaccessible or guarded.

1971: Newly passed OSHA act encompassed all industry standards, including those on guarding.

Contributory Factors

Many hazards remain dormant, especially hazards as multifaceted as pinch points, because the two magic words are never asked: *what if?* In this set of circumstances, it is shown that if a guard fails and the controls are accessible, someone is innocently invited into a danger zone that is not readily apparent, and a dangerous pinch point can occur. Examination of *any* equipment should look to the consequence of any component failure.

Hazard Identification and Prevention Matrix

	Eliminate the Hazard		Guard the Hazard		Provide a Safety Factor		Provide Redundancy		Provide Reliability
	Hazard	Safety	Hazard	Safety	Hazard	Safety	Hazard	Safety	
Natural	Human error			Warning signs					
Structural/ Mechanical	Pinch point			Guard				Guard the control	100%*
Electrical									
Chemical									
Radiant Energy									
Biological									
Automated Systems									

*Prevention of chain failure with guarding is near perfect; along with redundant safeguards, such as a way to reach the control from the ground, would create a control that is 100% reliable.

Applicable Standards and Literature

OSHA

OSHA 1910, Subpart O, .147, .212, .217, .219

Subpart P- H and Portable Powered Tools and other Hand Held Equipment

OSHA Fact Sheet #91-36; New OSHA Civil Penalties Policy; U.S. Labor Dept.

OSHA Bulletin #3067; Concepts & Techniques of Machine Guarding, 1981.

The Principles and Techniques of Mechanical Guarding, OSHA Booklet 2057.
Concepts and Techniques of Machine Guarding, OSHA Booklet 3067, 1980.

OSHA Subpart N, 1926.550(a)(9), requires: "Accessible areas within the swing radius of the rear of the rotating superstructure of the crane, either permanently or temporarily mounted, shall be barricaded in such a manner as to prevent an employee from being struck by the crane."

OSHA 1910.180(i)(6) requires that locomotive cranes should not be swung when the possibility exists that they might strike other railroad cars on adjacent tracks and defines the need to barricade the swing area effectively to avoid striking people with the crane cab and counterweight.

ANSI

ANSI B11 Series

Since 1921 ANSI Safety Standard for Mechanical Power Transmission Apparatus B15.1-1996 is the primary guide for protecting people from motion hazards. This standard has been continually revised and strengthened. The intent of this standard is to prevent bodily injury due to contact with moving parts by anyone walking into, reaching over, falling, slipping, tripping, or to be struck by flying objects set in motion by moving parts or the result of breakage of any component. The standard is the basic guideline applicable as necessary to all other ANSI safety standards and machines not specifically covered by ANSI standards.

Comment: These standards are the same as those shown in Section 5 of this chapter; however, these can be also specifically applied to pinch point hazards.

ISO

ISO 9000-2 (1993E) Quality Management

4.4 Design Control

Commentary

The absence of pinch point guarding still exists, either when guards are not provided or have been discarded. In some instances, the designer has presumed that sufficient safety is provided if moving parts are located behind a panel door. Unguarded moving parts within the engine compartment, where things may be stored, are an invitation for entrapment by a pinch point.

Pinch points and shear points can be avoided by modifying the design of some equipment. The following illustration shows how a rotating crane cab can be raised to a safe height at an angle. With this design suggestion, the body of the cab that could contribute to a pinch point is raised to an elevation taller than most construction workers. The possibility of pinch point on that piece of equipment is then eliminated.

DESIGN CONCEPT TO PREVENT CRANE-CAB/CARRIER-FRAME PINCH POINT INJURIE

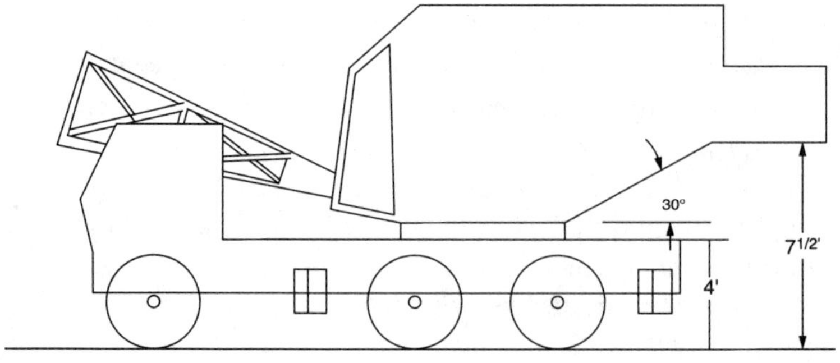

Recommendations

Designers need to provide guarding as standard equipment. Construction managers should barricade large rotating equipment to prevent access to inherent pinch points.

European workplace production machinery should bear the approval stamp of "CE," which means *compliance to European standards*. This standard should be extended to construction equipment in the United States. This is part of the International Standards Office (ISO) 12100-1 standard. This standard requires that the manufacturer certifies that the equipment complies with all of the design requirements including guarding. The United States has a voluntary program developed by Underwriters Laboratories (UL) that authorizes the display of the UL Seal of Approval if the manufacturer complies with the testing program. Adhering to such a standard would reduce the number of dangerous pinch points exposed at the job site.

SECTION 11

Compressed Gases

The Hazard Syndrome

Compressed-gas cylinders should never be transported in enclosed vans. This statement encompasses gases that support combustion (oxygen, chlorine), those that are flammable (methane, acetylene), and those that are inert (helium, fluorine, neon). Construction-service vans can travel into both business and residential areas on public roads, presenting a high hazard of explosion in a congested area. Welding gases such as oxygen transported in a closed van may appear to be *dormant,* but a very small leak *arms* the hazard. An oxygen-rich atmosphere with the presence of combustible materials will allow for an enhanced possibility of ignition, usually with explosive results. Combustion or explosion is an *active* hazard.

Scope

Compressed or liquefied gases are used extensively in production facilities, construction activities, maintenance tasks, transportation, and other conventional workplaces, as well as in many specialized support services and scientific endeavors. The hazards associated with the use of many gases can be very insidious, as gases are often invisible, silent, and give very few clues and little warning as to when a small leak causing a change of circumstance will result in disaster. Because they are so widely used, commercially available gases aren't thought of as being dangerous until someone is injured or dies. The dangers associated with their presence, particularly when they can accumulate in confined locations and create dangerous concentrations, are sometimes overlooked. Appropriate ventilation *must* be available when *any* compressed gas is stored or in use.

Peril from the Failure Mode

Leaking oxygen in a confined space can create a circumstance of explosive combustion. Leaking gases that are used as fuels, such as propane, acetylene, or hydrogen, are exceedingly explosive. These gases present a *chemical* hazard, for even naturally occurring gases such as methane will become dangerous if compressed and concentrated in an explosive mixture. Though explosion is the most common hazard, asphyxiation from inert gas is another frequent source of injury or death.

Brief Summary of the Occurrence

Oxygen gas does not need a total concentration in order to prompt an explosion. In Sierra Vista, Arizona, a 2.5 ton enclosed medical van blew up while making home deliveries of oxygen and other medical supplies to disabled people. It appears that a large oxygen leak occurred inside the van, either in the large liquid-oxygen tank with piping used to refill home-use oxygen cylinders, or in the other compressed-oxygen tanks. The oxygen-rich atmosphere was ignited by a source of unknown origin—perhaps by the rubbing of metal within the van causing friction or static, or possibly a small hole in the floor that allowed the oxygen to reach the hot exhaust pipe. Whatever the source, the cylinders of compressed and liquefied oxygen in the truck ruptured and exploded, demolishing two homes and damaging twenty-two others. Fortunately, the exploding cargo that sent numerous compressed-gas cylinders flying like unguided missiles seriously injured no one. It should be noted that commercial suppliers of tanks containing compressed industrial oxygen and other gases carry them in open-bed trucks where leaking gas cannot accumulate as in the case of this enclosed van. The following images are before and after pictures of the occurrence.

Date of Occurrence

March 5, 1997

Source of Information

The *Arizona Daily Star* in Tucson and *The Herald* in Sierra Vista documented the event. An out-of-court settlement was quickly reached.

History of the Hazard

December 19, 1910; New York City: A storage-battery house, where hydrocarbon gas and hydrogen were being compressed and stored in little cylinders used for lighting railroad cars and light buoys, exploded and wrecked several other buildings, killing ten persons and injuring more than a hundred others.

March 18, 1937; London, Texas: Surplus well gas from a nearby oil field was used to heat a school building. One wing of the schoolhouse exploded. It was believed that some of the rubber hoses carrying the gas to room heaters failed and leaked because of the moisture in the unrefined gas.

September 28, 1948; New York City, La Guardia Airport: An internal oxygen line in a passenger aircraft ruptured during preflight servicing, and an unknown source ignited the oxygen, destroying the plane.

January 24, 1956; London, England: The same type of occurrence as listed above took place on a Stratocruiser airliner.

December 1, 1962; Bombay, India: A similar occurrence on a Boeing 707.

November 16, 1963; Miami, Florida: A similar occurrence on a Boeing 727.

January 27, 1967; Cape Canaveral, FL: Three astronauts were incinerated in the *Apollo I* cockpit when a pure oxygen atmosphere ignited, burning the flammable material in the cockpit.

September 7, 1968; Rio de Janeiro, Brazil: A similar incident as in the 1948 LaGuardia example occurred on a Boeing 707.

January 25, 1969; Boeing Field; Seattle, WA: A similar occurrence on a Boeing 737.

June 21, 1970; Crescent City, Illinois: Sixteen railroad cars and ten of twelve 34,000-gallon liquid-propane tank cars exploded when the lead car was punctured. One of the cars exploded, broke in half, and traveled like a rocket for 800 feet. Sixteen businesses and twenty-three houses were destroyed, and eleven other buildings were damaged. This incdent was an early warning of the propensity of liquid-petroleum railroad tank cars to blow up. A system safety analysis of the cars should have been done and the design improvements should have evolved before the Kingman, Arizona occurrence.

December 31, 1970; Washington National Airport, Washington, DC: A similar occurrence again on a Boeing 737. *Most of the aircraft in these incidents were destroyed because pipes carrying oxygen were under pressure and fed the fire.*

July 5, 1973; Kingman, AZ: A leak in a fuel line transferring liquid propane from a railroad car to a storage tank was the source of a catastrophic explosion that killed eleven firefighters. This incident remains under study by fire departments and fire prevention specialists worldwide.

March 19, 1981; During the ground test of the Columbia shuttle, the use of pure nitrogen to prevent ignition from any type of spark provides an example of an excellent design feature for fuel tanks.

April, 1997; The discovery of banned oxygen generators on a Continental Airlines jet finally brought about installation of a much-needed backup safety system recommended in 1988 by the National Transportation Safety Board (NTSB).

As of May 1997, airlines have finally agreed to install fire-detecting and suppression systems in the cargo bays of all passenger aircraft.

Note: These historical occurrences were listed in *Fire in America* by Paul Lyons, published by NFPA.

Contributory Factors

The various disasters listed above illustrate the anti-safety mindset that ignores hazardous circumstances until a catastrophe occurs and becomes a widely publicized issue of public concern.

First, management often gives designers short-term priorities to complete the design without authority to pursue safer design alternatives that would reduce danger in the long run. Transporting compressed gases in a service van must entail a vehicle design accommodating increased airflow to minimize the presented hazard. Management that oversaw the design of the Apollo command module and the Hindenburg dirigible both assumed that design and operating procedures could overcome inherently dangerous gases, so they decided to forego much safer alternatives.

Second, safety expertise is usually not a high policy priority, and sometimes, to the leadership team, safety is considered an intrusion rather than an enhancement.

Third, it is assumed that safety is part of the regulatory role of government; and if the principal role of an agency is to support the development of an industry, safety often becomes a conflict of interest.

Fourth, people think that a hazard with which they must deal affects only their particular activity, and don't realize that the same hazard is also present in other industries, enterprises, services, and activities. When someone is hurt or a catastrophe occurs, it is often called a "freak accident." The reality is that the same hazard has been the culprit over and over again, only in different settings.

Fifth, reasonable safety controls are sometimes violated by the unscrupulous. According to the Department of Transportation's (DOT) Research and Special Programs Administration (RSPA), some compressed-gas cylinders are damaged when the test identification numbers are ground off and replaced with unauthorized numbers so old cylinders can be reused. Such a practice can cause a rupture.

Hazard Identification and Prevention Matrix

	Eliminate the Hazard		Guard the Hazard		Provide a Safety Factor		Provide Redundancy		Provide Reliability
	Hazard	Safety	Hazard	Safety	Hazard	Safety	Hazard	Safety	
Natural									
Structural/ Mechanical	Source of ignition (friction)			Open truck				Ventilation	1*
Electrical	Static electricity/ electrical spark			Grounded conductors				Explosion-proof electric motors	2*
Chemical	Combustion/ explosion			Gas detectors					3*
Radiant Energy									
Biological									
Automated Systems									

*With a combination of safeguards in 1, 2, & 3, the atmosphere is unable to support combustion, so reliability is 100%.

Applicable Standards and Literature

OSHA

OSHA 1910.101, Subpart H, General Requirements, Hazardous Materials lists:

1910.102 Acetylene

1910.103 Hydrogen

1910.104 Oxygen

1910.105 Nitrous Oxide

1910.110 Storage and Handling of Liquefied Petroleum Gases

1910.111 Storage and Handling of Anhydrous Ammonia

1910.134 Subpart I, Respiratory Protection, Personal Protective Equipment

1910.166-171 Subpart M, Compressed Gas and Compressed Air Equipment

1910.1200 Subpart Z, Hazard Communication, Toxic and Hazardous Substances

OSHA 1926.153 Subpart F, Fire Protection and Prevention, Liquefied Petroleum Gas

OSHA 1926.350 Subpart J, Welding and Cutting, Gas Welding and Cutting

MSHA

MSHA 56.4502 Battery Charging Station

MSHA 56.4600 through **56.4604** Welding/Cutting/Compressed Gas

MSHA 56.16005 Securing Gas Cylinders

MSHA 56.16006 Protection of Gas Cylinder Valves

MSHA 57.4463 Liquefied Petroleum Gas Underground

MSHA 57.4600 through **57.4660** Welding/Cutting/Compressed Gas

MSHA 57.801 through **57.907** Rotary Jet Piercing Surface (only concerns the use of oxygen)

MSHA 57.16005 Securing Gas Cylinders

MSHA 57.16006 Protection of Gas Cylinder Valves

MSHA 75.1106-2 Transportation of Liquefied and Non-Liquefied Gas Cylinders; Requirements

MSHA 75.1106-3 Storage of Liquefied and Non-Liquefied Compressed Gas Cylinders; Requirements

MSHA 75.1106-4 Use of Liquefied and Non-Liquefied Compressed Gas Cylinders; General Requirements

MSHA 75.1106-5 Maintenance and Tests of Liquefied and Non-Liquefied Compressed Gas Cylinders; Accessories and Equipment; Requirements

MSHA 75.1106-6 Exemption of Small Low Pressure Gas Cylinders Containing Nonflammable or Non-explosive Gas Mixtures

MSHA 77.208(d)(e) Storage of Materials

MSHA 77.1915 (a) Storage and Handling of Combustible Materials

MSHA 56/57.20012 Labeling of Toxic Materials

U.S. Department of Transportation (DOT)

See Title 49, Code of Federal Regulations, Part 173, Subpart B, 17S.32, a, b, and c. For detailed information on portable tanks, see 173.34 discussing maintenance of cylinders.

For additional information developed by the NTSB, check the following Web site: www.ntsb.gov.

U.S. Army Corps of Engineers

EM385-1-1, September 1996, *Safety and Health Requirements Manual*, Chapter 9, B, C & D "Fire Prevention and Protection,"; Chapter 10, C, D, E & F "Welding and Cutting,"; and Chapter 11, F & G "Electrical."

ANSI

ANSI B57.1-87 Compressed Gas Cylinder Valve Outlet and Inlet Connections

ANSI Z48-54 Marking Portable Compressed Gas Containers

ANSI Z293.1 (Rev. 1/1/59) - Fire Hazards in Oxygen Enriched Atmospheres

NFPA

NFPA Volume 4, 50 Bulk Oxygen Systems

NFPA Volume 4, 50A Gaseous Hydrogen Systems

NFPA Volume 4, 50B Liquefied Hydrogen Systems

NFPA Volume 4, 51 Welding and Cutting

NFPA Volume 13, 53M Fire Hazards in Oxygen Enriched Atmospheres

NFPA Volume 13, 56HM Home Respiratory Therapy

National Board of Boiler and Pressure Vessel Inspectors (NBBI)

This organization is building a $1.5 million training and conference center next to its Columbus, Ohio headquarters to accommodate the need for pressure vessel expertise.

Other References

- David Von Drehle, *Triangle: The Fire that Changed America,* Grove Press, New York, 2003.
- "Dangerous Compressed and Confined Gases" *Hazard Information Newsletter*, Volume 2, Issue 7, October 1997.

Commentary

It isn't hard to reason that a pure-oxygen space capsule is like a bomb ready to explode, and that placing a large liquid-oxygen refill tank with supporting piping inside an enclosed van (even if it has small vents) is asking for trouble. Normal truck travel will cause vibrations, which can easily loosen pipe joints and cause leaks. The truck now becomes a roving time bomb, and no one knows when it will go off, in what neighborhood, or whether in heavy traffic or on a crowded street filled with lots of people! Liquid-oxygen tanks and compressed-oxygen cylinders need to be in a separate, outside expanded-metal or wire-mesh locker so that it is impossible for an oxygen-enriched atmosphere to accumulate. Such trucks should not be parked inside an enclosed building or garage or driven through long tunnels.

Design philosophy needs to be broadened to include a "safety culture." In civil design, the licensed professional engineer signs and stamps his work product and bears full responsibility for a safe design. In the military hierarchy, the safety engineer advises the Commander on all matters concerning safety, and this support of safety is found in all staff relationships as well. If the safety engineer gives the Commander bad advice, either by misinformation or omission, it is the safety engineer's "hot potato." If the Commander fails to consult with or overrules the engineer, the Commander has the "hot potato." This one-on-one approach to high-loss hazards has minimized high-tech snafus. System safety should be as important in management's hierarchy as sales, production, and other administrative departments. Those with safety expertise must have the undisputed authority to say "go" or "no go," and need to be included in the development of all management policies. However extreme this concept may appear to many management groups, safety is a necessary member of the management team and, if given a chance, could soon prove its worth by keeping hazards at bay and increasing profitability.

OSHA, MSHA, and the Product Safety Commission were all established to reduce injuries, but because their rule making is based upon consensus standards, their requirements are sometimes very weak and inadequate, and rarely discuss hazards or how to prevent them. It is very difficult for any governmental agency to come up with adequately strong rules when it must deal with so much political pressure from all sides. Perhaps governmental safety agencies could be more effective if their appointed leaders would focus on hazards from a bully pulpit and get the public involved. For example, the "History" part of this section shows that oxygen-enriched atmospheres have been known for many years to be a critical hazard in the airline industry, as some seven large aircraft were destroyed between 1948 and 1970 during on-the-ground testing of oxygen systems necessary for high-altitude flying; however, publicity about this hazard remains relatively scarce. The promotion of the aviation industry has often seemed to take priority over flight safety.

The OSHA HAZMAT programs to identify dangerous substances are a step in the right direction, but they often fail to reach operating personnel with explicit information about a particular hazard and its failure modes. For example, there is little said about the dangerous oxygen-enriched atmosphere that may be created when a person with an oxygen-assist portable system rides in an enclosed automobile for an extended period of time. When they carry these bottles in the trunk of a car, a disaster can ensue if there is a small leak or valve failure; compare this practice to the commercial delivery of industrial oxygen and welding gases that are always transported on open delivery trucks.

Recommendations

Safety must become an engineering function:

- Vehicles designed to transport cylinders of liquefied or compressed gases should have open beds to prevent the accumulation of flammable or toxic gases. The illustration below depicts an example of an open transport cage for compressed gases.

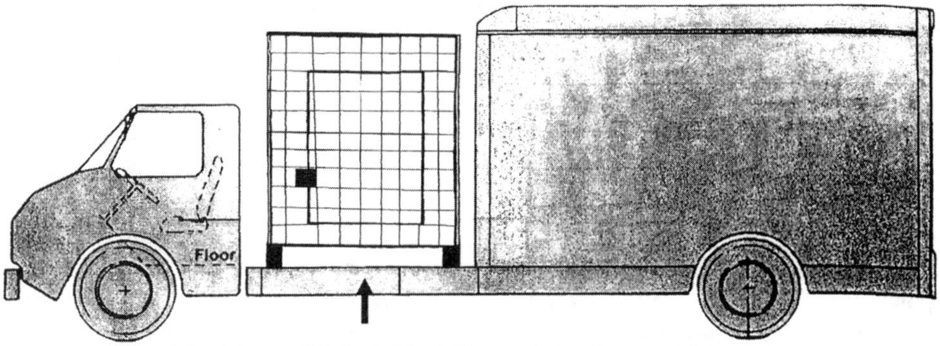

Expanded metal open grill locker to hold liquid oxygen tank and compressed oxygen cylinders away from the enclosed section of the van.

- Designs that require poisonous gas under high pressure require corresponding safety features. A licensed professional engineer who is eminently qualified in this area should design such systems and specifically sign and seal the drawings and specifications.
- Tanks holding gases should be designed to continuously self-purge the hazardous gas so it cannot collect in a confined space. Providing safe access and restricting entrance into such spaces until these gases have been purged are hazard-prevention measures that need implementation.
- Devices that will detect excess flow beyond the needs of the appliance by applying the same safe-design concepts for gas flow that are in use for electrical current flow need to be designed. The addition of odorizers does not offer complete protection from leaking gas. This improved design would make great strides in controlling the hazards of asphyxiation, combustion, and explosion.
- The structural integrity of a compressed gas cylinder must not been compromised by the unauthorized grinding to remove the Retested Identification Numbers (RIN), making it potentially vulnerable to rupture. To ensure reliable ID numbers, call DOT's Office of Hazardous Materials at (909) 483-5624. Expiration dates can also be checked with the DOT.

SECTION 12

Hazardous Secondary Voltages

The Hazard Syndrome

All electrical systems in construction need to be wired for ground-fault detection and automatic de-energizing. People who work regularly with lower-voltage tools and equipment sometimes become desensitized to the *dormant* hazard of the potential for low-voltage current flow that becomes *armed* when energized, and *active* by a change of circumstance that allows the current to flow silently into an extremely harmful or deadly situation. Low voltages of 440v and 110v comprise the backup electrical system on construction sites. When undertaking low-voltage "hot" work, the hazard is armed, and it is hard to predict the outcome of the shock the worker will receive if the hazard becomes active.

Modern construction-facility operation and maintenance depend on the availability of electric power for tools and equipment that use 600 volts or less. The frequency of electrocutions from this source suggests that many individuals in management and the workforce fail to recognize the magnitude of this life-threatening hazard. Construction managers and safety planners tend to focus on the danger of high-voltage powerlines and forget how dangerous lower voltages can be. Heavy-duty table saws and similar electrically-powered equipment still create the possibility for electric shock, which has the potential to be fatal. The use of inherently safe battery-powered tools has changed the complexion of portable power tools and dramatically reduced the hazard of low-voltage electrocution from this source. However, such electrocutions still present a viable hazard on modern construction sites.

Scope

Electrically powered tools such as drills, saws, and sanders are common on construction sites. Lower operating voltages of 110–440 are killers as efficient as unintentional contact with high voltages. Sometimes those who have received secondary-voltage electrical shocks have died a few hours later. Autopsies have revealed that the electrical current flow had caused serious internal hemorrhaging. Anyone who feels more than a tingle from an electrical current should seek medical assistance immediately.

Peril from the Failure Mode

An easy rule of thumb to remember is that the greater the current flow, the greater the injury. However, a number of electrically related injuries have not been caused by the shock itself, but from the fall or injury that ensued as a result of the shock. Though secondary voltage presents a primarily *electrical* hazard, injuries can occur from almost any other type of source when the worker is immobilized due to the effects of electrical shock.

As might be expected, due to the troubleshooting nature of their work, those most in danger from electrical shock and burns are electricians and maintenance personnel. Electricians experience over 50% of reported shocks and burns. Mechanics/repairmen, welders, laborers/helpers, machinists, those who use portable electric tools, and other employees in other occupations are also in danger from electrically initiated fires and explosions. In addition, firefighters, janitors, and other personnel in occupations that require standing in water while performing their tasks tend to experience more shocks than do others. Electrical shorts or indirect electrical hazards cause numerous fires and explosions, and account for millions of dollars in damages.

The effect electricity has on an individual depends upon four factors:

- The voltage and current the circuit carries
- Skin condition and internal body resistance
- The amount of current that flows through the body
- The path electricity takes through the body

Table 9–3
Estimates of Human Response to Exposure to U.S. Electrical 60-cycle Current.

Exposure	Milliamperes
Slight sensation on hand	1
Mild shock, tingle	2
Pain	9
Chest contraction	23
Ventricular fibrillation	40–100
Tissue damage/hemorrhages	300+

Even secondary-voltage current flow in series-type Christmas tree lights has been known to cause fibrillation of the heart, which can cause death. The following chart shows the time it takes to cause bodily damage depending on the strength of the electrical current. Exposure to 60 milliamperes of current can lead to severe damage or death in under three minutes.

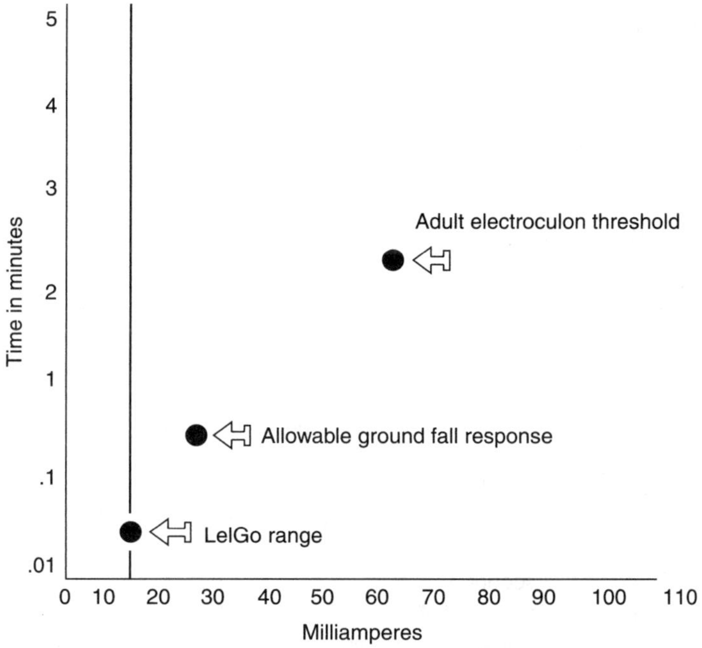

Contact with electrical energy causes harm in a number of ways:

- Burns are caused by the heat generated by current flowing through body tissue, because flesh is an excellent conductor but cannot carry a current load, and the flesh literally blows up.
- Involuntary muscle contractions prevent releasing the grip on an energized conductor.

- Ventricular fibrillation occurs when the electrical impulses controlling the heart muscle are disrupted by a foreign current and the heart muscles function in an uncoordinated, rapid pulsing that does not pump blood.
- Cardiac arrest results when nerve centers controlling heart rhythm are paralyzed.
- Pulmonary arrest is the inability to breathe, resulting from paralysis of the nerve centers that control breathing.

Brief Summary of the Occurrence

The injured received a shock that caused fibrillation of the heart while working on a stepladder, from which he fell. The injured was correcting permanent wiring conductors in an electrical junction box and assumed that they were de-energized. The new construction involved mixing temporary wiring with permanent wiring, which led to the error of allowing workers to energize segments of the permanent wiring being installed in order to run the operating equipment.

Date of Occurrence

2000

Source of Information

Idaho Workers' Compensation claim of subrogation

History of the Hazard

1965: 1,071 electrocutions reported.

1995: Reported electrocutions reduced to 550.

This 50% reduction in thirty years is an example of the remarkable progress that can be achieved by the use of available safeguards as they are developed. The reduction is even more dramatic considering the tremendous proliferation of electrically-powered portable tools that has occurred since the 1950s. Portable electric tools are safer today because of the evolution from ungrounded tools to

- grounded electrical tools,
- provisions for ground-fault-circuit interrupters (GFCI), which are sensitive devices intended to prevent electrical shock by de-energizing an electrical circuit within a fraction of a second when leakage to the ground begins to exceed a trip threshold of 5 ma = +/–, as specified in Underwriters Laboratories (UL) Standard 943, and consistent with National Electrical Code (NEC) 150-81 through 150-154,
- the adoption of double-insulated tools (double insulation means the power tool has two separate layers of insulation), and
- the availability and convenience of portable battery-powered tools.

Contributory Factors

A list of possible contributors to the continued occurrence of low-voltage electrocution on construction sites includes

- Provision of a temporary electrical system to power production equipment while the permanent wiring system is being built or replaced
- The underestimation by many people of the dangerous nature of secondary voltages
- Lack of a conspicuously posted diagram of the wiring circuit showing what goes where
- Lack of labels on each switch box showing what is on the circuit it controls
- Random placement of control panels and switch boxes throughout the facility with little consideration for convenience or easy access
- Absence of a lockout type of switch on each individual piece of production equipment
- Electrically defective power tools, machines, welders, and other equipment
- Faulty or incorrect plugs and receptacles for various voltages
- Use of improper plugs and receptacles that allow lower voltage equipment to be connected to higher voltages
- Improper polarity of wiring or equipment
- Defective or aging appliances
- Failure to implement lockout/tagout procedures and to use adequate protective equipment

Hazard Identification and Prevention Matrix

	Eliminate the Hazard		**Guard the Hazard**		**Provide a Safety Factor**		**Provide Redundancy**		**Provide Reliability**
	Hazard	Safety	Hazard	Safety	Hazard	Safety	Hazard	Safety	
Natural	Gravity/ fall from ladder			Staging					
Structural/ Mechanical									
Electrical	Current: mixing permanent wiring with temporary instal-lation	Tem-porary power system	Unident-ified conduc-tors	Separation and labeling of temporary system		Separate the two systems		Battery-powered hand tools	
Chemical									
Radiant Energy									
Biological									
Automated Systems									

Applicable Standards and Literature

OSHA

1910.137 Electrical protective devices

1920.138 Hand Protection, Subpart S 301 Introduction

1910.304 Wiring Design and Protection

1910.305 Wiring Methods, Components and Equipment for General Use.

1910.306 Specific Purpose Equipment and Installations

1910.307 Hazardous (Classified) Location

1910.308 Special Systems

1910.331 Scope

1910.332 Training

1910.333 Selection and use of Work Practices

1910.334 Use of equipment

1910.335 Safeguards for Personal Protection

1926 Subpart K:

1926.400 Introduction

1926.402 Applicability

1926.403 General Requirements

1926.404 Wiring Design and Protection

1926.405 Wiring Methods, Components and Equipment for General Use

1926.406 Specific Purpose Equipment and Installations

1926.407 Hazardous (Classified) Locations

1926.408 Special Systems

1926.416 General Requirements

1926.417 Lockout and Tagging of Circuits

1926.418- 1926.430 Reserves

1926.431 Maintenance of Equipment

1926.432 Environmental Deterioration of Equipment

1926.441 Batteries and Battery Charging

1926.449 Definitions Applicable to this Subpart

1926.950 through **1926.960 Subpart V** Power Transmission and Distribution (mostly high voltage safety)

MSHA

MSHA covers electricity in CFR 30, Subpart K, 56.12001 through 56.13030 for Surface Metal and Non-Metal Mines, and Subpart K, 57.12088 through 57.1303 for Underground Metal and Non-Metal Mines.

National Electrical Safety Code (NESC)/ANSI C2

This code is written by the Institute of Electrical and Electronics Engineers, Inc. (IEEE), and contains standards for electrical transmission and distribution from the source of power generation to the meter, usually high-voltage lines, except for the 440 volt lines from the step-down transformer to the meter.

National Electrical Code (NEC)

This code is published by the National Fire Protection Association (NFPA) and contains standards from the meter to the energy-consuming device.

The NEC addresses proper electrical systems and equipment installation to protect people and property from hazards arising from the use of electricity in buildings and structures, generally addresses secondary voltage systems, and is the standard that must be followed for both temporary and permanent installations.

U.S. Army Corps of Engineers

EM385-1-1, (November 3, 2003) *Safety and Health Requirements Manual* has a complete overview of secondary electrical hazards in Section 11. The following excerpts highlight the key safety issues.

Section 11.A, General

11.A.01 Approval and qualification

(a) All electrical wiring and equipment shall be a type listed by a nationally recognized testing laboratory for the specific application for which it is to be used.

(b) All *electrical work* shall comply with applicable National Electrical Safety Code (NESC), NEC, and for marine operations (USCG regulations).

(c) All work shall be performed by *qualified* personnel familiar with *applicable* code requirements.

11.A.02 Isolation

(a) Before work is begun, the person in charge shall ascertain by inquiry, by direct observation, or by instruments, whether any part of an electric power circuit—exposed or concealed—is located such that the performance of work could bring any person, tool, or machine into physical or electrical contact with it.

(b) Whenever possible, all equipment and circuits to be worked on shall be de-energized before work is started and personnel protected by clearance procedures and grounding.

(c) Live parts of wiring or equipment shall be guarded to protect all person or objects from harm.

11.A.03 Flexible cords

(a) Flexible cord sets used on construction sites or in damp locations shall contain the number of conductors required for the service plus an equipment ground wire: the cords shall be hard usage or extra hard usage as specified (see Table 9-4).

(b) Electric wire and flexible cord passing through work areas shall be protected from damage (including that caused by foot traffic, vehicles, sharp corners, projections, and pinching); flexible cords and cables passing through holes shall be protected by bushings or fittings.

Flexible Cord and Cable Usage

The following initials indicate the type of electric cable that is suitable for particular types of use.

Table 9–4 Ground requirements.

Trade Name	Identifying Code	Usage
Portable power cable	G,W	Portable-extra hard usage
Termoset-jacketed hard cord	HS, HSO	Portable or portable heaters, damp locations, extra hard usage
	HSJ, HSJO	Portable or portable heaters, damp locations, hard usage
Hard service cord	S, SW, SEO, SO, SOO, ST, STO, STOO	Pendant or portable, damp locations, extra hard usage
Junior hard service cord	SJ, SJE, SJEO, SHO, SJOO, SJT, SHTO, SHTOO	Pendant or portable, damp locations, hard usage

Exemptions for Grounding

Portable and Vehicle-Mounted Generators (from NEC 250-6)

Portable Generators

Under the following conditions the frame of a portable generator is not required to be grounded and shall be permitted to serve as the grounding electrode for a system supplied by the generator.

(a) The generator supplies only equipment mounted on the generator and/or cord- and plug-connected equipment connected through receptacles mounted on the generator.

(b) The noncurrent carrying metal parts of equipment and the equipment grounding conductor terminals of the receptacles are bounded to the generator frame.

Vehicle-Mounted Generators

Under the following conditions the frame of a vehicle is permitted to serve as the grounding electrode for a system supplied by a generator located on the vehicle (all conditions must be satisfied).

(a) The frame of the generator is bonded to the vehicle frame.

(b) The generator supplies only equipment located on the vehicle and/or cord- and plug-connected equipment through receptacles mounted on the vehicle or on the generator.

(c) The noncurrent carrying metal parts of equipment and the equipment grounding conductor terminals of the receptacles are bonded to the generator frame.

(d) The system complies with all other NEC grounding requirements.

Neutral Conductor Bonding

A neutral conductor shall be bonded to the generator frame when the generator is a component of a separately derived system; the bonding of any conductor other than a neutral within the generator to its frame shall not be required.

Receptacle outlets that are not part of the permanent wiring of the building or structure shall be GFCI protected by one of the following means:

(a) a receptacle outlet with integral GFCI protection;
(b) a standard receptacle outlet connected downstream of a receptacle outlet with integral GFCI protection; or
(c) receptacles protected by a GFCI-type circuit breaker.

Receptacle outlets that are part of the permanent wiring of the building or structure (including portable generators) shall use a portable GFCI device if the receptacle outlets are not already GFCI protected. The portable GFCI device shall be as near as practicable to the receptacle outlet.

Electric tool circuits that are "hard-wired" directly to an electrical source of power shall be GFCI protected by a GFCI-type circuit breaker.

11.D Temporary Wiring and Lighting

11.D.01—A sketch of proposed temporary power distribution systems shall be submitted to the designated authority and accepted for use before temporary power is installed: the sketch shall indicate the location, voltages, and means of protection of all circuits, including receptacles, disconnecting means, grounding, GFCIs and lighting circuits.

11.D.03—The vertical clearance of temporary wiring for circuits carrying 600 volts or less shall be:
(a) 3 m (10 ft) above finished grade, sidewalks, or from any platform;
(b) 3.6 m (12 ft) over areas subject to vehicular traffic other than truck traffic;
(c) 5.4 m (18 ft) over public street, alleys, roads, and driveways;
(d) 4.5 m (15 ft) over areas other than those specified in 11.D.03c that are subject to truck traffic.

11.D.05—Wires shall be insulated from their supports.

11.F.01—Batteries of the non-sealed type shall be stored in enclosures with outside vents or in well-ventilated rooms and be so arranged as to prevent the escape of fumes, gases, or electrolyte spray into other areas.

11.F.02—Ventilation shall be provided to ensure diffusion of battery gases to prevent the accumulation of an explosive mixture.

11.G.01—Hazardous (classified) Locations

Locations of electrical equipment and wiring shall be classified on the properties of the flammable vapors, liquids or gases, or combustible dusts or fibers that may be present therein and the likelihood that a flammable or combustible concentration or quantity is present.

11.I.01 Underground Electrical Installations

Guarding underground openings

(a) Warning signs and rigid barricades shall be promptly placed when covers of manholes, handholds, or vaults are removed.
(b) When an employee enters an underground opening the opening shall be protected with a barricade, temporary cover, or other guard appropriate for the hazard.
(c) Underground opening guards and warning signs shall be lighted at night.

11.I.02—Maintenance holes and unvented vaults shall be treated as, and subjected to the requirements of confined spaces.

Underwriters Laboratories (UL)

This organization has developed electrical-safety standards for many of the electrical tools, machines, and devices used in both industrial and consumer applications.

National Electrical Manufacturers Association (NEMA)

NEMA has developed standards for a wide range of electrical devices used for wiring systems.

Factory Mutual (FM)

FM also provides excellent references and standards on protective grounding, electrical maintenance, electrical equipment in hazardous locations, and special electrical installations.

Other References

- David V. MacCollum, "Phase Planning for Electrical Systems," *Construction Safety Planning*, Chapter 27, Van Nostrand Reinhold, 1995.
- *Supervisor's Safety Manual* 7th Edition, National Safety Council.
- "Hazardous Secondary Voltages," *Hazard Information Newsletter*, Volume 2, Issue 6, September 1997.

Commentary

Fail-safe systems can and should be developed and implemented. When people are engrossed in a project, they will predictably forget about everything else, including safety.

It is essential that we use tools and equipment that are fail-safe to help overcome this known trait. When activities become routine, it is even more crucial to ensure that fail-safe secondary electrical systems are designed and in place, and that operating procedures are monitored. The long-term benefits of completely separate and isolated permanent wiring systems in construction wiring ensures no unused or hot circuits when the construction is completed.

The landowner and contractor should not attempt to operate power machinery on a new electrical system until it is completely installed. Temporary power sources for existing production equipment should be planned at the time of design and installed before rewiring work begins.

Recommendations

Use of battery-powered tools avoids the necessity of temporary wiring. The electrical contractor, if different from the general contractor, should be contacted during the safety-planning phase of the construction, and this person's input should be written into the plan.

The use of the partially installed electrical system leads to unauthorized energizing of conductors. Planning should include temporary circuits to power operating equipment while the new system is being installed.

10 Access Hazards

INTRODUCTION

This chapter discusses the major causes of construction injuries related to access to the job site. The remedies for these hazards are usually accomplished with provisions that fall within existing standards and requirements. The following section contains seven examples:

1. Fall prevention
2. Fall protection
3. Falling objects
4. Ladders
5. Dangerous access to equipment
6. Traffic control
7. Haul roads

SECTION 1

Fall Prevention

The Hazard Syndrome

Falls can be categorized into three basic hazards:

- Slips for the lack of traction
- Trips in the path of travel or due to uneven walkway
- Falls to a lower level

Most walking surfaces present varying degrees of a *dormant* hazard. The hazard of slips is *armed* when the walking surface lacks traction and handrails. A decrease of friction creates a slippery surface, and the hazard becomes *active* when someone slips, loses balance, and falls. Fall prevention includes those measures to avoid a fall of any type.

Scope

Falls are a problem in virtually every construction, industrial, and public area. The design of walking surfaces such as floors, ramps, stairs, ladders, platforms, and scaffolding, whether fixed or as part of mobile equipment, must take into consideration the environmental conditions that may be expected or foreseen, which might make surfaces slippery. Walkway features that will provide adequate traction in a particular environment should be specified, designed, constructed, and maintained.

Absence of a walkway over a freshly poured concrete footing that allows access in and out of a home during construction is a common hazard encountered by laborers. In many cases, such concrete footings contain protruding steel rebar, which presents an additional hazard of impalement.

Peril from the Failure Mode

An uneven walkway with long steps to avoid wet concrete and unprotected rebar can cause trips that result in severe injury. Because falls have the potential for serious injury or death, every fall should be considered life threatening. Though they are caused by the *natural* hazard of gravity, falls can be *mechanically* controlled through means such as access points and traction.

Table 10–1 Construction locations where falls are most likely to occur.[1]

Percentage	Where Falls Occur
27%	Roofs
17%	Scaffolds
10%	Steel erection
8%	Ladders
7%	Floor openings
4%	Open-sided floors
3%	Concrete erection
3%	Aerial lift/basket
3%	Suspended platform/vehicle
16%	Not otherwise identified

Note: Approximations. % error = ± 2%.

[1]Culver and Connolly, "Prevent Fatal Falls in Construction," *Safety and Health,* National Safety Counscil, September 1994.

Fatalities: The National Safety Council's (NSC) figures for 2003 list a conservative figure of 13,300 deaths from falls on the same level, or from one level to another. The NSC tabulates falls while boarding or alighting from transport vehicles under another category, so a percentage of the 2500 deaths reported from that category would increase this number of fall fatalities to well in excess of 14,000.

Falls from an elevation accounted for 33% of construction fatalities for the period 1985–1993, and this type of fall accounts for over half the deaths experienced by ironworkers.

When examining these occurrences, it is easy to get caught up in calculating the distance someone fell because of a floor opening or open edge, and overlook the initial cause: the slippery surface, loose planks, uneven or inadequate walkways, or lack of guard rails. Slips and trips as the initial cause of a fall from an elevation are vastly under-reported.

Table 10–2 The percentages of deaths from various heights.[2]

Percentage	Height
36%	Higher than 30'
22%	Between 11'–20'
20%	Between 21'–30'
8%	Between 6'–10'
2%	Lower than 6'
12%	Not given

Injuries: Injuries from falls on the same level or to another level account for almost 16% of *all* occupational injuries. According to the Consumer Product Safety Commission's National Electronic Injury Surveillance System (NEISS), a nationwide electronic data collection system that compiles a database from computerized records in 119 selected U.S. hospitals, falls are the leading cause of injuries requiring hospital treatment, with the exception of automobile collisions. It can be estimated that in addition to the injuries included in the NEISS statistics, millions of other falls occur that result in minor or temporary disabilities that do not require emergency room treatment, and thus are never counted in injury statistics.

Brief Summary of the Occurrence

A homebuilding project required concrete footings around the perimeter of the entire house to be built. The worker tripped because there was no provision for a designated

[2]Culver & Connolly, "Prevent Fatal Falls in Construction," *Safety and Health,* National Safety Counscil, September 1994.

walkway or stile over the concrete footings with protruding rebar. He fell and impaled his scrotum on the unguarded rebar. Though the rebar depicted below is protected with guards, the worker in the case example fell onto rebar that was not protected.

Date of Occurrence

1990s

Source of Information

Workers' compensation claim

History of the Hazard

1920: Serious interest arose in the standards community to develop a quantitative safety code for walking surfaces.

1929: R.B. Hunter developed an articulated-strut machine for measuring frictional coefficients of walking surfaces. Later, Percy Sigler found that almost all dry surfaces are not slippery, but water, oil, or other contaminants cause a loss of traction.

1974: Dr. Robert Brungraber developed an elaborate static slipmeter.

1988: Harris & Shaw, "Slip Resistance of Floors; Users Opinions," *Journal of Occupational Accidents*, Amsterdam, pp. 287–298, showed the "Tortus" to be a cleverly disguised dragsled device.

1991: Robert Brungraber invented the Mk II for measuring both wet and dry surfaces.

1992: William English invented and filed for a patent on the *XL* for measuring both dry and wet surfaces.

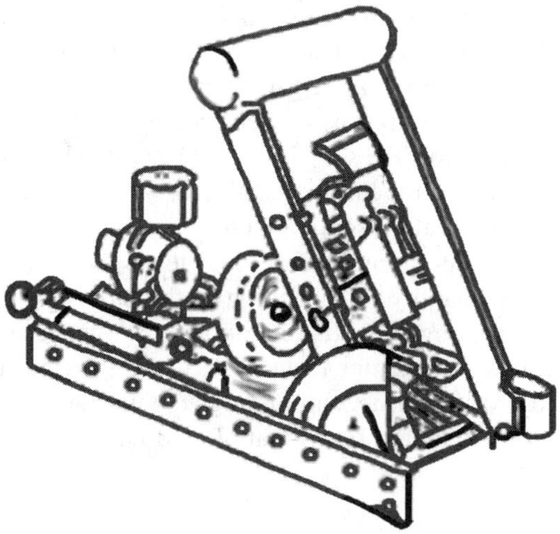

1995: William English demonstrated the *XST* for measuring both dry and wet shoe traction. It was the only device that mimics the pressures created when walking.

Contributory Factors

Most people feel they are invincible and can handle most situations, especially remembering to step over dangerous conditions such as those caused by freshly poured concrete and unguarded rebar. Such an attitude can lead to devastating results when a slip or trip occurs in an unfortunate circumstance.

Fall prevention has not been emphasized as the lifesaver it truly is. Slippery, uneven, and obstacle-strewn surfaces require immediate attention, as slips and trips are no laughing matter. Management must recognize that on-the-job awareness of the unsafe walkway across a floor opening six stories up, the open edge without a railing, or any other unguarded drop-off is not enough. Falls occur because fall prevention is not given a *now* priority for immediate remedy!

One of the main reasons falls remain a low-priority hazard on many construction sites is the epidemic of under-reporting or incorrectly reporting the frequency and severity of falls. When an injured worker dies from complications several months after a fall, this death generally goes unreported as a fatality as the result of a fall, so statistics don't always show the severity of the hazard. Also, most industrial databases are derived from workers' compensation analyses, which are generally based on reporting by supervisory and medical personnel. Unfortunately, this personnel pool typically has had little training in safety engineering and tends to classify events by injury type and body part, rather than by a description of the hazardous condition. Another void in the reporting of meaningful statistical data is the vast area of city, county, and state governments, which are exempt from OSHA and thus not required to keep injury/fatality records. Railroad, marine, and mining activities are also outside the authority of OSHA and use different data formats that often do not focus on specific hazards.

Table 10–3 Distribution of fatalities and injuries by work surface.[3]

Work Surface Involved	Estimated Annual		Estimated Annual	
	Injuries	Percent	Fatalities	Percent
Floor	53,970	51.4	29	22.0
Ramp	735	.7	1	.6
Roof	420	.4	15	11.0
Platform	8,715	8.3	9	7.0
Walkway	5,355	5.1	6	4.3
Stairs	13,965	13.3	15	11.2
Ladder	11,025	10.5	25	19.0
Scaffold	8,400	8.0	25	18.6
Other	2,415	2.3	8	6.3
Total	105,000	100	132	100.0

[3]Source: U.S. Department of Labor

Hazard Identification and Prevention Matrix

	Eliminate the Hazard		Guard the Hazard		Provide a Safety Factor		Provide Redundancy		Provide Reliability
	Hazard	Safety	Hazard	Safety	Hazard	Safety	Hazard	Safety	
Natural	Gravity								
Structural/ Mechanical	Unsafe access	Safe walkway/ guards on protruding objects	Lack of traction	Nonslip surface/ Handrails		Measuring device for traction			95%
Electrical									
Chemical	Slippery sub-stances	Remove contami-nation							85%
Radiant Energy									
Biological									
Automated Systems									

Applicable Standards and Literature

OSHA

The proposed standard for fall prevention was published August 9, 1994 and specifies a Slip Coefficient of Friction (SCOF) of 0.5.

Comment: It appears that final acceptance of the Steel Erectors Negstiated Rule Advisory Committee (SENRAC) may address wording that slip resistant paint will be equal to slip resistance of unpainted steel or better than 0.5. The averages of the slip indices when wet are as follows:

Material	Value
Steel	0.5
Rusty steel above	0.5
Ice	0.2
Terrazzo, smooth tile, plate glass	0.15

Table 10–3 Wet values.

The final standard issued by OSHA and published in the *Federal Register*, Friday, August 30, 1996, 29CFR Part 1926, "Safety Standard for Scaffolds used in the Construction Industry," Subpart L Scaffolds 1926.450 contains a requirement for safe scaffolding that, when erected, will provide fall prevention.

ANSI

ANSI A117.1-1992—Accessible and Usable Buildings and Facilities is a voluntary consensus standard that sets forth specifications for making buildings and facilities accessible not only by all people, but also by those with physical disabilities. The 1980 version was adopted in part and included in the Americans with Disabilities Act (ADA), which is now in effect.

American Society for Testing Materials (ASTM)

C1028-89—Test method for determining the static coefficient of friction of ceramic tile and other like surfaces by the horizontal dynamometer pull meter method.

D2047-93—Test Method for Static Coefficient of Friction of Polish-Coated Surfaces as Measured by the James Machine.

D5859-96—Standard test method for determining the traction of footwear on painted surfaces using the variable incidence tester (English XL).

F1679-96—Test Methods for using a VIT (English XL).

F802-88—Guide for selection of certain walkway surfaces when considering footwear traction.

Proposed ASTM Standards

New STD Z3403Z—Test Method for determining the static coefficient of friction and slip resistance of ceramic tile and other surfaces by the Mark II slip tester.

New STD Z4374Z-94—Test Method for determining the slip-resistance (static coefficient of friction) of footwear sole, heel, or related materials by the English XL slip resistance tester.

New STD Z4375Z-95—Test method for using a portable inclinable articulated strut slip tester (PIAST).

New STD Z4594Z-95—Test method for using a portable articulated strut slip tester (PAST).

Other Standards

National Fire Protection Association (NFPA)-1994; *Life Safety Code* ANSI/NFPS 101 1994.

Underwriters Laboratories-1992; UL 410; *Standard for Slip Resistance of Floor Surface Materials*.

Americans with Disabilities Act, Public Law 101-336, 7/26/90.

Other References

- William English, *Pedestrian Slip Resistance, How to Measure It and How to Improve It*, 1996.
- William English, and William Marletta, PhD "The Right Paint Can Prevent Slips and Falls," *Safety & Health*, November 1995.

- William English, "The Validation of Slipmeters," *Proceedings of the Annual Conference of Ergonomics Society*, University of Warwick, Coventry, England, April 1994.
- National Safety Council, *Accident Facts*, 1994 Edition.
- T.D., Proctor, " Slipping Accidents in Great Britain—an Update," *Safety Science*, pp. 367–378, 1993.
- William English, "Progress Toward Valid Walkway Slip-Resistance Standards," *Proceedings of the Sixth Southeastern Safety and Health Conference*, pp. 33–44, 1992.
- Robert J. Brungraber, et al., "Bucknell University F-13 Workshop to Evaluate Various Slip Resistance Measuring Devices," *ASTM Standardization News*, American Society for Testing and Materials, pp. 21–24, May 1992.
- William English and Robert Kohr, "A Study of the Comparative Slipperiness of Floor Cleaning Chemicals," *ASTM Standardization News*, American Society of Testing and Materials, pp. 62–65, 1991.
- Robert Brungraber and John Templer, "Controlled Slip Resistance," *Progressive Architecture*, pp. 112–116, March 1991.
- David A. Winter, *The Biomechanics and Motor Control of Human Gait*, Second Edition, University of Waterloo Press, Waterloo, Ontario, Canada, 1991.
- Raoul Gronqvist, J. Roine, E. Korhonen, and A. Rahikainen, "Slip Resistance Versus Surface Roughness of Deck and Other Underfoot Surfaces on Ships," *Journal of Occupational Accidents*, Amsterdam, pp. 291–302, 1990.
- William English, *Slips, Trips and Falls: Safety Engineering Guidelines for the Prevention of Slip, Trip and Fall Occurrences*, Hanrow Press, 1989.
- G.W. Harris and S.R. Shaw, "Slip-Resistance of Floors: Users' Opinions, Tortus Instrument Readings and Roughness Measurement," *Journal of Occupational Accidents*, Amsterdam, pp. 287–298, 1988.
- Robert O. Andres, *Essential Design Criteria for an Ergonomically Sound Portable Slip-Resistance Tester, and Other Field Measurement Considerations*, University of Massachusetts Department of Exercise Science, Amherst, 1985.
- R. Harrison and F. Malkin, "On-Site Testing of Shoe and Floor Combinations," *Ergonomics*, Vol. 26, No. 1, 101–108, 1983.
- P.J. Perkins and M.P. Wilson, "Slip Resistance Testing of Shoes—New Developments," *Ergonomics*, Vol. 26, No. 1, pp. 73–82, 1983.
- Sanford C. Adler and Brian C. Peirman, A *History of Walkway Slip-Resistance Research at the National Bureau of Standards*, NBS Special Publication 565, 1979.
- P.L. Armstrong and S.G. Lansing, "Slip-Resistance Testing: Deriving Guidance from the National Electronic Injury Surveillance System (NEISS)," *Walkway Surfaces: Measurement of Slip Resistance, ASTM STP 649*, Carl Anderson and John Senne, Eds., American Society for Testing and Materials, pp. 3–10, 1978.
- P.J. Perkins, "Measurement of Slip Between the Shoe and Ground During Walking," *Walkway Surfaces: Measurement of Slip Resistance, ASTM STP 649*, Carl Anderson and John Senne, Eds., American Society for Testing and Materials, pp. 71–78, 1978.
- Robert J. Brungraber, *A New Portable Tester for the Evaluation of the Slip-Resistance of Walkway Surfaces*, NBS Technical Note 953, U.S. National Bureau of Standards, Washington, 1977.

- Robert J. Brungraber, *An Overview of Floor Slip-Resistance Research with Annotated Bibliography*, NBS Technical Note 895, U.S. National Bureau of Standards, Washington, DC, 1976.
- National Safety Council, *Floor Mats and Runners*, Data Sheet 595, 1967.
- National Safety Council, *Falls on Floors*, Data Sheet 495, 1960, Rev. 1991.
- David V. MacCollum, "How Proper Scaffolding Cuts Costs," *Western Construction*, September 1957.
- J.B. Derieux, "The Coefficient of Friction of Rubber," *Journal of the Elisha Mitchell Scientific Society*, 50, pp. 53–55, 1934.
- "Fall Prevention," *Hazard Information Newsletter*, Volume 1, Issue 8, November 1996.

Commentary

The principle focus of fall prevention is to create walking surfaces with two characteristics: easy access and slip resistance. Particular emphasis needs to be given to those who walk on elevated surfaces. For example, the Ironworkers International Union realized that their members were being killed because of slippery surfaces on structural steel being erected at high elevations, so the Steel Erectors Negotiated Rule Advisory Committee, known as SENRAC, funded by the U.S. Department of Labor, proposed that painted steel will not be slipperier than unpainted steel, with the exception of some decking. Such a standard for slip resistance on structural steel is an important step towards improving the safety of workers in the erection industry. In this study, experienced ironworkers also contributed to this evaluation; and to support their observations, slipmeter measurements were made with the XL Slip Resistant Tester and the MARK II SLIP TESTER. Smooth walking surfaces should be measured to determine the traction in both wet and dry situations. To that end, great credit should be given to Bill English PE, CSP, who invented the XL Slip Resistant Tester and whose work extends over nearly fifty years in researching the nature of slipperiness and how it can be measured. This revolutionary device is an important step in the evolution of fall prevention and is very handy on a construction site. Portable slipmeters can be laid on a tabletop to demonstrate actual conditions of the floor surface when *contaminating materials* are present. Such devices facilitate research into the relative safety of various flooring materials under real-world-use conditions, and can be used to help determine which slip/fall remedies are effective in actual walking environments.

Recommendations

Construction management and supervision has an inalienable duty to ensure reasonably slip-resistant conditions on all parts of the job site. Injury-free worker performance should never be based upon the agility of workers to step over hazardous barriers.

To prevent falls, look for the following on walkways:

- Spills, water, grease spots, dust, small roller-bearing-like pellets, or loose gravel
- Projecting parts, discarded tools, or trash
- Liquids from leaking pipes or tracked in on rainy or snowy days
- Other materials or dusts from manufacturing processes that have lubricating properties that can be tracked to other locations

Avoiding Slips

The following steps can help avoid slips:

- Design and build facilities with slip-resistant floors
- Keep floors clean and dry
- Repair leaky pipes
- Provide mats to reduce tracking of liquids
- Audit the janitorial process
- Consider slips and falls when you analyze tasks
- Provide visual marking of steps leading to a different level
- Provide proper lighting
- Specify or provide slip-resistant shoes when possible

SECTION 2

Fall Protection

The Hazard Syndrome

Work needing to be performed on an unguarded elevation above ground level is a particularly broad *dormant* hazard. When employees work on any unenclosed elevated location without guardrails, the hazard of falling from that elevation is *armed*. The hazard is *active* when the worker falls.

Fall protection is specifically designed to guard against falls from elevation. Fall protection is necessary when all other options to prevent falls from an elevated location are not feasible. There are many examples of where fall prevention and fall protection are not considered by the designers. For example, designers of support equipment, such as formwork for concrete structures, traditionally ignore the need for safe access to aerial structures. As shown in Section 5, safe access is an integral concept in fall protection.

Scope

Fall from elevation is the cause of nearly 50% of fatal construction injuries. Such falls are the number one killer in construction today. Falls from elevation often occur when the worker has no fall-prevention system, such as a work scaffold, in place. Wood-frame construction, steel erection, roofing, and maintenance workers on finished buildings all work in trades that offer little fall protection; therefore, pains must be taken to ensure that fall-protection systems such as lanyards and working harnesses are in place.

Highway construction can be an example of a blatant disregard for construction-worker safety: the design of forms for overpass columns rarely includes a fall-protection system. These reusable steel forms developed by formwork manufacturers to cast tall concrete columns are dangerous, as they include no safe access ladder or lanyard anchor points, and limit the workers' fall-protection options. Worker ability and agility must not be relied upon to prevent potentially deadly falls. The highway overpass columns depicted below are a hazard to anyone.

Peril from the Failure Mode

Not only is an elevated platform difficult to climb, but the hazard of a serious fall from a high elevated area is always present. Falls from elevations are the principle source of construction fatalities. The hazard of falls is a *natural* hazard compounded by *structural/ mechanical* factors.

No Ladder or Place for Fall Protection Systems Creates Unsafe Access to Highway Overpass Columns

Brief Summary of the Occurrence

A worker was assigned the task of removing the forms from around a building pillar and had no place to attach his lanyard. He fell approximately ten feet and was seriously injured. The design of the six-foot-high semicirclular form panel included only two-inch wide ribbing to support the weight of the wet concrete against the panel itself. It was not wide enough to provide a foothold for the worker. As an additional problem, many of these column form panels are completely devoid of any fall protection opportunities, such as a ladder or fall-protection system. The only consideration in the design was how the semicircular panel could be bolted together horizontally or laterally, with no thought to accessibility for the worker. The worker was left with no place to attach the lanyard or safety belt on a panel completely devoid of handholds and footholds. The manufacturer was negligent in assuming workers would have ladders, aerial lifts, or other access to assemble and disassemble these form panels.

Date of Occurrence

December 29, 1989

Source of Information

Circuit Court, City of St. Louis, MO, # 942-01694 Div.

History of the Hazard

Since about **1850** workers have used protective devices while installing, maintaining, and working on telegraph poles. Today's electrical-utility workers use aerial lifts with lanyards to ensure their safety.

1880s: The first steel skyscraper was built in Chicago, and rudimentary fall-protection devices were introduced to protect workers who climbed the steel framework.

1937: While the Golden Gate Bridge was built, safety nets were used, and nineteen lives were saved.

1946: The U.S. Coast Guard rehabilitated a captured German naval training vessel renamed *Eagle*, and included a safety tie-off system for cadets who climbed the masts and yardarms to set the sails.

1954: Shock-absorbing devices were not available. The American Society of Safety Engineers (ASSE) and the NSC published a study that identified the severe internal injury-producing forces of 4,000 lbs. that occur when 6-foot lanyards are used with a belt.

1967: The Boeing Company published a report demonstrating the advantages of safety harnesses as compared with belts.

Contributory Factors

There are a lot of dangerous myths that contribute to inadequate fall protection. A few of them include

Belts are good enough protection!

Truth: The era of the belt for fall arrest has passed. The reality is that internal injuries to body organs can be severe, and asphyxiation can occur when totally suspended. Also, the individual can fall through a loose belt.

Hooking up below your feet is the best we can do!

Truth: Safety planning must be done to ensure that lanyards are attached to a safety line, which is positioned at waist level and preferably above the head.

Guardrails are OK to use for an anchorage point.

Truth: OSHA states that guardrails will not be used for fall arrest. If the qualified person is an engineer and can determine that part of a railing can withstand the load requirements for the specific force and is adequate for safe use, then an engineered guardrail system may be practical for use.

It's OK to tie-off to a handline.

Truth: A handline is used only to lift tools; therefore, it may not be able to sustain the forces of the fall. The term "tie-off" has been replaced by "anchorage point." Anchorage points have to meet specific structural requirements.

Harnesses with belts are OK.

Truth: A harness with a body-encircling belt is still a belt (belts became illegal for fall arrest after December 1997). However, a harness with a circle- or D-ring is excellent for positioning when working at heights.

Double-locking hooks are OK to attach to anything solid.

Truth: Locking snaphooks and anchorage points must be compatible.

Double lanyards (Y-lanyards) are the greatest things since sliced bread!

Truth: If misused, a Y-lanyard can allow a fall up to 31 feet. Sooner or later this will happen. It is better to use the engineered HLL and have the anchorages approved. There is less hooking and unhooking when using an HLL that the lanyard can slide along.

Hazard Identification and Prevention Matrix

The following combination of safeguards can be exceedingly effective and offer near total protection from falls from an elevation.

	Eliminate the Hazard		Guard the Hazard		Provide a Safety Factor		Provide Redundancy		Provide Reliability
	Hazard	Safety	Hazard	Safety	Hazard	Safety	Hazard	Safety	
Natural	Gravity								
Structural/ Mechanical	No safe access	Provide safe access		Guardrails		Horizontal handholds, harnesses		Safety line anchor points	99%
Electrical									
Chemical									
Radiant Energy									
Biological									
Automated Systems									

Applicable Standards and Literature

OSHA

1910.269 Electric Power Generation, Transmission, and Distribution (9)(2)(i) Personal Fall Arrest Equipment shall meet the requirements of Subpart M of Part 1926 of this chapter.

1915.159 Subpart I, Shipyards - Personal Fall Arrest Systems (PFAS).

1926.250 General Requirements for Storage (b)(2) Each employee required to work with stored materials in silos, hoppers, tanks, and similar storage areas shall be equipped with personal fall arrest equipment meeting the requirements of Subpart M of this part.

1926.451 Subpart L, General Requirements, Scaffolds

1926.500 Subpart M, Fall Protection
(a) Scope of Application
(b) Definitions

1926.501 Duty to have Fall Protection

1926.502 Fall Protection Systems Criteria and Practices

1926.503 Training Requirements, Subpart M, Appendix:
(a) Determining Roof Widths
(b) Guardrail Systems
(c) Personal Fall Arrest Systems
(d) Positioning Device Systems
(e) Fall Protection Plan
  (i) Statement of Company Policy
  (ii) Fall Protection Systems to be Used on this Project
    - Safety Monitoring System
    - Control Zone System
    - Holes
  (iii) Implementation of Fall Protection Plan
    - Columns
    - Inverted Tee Beams
    - Spandrel Beams
    - Floor & Roof Members
    - Detailing
  (iv). Conventional Fall Protection Considered for Point of Erection or Leading Edge Erection Operations
    - Personal Fall Arrest Systems
    - Safety Net Systems
    - Guard Rail Systems
  (v) Other Fall Protection Measures Considered for This Job
  (vi) Enforcement
  (vii) Accident Investigation
  (viii) Changes to Plan

1926.550 Cranes & Derricks, Subpart N, Cranes, Derricks, Hoists, Elevators and Conveyors

1926.753 Safety Nets, Subpart R Steel Erection is to be added to fall protection requirement for steel erection.

Note: Sloped roofs slide guards (2 × 6 inch boards held on edge by brackets) became past of the regulations for fall protection in December 1995.

1926.951 Tools and protective equipment, Subpart V, Power Transmission and Distribution.

1926.1050-1053, Subpart X, Stairways and Ladders - 1926.1053(a)(19)(i) Ladder Safety Devices.

Comment: Use warning lines positioned 6–10 feet from flat edges of roofs. Some federal OSHA-allowed systems include roofing slide guards for pitched roofs; control access zones for brick and block laying; monitoring systems for roofs; and procedures developed in an OSHA-style fall protection plan.

NIOSH

Training should be provided minimally every 6 months per NIOSH Criteria Document 76–128 for egress from elevated workstations.

MSHA

56.15005 Safety Belts and Lines

ANSI

A10.6-90 Safety Requirement for Demolition Operations, General Protection, Chapter 5, Section 9

A10.8-88 Scaffolding Safety Requirements, Section 6.5, Body Belts and Drop Lines

A10.11-89 Construction and Demolition Operations—Personnel and Debris Nets

A10.14-91 Construction and Demolition Operations—Requirements for Safety Belts, Harnesses, Lanyards and Lifelines for Construction and Demolition Use

A10.28-90 Work Platforms Suspended from Cranes or Derricks for Construction and Demolition Operation, Section 10–Personal Protection Equipment

A10.32 Standards for Demolition and Construction, Dec. 2004

A14.3-92 Safety Requirement for Fixed Ladders, Section 6.3—Ladder Safety Devices

A90.1-92 Safety Standard for Belt Manlifts

A92.3-90 Manually Propelled Elevating Aerial Platforms, Section 4.9.5—Attachments points

A120.1-96 Safety Requirements Powered Platforms Exterior Building, Section 28—Safety Belts and LifeLines

Z359.1-92 Safety Requirements for Personal Fall Arrest Systems, Subsystems and Component

U.S. Army Corps of Engineers

EM 385-1-1, (3 Sept. 1996) *Safety and Health Requirements Manual*

Comment: The Corps of Engineers' requirements meet or exceed OSHA standards.

05.F Body Belts, Harnesses, Lanyards, and Lifelines—Selection of Components

05.F.01 Personal fall arrest and positioning device systems

(a) Personal fall arrest systems require the use of body harnesses: body belts are not acceptable as part of personal fall arrest systems.

(b) The use of a body belt is permitted in a positioning device system.

05.F.02 Body belts and harnesses

(a) Connectors shall be drop forged, pressed or formed steel, or made of equivalent materials: shall have corrosion resistant finish: and all surfaces and edges shall be smooth to prevent damage to interfacing parts of the system.

(b) D-rings, snaphooks, and other connectors shall have a minimum tensile strength of 2.270 kg (5000 lbs.): D-rings and snaphooks shall be proof-tested to a minimum tensile load of 1600 kg (3600 lbs.) without cracking, breaking, or taking permanent deformation. The manufacturer typically conducts proof testing, and a specification of proof testing supplied with the manufactured goods.

(c) Body belt/harness systems shall decelerate and bring the employee to a complete stop within 1 m (42 in), excluding lifeline elongation, after free fall distance.

(d) Body belt/harness systems, when stopping or preventing a fall, shall not produce an arresting force on an employee of more than 10 times the employee's weight or 800 kg (1800 lbs.), whichever is lower.

(e) Body belts shall be at least 4 cm (1-5/8 in.) wide.

05.F.03 Lifelines and lanyards

(a) Lanyards and vertical lifelines shall have a minimum tensile strength of 2,270 kg (5000 lbs.).

(b) Horizontal lifelines shall be designed, installed, and used, under the supervision of a qualified person, as part of a complete personal fall arrest system, which maintains a factor of safety of at least two.

(c) Self-retracting lifelines and lanyards that automatically limit free fall distance to 60 cm (2 feet) or less shall be capable of sustaining a minimum tensile load of 1360 kg (3000 lbs.) applied to the device with the lifeline or lanyard in the fully extended position. Self-retracting lifelines and lanyards that do not limit free fall distance to 60 cm (2 feet) or less, ripstitch lanyards, and tearing and deforming lanyards shall be capable of sustaining a minimum tensile load of 2,270 kg (5000 lbs.) applied to the device with the lifeline or lanyard in the fully extended position.

(d) Ropes and straps (webbing) used in lanyards, lifelines, and strength components of body belts and body harnesses shall be made from synthetic fibers.

05.F.04 Lineman's equipment

(a) All fabric for safety straps shall be capable of withstanding an alternating current dielectric test of not less than 2500 volts per foot "dry" for 3 minutes, without visible deterioration.

(b) All fabric and leather used shall be capable of being tested for leakage current and not exceed 1 milliampere when a potential of 3000 volts is applied to the electrodes 30 cm (12 in.) apart.

(c) Direct current testing may be permitted in lieu of alternating current testing.

05.G Safety and Debris nets—Design and Testing

06.G.01 Design and safety nets

(a) The minimum size of each safety net mesh opening shall not exceed 90 cm (36 in.) square, nor be longer than 15 cm (6 in.) on any side, and the opening - measured center-to-center of mesh ropes or webbing - shall not be longer than 15 cm (6 in.). All mesh crossings shall be secured to prevent enlargement of the mesh opening.

(b) All new nets shall meet accepted performance standards of 17,500 foot-pounds minimum impact resistance as determined and certified by the manufacturer and shall bear a label of proof test.

(c) Each safety net (or section) shall have a border rope for webbing with a minimum breaking strength of 2,270 kg (5000 lbs.).

(d) Connections between safety net panels shall be as strong as integral net components and spaced not more than 15 cm (6 in.) apart.

05.G.02 Testing of safety nets

(a) Safety nets and safety net installations shall be tested in the suspended position immediately after installation and before being used as a fall protection system, whenever relocated, after major repair, and, when left at one location, at not more than six-month intervals.

(b) The test shall consist of dropping into the net a 180 kg (400 lb.) bag of sand, not more than 75 cm +/–5 cm (30 in +/– 2 in.) in diameter, at least 1 m (42 in.) above the highest working/walking surface at which employees are exposed to fall hazards.

(c) Defective nets shall not be used.

05.G.03 Design of debris nets

(a) Debris nets shall be constructed of wire or synthetic netting of not more than 25 mm (1 in.) mesh.

(b) Wire mesh shall be made of not less than 22-gauge wire and synthetic mesh of not less than Number 18 twine.

Other Government Specs

MIL STD H-24460 Harness, Safety; and Lanyards, Safety and Working, 8/30/93.

Air Force OS Health 91-2 Vehicle-Mounted Elevating and Rotating Work Platforms, Manually Propelled and Self-Propelled Mobile Work Platforms, and Scaffolds, 12/28/93.

Federal Specs 58005, Truck, Fire Fighting (Telescoping Aerial Ladder and Work Platform), 4/2/92.

MIL-STD 2151 Inclined Ladder Tread Test Methods and Equipment for Wear, Slip-Resistance and Impact, 3/21/91.

MIL STD P-23210 Platforms, Aluminum and Steel; and Accessories (for ships' prows), 2/5/91.

Federal Spec 50120 Harness, Climbing, Full-Body, 2/21/90.

MIL-STD H-24626/1 Harness Assemblies, Cable, Pressure-Proof, Fiber, 11/27/87.

MIL-STD H-24626 Harness Assemblies, Cable, Pressure-Proof Fiber, 11/27/87.

Federal Spec JJ-B160 Belt, Safety, Industrial, Safety-Body-Pad Harness, 4/1/77.

European Specs

These European Standards came into effect in 1995 linked to ISO 9000, which requires approved suppliers for fall equipment.

5973 (British)-Code of Practice for Access and Working Scaffolds and Special Scaffold Structures in Steel, 1/1/93.

7478 (German) Safety Harnesses; Upper Body Harness for Mining, 4/1/93.

7472 P2 (German) Safety Harness; Steel Fittings; Connecting Dimensions, Safety Requirements and Testing, 1/1/91.

280 (Preliminary Draft) Mobile Elevating Work Platforms-Design Calculation

Stability Criteria-Construction-Safety -Examination and Tests, 9/1/91.

Service and Working Scaffolds; Ladder Scaffolds; Safety Requirements, 12/1/90.

7171 (British) Specification for Mobile Elevating Work Platforms, 1/1/89.

6289-P1 (British) Work Platforms-Code of Practice for Mobile Scissor Operated Work Platforms, 1/1/82.

Other References

- Standards updates are located on the web at *http://www.fallsafety.com.*
- Institute of Electrical and Electronics Engineers (IEEE) 1307, *Trial Use Guide for Fall Protection for the Utility Industry,* 6/20/96.
- *Introduction to Fall Protection* by J. Nigel Ellis, PhD, CSP, PE, published by ASSE, 1988, 1993, & 2001.
- "Fall Protection," *Hazard Information Newsletter*, Volume 2, Issue 4, July 1997.
- David MacCollum, *Construction Safety Planning,* Van Nostrand Reinhold, 1995 (now by John Wiley).
  - Fall Prevention and Protection, *Fall Prevention*, Chapter 15, pp. 101–104.
  - *Phase Planning for Steel Erection*, Chapter 25, pp. 187–189.
  - *Phase Planning for Exteriors*, Chapter 29, p. 218, Figure 29-1.
  - *Phase Planning for Roofs*, Chapter 30, p. 230, Figure 30-1.

Commentary

Hazards from a fall from elevations require special methods of control. As the location and attributes of elevated locations differ, the most effective method of fall protection is specific to the site. Much credit goes to Nigel Ellis, the foremost expert on fall protection and noted author on fall safety issues. His research and study of providing safe fall protection speaks well of his dedication to the safety of the construction worker. Nigel can be contacted at

J Nigel Ellis, PhD, CSP, CPE, PE
President, CEO: Fall Safety Solutions

306 Country Club Dr.
Wilmington, DE 19803
(800) 372-7775
(302) 571-0767
fax (302) 571-0756
Nigel@fallsafety.com

Recommendations

Manufacturers of steel formwork for columns need to include safety-line anchor attachments and a permanent ladder system for access on form panels. Similar elevated work places must contain designated anchor attachments in the design of the piece. Architects should design permanent attachment opportunities into elevated locations for both builders and maintenance workers. A design calling for a permanent guardrail around the rooftop shown below would eliminate the possibility of falls during construction or maintenance activities.

SECTION 3

Falling Objects

The Hazard Syndrome

A great number of injuries and deaths are the result of pedestrians being struck with falling objects. In addition to the specific source, falling objects have been mentioned in Chapter 7, (Two-blocking, Jib Storage, Latticework-Boom Collapse) and Chapter 9 (Killer Hooks, Nuts, Bolts, and Connectors). Since it is working against the natural force of gravity, the hazard of falling objects is always *armed* when the object is unsecured, and becomes *active* when an object falls. The hazard of falling objects continues to create problems on many job sites, and is responsible for as many as 10% of construction-site injuries. This hazard must be carefully considered in the initial design of building projects. If overlooked, very expensive corrective methods may become necessary to avoid injury.

Scope

The danger of falling objects appears every time an object is supported or suspended in an overhead position. Many objects fall due to exposure to factors that create a change of circumstance that activates the hazard and causes the object to lose stability and subsequently fall. Some falling-object hazards are caused by objects that have been improperly anchored in their overhead position; others are caused by circumstances that were not factored into design specifications. The following two examples illustrate that falling objects can strike at any time, catching unsuspecting workers and even passersby off-guard.

A number of years ago a unique, attractive, and totally glass-enclosed skyscraper was erected in Boston. Not too long after its completion, some of the large glass panels began to fall from the building during temperature changes or high winds. The surrounding area had to be barricaded until the glass panels and frames could be completely redesigned to withstand the elements and the building was resheathed.

Visitors to Moscow, Russia often see eight-foot pieces of wood with wire netting extending from the second-story level of many buildings. These debris nets have been installed to catch the many building tiles that continually fall from the face of building walls. Apparently the adhesive used to secure these tiles cannot resist the effects of the severe winter climate.

Substantial monetary savings would have resulted in both of these instances had the initial design taken into consideration the effects of severe weather conditions on the materials chosen for these buildings.

Safety procedures, standards, and requirements abound for preventing falling objects from injuring or killing people. Each situation varies, and different and creative approaches are necessary to ensure the prevention of falling objects.

Peril from the Failure Mode

What goes up must come down. An object in an elevated location must be constrained in order to prevent its being dislodged and falling, for gravity is a *natural* hazard that is always with us. However, this hazard can be mostly overcome by engineering controls. Many times a falling object can be traced back to a *mechanical/ structural* failure.

Hard hats are not enough. One of the most common falling objects is an unbraced joist upon which pallets of decking are placed during construction. These joists often tip and cause the materials placed on them to fall through the joists, crushing those below.

Brief Summary of the Occurrence

A crane was used to place a pallet of decking on unbraced joists. The joists fell near a group of workers like dominoes. The bracing attachment between the joists was defective, causing the joists to tip. A pallet fell through and landed on a worker, crushing him to death.

Date of Occurrence

March 18, 1985

Source of Information

Circuit Court, Wayne County, MI, # 85-508-650-NO

History of the Hazard

People have been plagued by this hazard since Biblical times. In Today's English Version of the Good News Bible, Ezekiel 13, Verses 10–12 in the Prophecy against False Male Prophets, is the following quote: "The Prophets mislead my people by saying that all is well. All is certainly not well! My people have put up a wall of loose stones and then the prophets have come and covered it with whitewash. Tell the prophets that their wall is going to fall down. I will send a pouring rain. Hailstones will fall on it, and a strong wind will blow against it. The wall will collapse, and everyone will ask you what good the whitewash did." Today, whitewashing over any hazard doesn't cut it!

Post–World War I: Protective headgear was introduced for miners.

World War II: Shipyard workers used aluminum hard hats as protection from small falling objects. After World War II, loggers and construction workers soon adopted hard hats.

Late 1940s: At logging camps, towering Douglas fir trees being felled or tree limbs falling during a windstorm were labeled as "widowmakers," and the only protection was to avoid these locations when possible. Since then, tractors, forklifts, dump trucks, and other vehicles have included FOPS/ROPS to alleviate widowmaker conditions.

For over 50 years falling objects have been recognized as a significant hazard, as shown by *American Standard Safety Code for Building Construction*, ASA A10.2-1944, pp.12–13 and 80–82, (published June 7, 1944).

Contributory Factors

Human nature has some odd quirks, as all of us can recall times that a temporary fix to prop up an item didn't work. One common myth is that "It will only take a moment to fix. This temporary support will hold until I'm finished." The truth of the matter is that any suspended load or jacked-up structure presents an armed hazard and is not safe to work underneath, unless it is securely supported with cribbing, blocking, or other solid and strong material.

In construction, there is little exchange of hazard information at the time of design. For instance, the architect or engineer should incorporate a bracing system between joists to prevent tipping during construction. Such a system could work so the brace easily folds out when the joist is first set.

Hazard Identification and Prevention Matrix

Simple requirements make the prevention of falling objects a realistic goal.

	Eliminate the Hazard		Guard the Hazard		Provide a Safety Factor		Provide Redundancy		Provide Reliability
	Hazard	Safety	Hazard	Safety	Hazard	Safety	Hazard	Safety	
Natural	Gravity								
Structural/ Mechanical	Com-ponent failure	Joists equipped with brace mechanisms				Locking point (joist attachment)		Duplicate joist-securing device	100%
Electrical									
Chemical									
Radiant Energy									
Biological									
Automated Systems									

Applicable Standards and Literature

OSHA

Code of Federal Regulations (CFR) - *Labor Part* 1910 (Rev. July 1, 1988).

1910.28 Safety Requirements for Scaffolding (a)(3)(ii) "Guardrails should all be 2 × 4 inches or the equivalent, installed no less than 36 inches or not more than 42 inches high, with a midrail, when required, of 1 × 4-inch lumber or equivalent. Supports should be at intervals not to exceed ten feet. Toeboards shall be a minimum of 4 inches in height."

1910.135 Occupational head protection. Helmets for the protection of heads of occupational workers from impact and penetration from falling and flying objects and from limited electric shock and burn shall meet the requirements and specifications established in American National Standard Safety Requirements for Industrial Head Protection, Z89.1-1969.

1910.176 Handling materials - general. (b) Storage of material shall not create a hazard. Bags, containers, bundles, etc., stored in tiers shall be stacked, blocked, interlocked and limited in height so that they are stable and secure against sliding or collapse.

1910.244 Other portable tools and equipment. (a) Jacks - (1) Loading and marking. (i) The operator shall make sure that the jack used has a rating sufficient to lift and sustain the load. (2) Operation and maintenance. (i) In the absence of a firm foundation, the base of the jack shall be blocked. If there is a possibility of slippage of the cap, a block shall be placed in between the cap and the load. (iii) After the load has been raised, it shall be cribbed, blocked, or otherwise secured at once.

1910.266 Pulpwood logging. (d)(2) Protective Canopy. A protective canopy shall be provided for the operator of mobile equipment. It shall be so constructed as to protect the operator from injury due to falling trees or limbs . . .

Code of Federal Regulations (CFR) - Labor Part 1926 (Rev. July 1, 1993).

1926.100 Head Protection. (a) Employees working in areas where there is a possible danger of head injury from impact, or from falling of flying objects, or from electrical shocks and burns, shall be protected by protective helmets. (b) Helmets for protection of employees against impact and penetration of falling and flying objects shall meet the specifications contained in American National Standards Institute, Z89.1-1969, Safety Requirements for Industrial Head Protection.

1926.250 General requirements for storage. (a) General. (1) All materials stored in tiers shall be stacked, racked, blocked, interlocked, or otherwise secured to prevent sliding, falling or collapse. (b) Material storage. (1) Material stored inside buildings under construction shall not be placed within 6 feet of any hoist way or inside floor openings, nor within 10 feet of an exterior wall which does not extend above the top of the material stored.

(b)(4) Bagged materials shall be stacked by stepping back the layers and cross-keying the bags at least every 10 bags high.

(b)(5) Materials shall not be stored on scaffolds or runways in excess of supplies needed for immediate operations.

(b)(6) Brick stacks shall not be more than 7 feet in height. When a loose brick stack reaches a height of 4 feet, it shall be tapered back 2 inches in every foot of height above the 4-foot level.

(b)(7) When masonry blocks are stacked higher than 6 feet, the stack shall be tapered back one-half block per tier above the 6-foot level.

(b)(8) Lumber:

(ii) Lumber shall be stacked on level and solidly supported sills.

(iii) Lumber shall be so stacked as to be stable and self-supporting.

(iv) Lumber piles shall not exceed 20 feet in height provided that lumber to be handled manually shall not be stacked more than 16 feet high.

(b)(9) Structural steel, poles, pipe bar stock, and other cylindrical materials, unless racked, shall not be stacked and blocked so as to prevent spreading or tilting.

1926.600 Equipment

(a)(3)(I) Heavy machinery, equipment, or parts thereof, which are suspended or held aloft by use of slings, hoists or jacks shall be substantially blocked or cribbed to prevent falling or shifting before employees are permitted to work under or between them. Bulldozer and scraper blades, end-loader buckets, dump bodies, and similar equipment, shall be either fully lowered or blocked when being repaired or when not in use. All controls shall be in neutral position with the motors stopped and brakes set, unless work being performed requires otherwise.

1926.603 Pile driving equipment

(a)(3) Overhead protection, which will not obscure the vision of the operator and which meets the requirements of Subpart N of this part, shall be provided. Protection shall be the equivalent of 2-inch planking or other solid material of equivalent strength.

(a)(5) A blocking device, capable of safely supporting the weight of the hammer, shall be provided for placement in the leads under the hammer at all times while employees are working under the hammer.

(a)(7) When leads must be inclined in the driving of batter piles, provisions shall be made to stabilize the leads.

1926.604 Site clearing

(a)(2) All equipment used in site clearing operations shall be equipped with rollover guards meeting the requirements of this Subpart. In addition, rider-operated equipment shall be equipped with an overhead and rear canopy guard meeting the following requirements: (I) The overhead covering on this canopy structure shall be of not less than 1/8-inch steel plate or 1/4-inch woven wire mesh with openings no greater than 1 inch, or equivalent.

1926.651 General requirements

(e) Exposure to falling loads. No employee shall be permitted underneath loads handled by lifting or digging equipment. Employees shall be required to stand away from any vehicle being loaded or unloaded to avoid being struck by any spillage or falling materials. Operators may remain in the cabs of vehicles being loaded or unloaded when the vehicles are equipped, in accordance with 1926.601(b)(6), to provide adequate protection for the operator during the loading and unloading operations.

1926.1003 Overhead protection for operators of agricultural and industrial tractors.

(a) General-(1) Purpose. When overhead protection is provided on wheel-type agricultural and industrial tractors, the overhead protection shall be designed and installed according to the requirements contained in this section. The provisions of 1926.1001 for rubber tired dozers and rubber-tired loaders may be used in lieu of the standards contained in this section. The purpose of this standard is to minimize the possibility of operator injury resulting from overhead hazards such as flying and falling objects, and at the same time to minimize the possibility of operator injury from the cover itself in the event of accidental upset.

ANSI

A1264.1-1995 American National Standard, Safety Requirements for Workplace Floor and Wall Openings, Stairs & Railings Systems, which superseded ANSI A12.1-1989 Safety Requirements for Floor and Wall Openings, Railing and Toeboards, (published since 1967).

A10.6-1990 Safety Requirements for Demolition Operations has many requirements that provide effective protection against falling objects.
Comment: It would be a significant improvement to address the hazard of falling objects to both workers and the public to ensure for management priority to overcome this hazard.

A10.11-1989 Minimum Requirement for Safety Nets (published since 1971). **Summary:** Safety nets need sufficient clearance to prevent contact with objects below. Install nets that can absorb the impact force of a drop test, which should be performed and documented at the jobsite. The maximum opening of the safety net should not exceed 36 square inches and be no longer than 6 inches on any side. When measuring center to center, the rope or webbing should not exceed 6 inches. The breaking strength of the border rope should be 5000 pounds minimum. Connectors need to be as strong as the integral net and spaced not more than 6 inches apart.

A10.13-1989 Safety Requirements for Steel Erection also covers safety nets (published since 1972).

A10.34 Public Protection (under development).

A10.37 Debris Nets (under development)

Z89.1-1986 Protective Headwear for Industrial Workers - a standard that provides guidelines for "level 1" helmets, which protect against objects striking from above, and a "level 2" helmet, which protects against impact from the sides, front and back.

Society of Automotive Engineers (SAE)

SAE J1356 Performance Criteria for Falling Objects Guards for Excavators, February 1, 1988.

SAE J1119 Steel Products for Rollover Protective Structures (ROPS) and Falling Object Protective Structures (FOPS), December 1, 1988.

SAE J231 Minimum Performance Criteria for Falling Object, January 1, 1981.

SAE J1043 Performance Criteria for Falling Object Protective Structure (FOPS) for Industrial Machines, September 1, 1987.

U.S Army Corps of Engineers

Safety and Health Requirements Manual, EM 385-1-1, U.S. Army Corps of Engineers, U.S. Government Printing Office, Washington, D.C., November 3, 2003. "Overhead Protection shall be provided where the public or workers are subject to injury from falling objects" (published since 1946).

National Safety Council (NSC)

Safety Nets for Construction Projects, Data Sheet 608, National Safety Council, 1983, (published since 1971).

Society of Automotive Engineers (SAE)

SAE J1356 Performance Criteria for Falling Objects Guards for Excavators, February 1, 1988.

SAE J1119 Steel Products for Rollover Protective Structures (ROPS) and Falling Object Protective Structures (FOPS), December 1, 1988.

SAE J231 Minimum Performance Criteria for Falling Object, January 1, 1981.

SAE J1043 Performance Criteria for Falling Object Protective Structure (FOPS) for Industrial Machines, September 1, 1987.

International Organization for Standardization (ISO)

ISO 3449 Earth Moving Machinery-Falling Object Protective Structures - Laboratory Tests and Performance Requirements, May 15, 1992.

ISO 3164 Earth-Moving Machinery - Laboratory Evaluations for Roll-over and Falling-Object Protective Structures - Specifications for Deflection - Limiting Volume, November 15, 1995.

ISO 8083 Machinery for Forestry - Falling Object Protective Structures - Laboratory Tests and Performance Requirements, December 15, 1989.

Other References

- Planning is essential to avoid injury from falling objects. The book *Construction Safety Planning* by David V. MacCollum, PE, CSP, Van Nostrand Reinhold 1993, (now John Wiley) provides many examples of how to recognize those objects that are prone to fall unless special measures are taken.
- "Self-Climbing Safety Screens Catch Construction Debris," *Engineering News Record,* p. 16, August 27, 1987.
- "Safety Net Result: Money Spent, Lives Saved," *Engineering News Record*, July 8, 1976.
- "How to Rig Perimeter Nets," *National Safety News*, pp.24 and 83–84, November, 1964.
- "Falling Objects," *Hazard Information Newsletter*, Volume 1, Issue 11, February, 1997.

Commentary

Without the design to incorporate methods to prevent falling objects, current construction supervision can foreseeably overlook circumstances that create conditions that allow falling objects. The lesson in the described occurrence highlights the need for prepacked joist systems that will provide bracing that folds out when joists or trusses are set in place. This safeguard prevents tipping and collapsing first.

In addition, the design of staging always needs to include toeboards to prevent tools, materials, or other items from being kicked from the elevated staging when someone stumbles over these tripping hazards. Equipment such as dump trucks and front-end loaders should have safety struts to lock the raised dump bed or back loader in a raised position when service is being performed to prevent them from falling down and striking someone underneath. These raised devices can become a trap without the proper protection.

Recommendations

There is a great need for architects, engineers, and construction managers to have an opportunity to mutually review the design of structural components so it does not conflict with the construction methods. By applying technology specific to one hazard into the prevention of another hazard and sharing ideas and knowledge, universal hazards such as falling objects can be more effectively controlled.

SECTION 4

Ladders

The Hazard Syndrome

The primary purpose of a portable ladder is to provide access by ascending to or descending from an elevated area. This action presents the *dormant* hazard of falling from an elevation. Ladders are not intended to provide a work platform, but are used as an expediency device to perform infrequent maintenance such as replacing burnt-out light bulbs, painting, or hanging wallpaper. A ladder used as a work platform, a ladder with inadequate handholds, or too much space between rungs all create conditions that *arm* the hazard. Falls from a ladder occur in the *active* mode.

Scope

If properly designed, ladders can be used infrequently as a means of access. Portable ladders are exceedingly convenient. In numerous instances, though, portable ladders are misused on construction sites and result in serious injuries such as slipping or falling from elevation. In addition to the temptation towards improper use of the ladder as a work platform, many people will attempt to carry materials or tools up and down a portable ladder. Hauling objects up or down a ladder compromises the individual's grip on the ladder, increasing the probability of a fall. Ladders with a slightly damaged rung or side rail are particularly dangerous, because they present a temptation for continued use. Defunct or compromised ladders should be fixed or destroyed.

The seriousness of injuries from ladders is directly related to the height of the ladder. The greater the distance of the fall, the more serious the injury. Knee injuries from overuse are also common. Every ascent and descent on a ladder increases a worker's chance for injury or death.

Peril from the Failure Mode

A worker on a ladder being used as a work platform can topple from an unanchored position and sustain a very serious injury. Ladders present the *natural* hazard of gravity exacerbated by *mechanical* hazards of instability. According to the National Safety Council's Accident Facts, 1997 Edition, deaths occurring from falls from ladders and scaffolding average more than 300 each year. However, this data does not break down the figures into those just for ladders. Injury tracking of hospital emergency visits shows some 238,000 injuries per year as a result of falls from ladders. Unfortunately, OSHA reporting does not focus on ladders in particular. It states in its 1993 OSHA Booklet #3124, *Stairways and Ladders*, "OSHA estimates that there are 24,882 injuries and as many as 36 fatalities per year due to falls from stairways and ladders used in construction. Nearly half of these injuries are serious enough to require time off the job." As long as nonspecific statistics are published and mix ladders with scaffolds or stairs with no direct emphasis on the type of hazard that caused the failure mode, operating personnel often do not suspect how dangerous ladder use really is and rarely take proper measures to ensure safer ladder use or provide, safer alternate access.

Brief Summary of the Occurrence

A worker toppled from a ladder while trying to attach a building-cladding panel. The design of the panel did not allow the panel to be fastened when standing inside the building; it required the use of a ladder from outside the building. The worker fell from the ladder and suffered brain damage.

Date of Occurrence

November 3, 1988

Source of Information

Circuit Court, Cook County, Law Division, IL, # 88-L-20470

History of the Hazard

Portable ladders have been with us since the earliest times. The Chinese were among the earliest users, as were many Native American tribes in pueblos in the Southwest.

Portable ladders in European medieval times were often used to scale castle walls during sieges.

1880s: Ladders were developed by fire departments and quickly popularized.

1923: American Standards Association (now known as American National Standards Institute or ANSI) began to develop standards for all types of ladders.

2003: Nigel Ellis' book on fall protection and prevention, mentioned in Chapter 9, discussed safer access at the landing where the ladder provides access. It has been found that a horizontal handhold affords a more efficient grip than does a vertical handrail.

Contributory Factors

As an inexpensive expediency, use of portable ladders instead of a work platform, staging, scaffolding, aerial lifts, or other safer means of access is a dangerous temptation. A scenario as common as the practice of someone standing on a 20-foot lightweight aluminum stepladder to replace burnt-out light bulbs in a church sanctuary, rather than using stable scaffolding that would straddle the pews, is a hazardous practice. A fall from the ladder could result in lifetime disability or death.

Many enterprises rely upon people who are unaccustomed to using ladders to perform tasks from an elevated location. These workers probably do not know that the particular use of the ladder is not the safest choice. Another ladder failure mode is the collapse of a side rail below the bottom rung. This situation may arise when one side rail of an aluminum stepladder sinks into the ground and exerts a bending force that causes the ladder to tip sideways and fall when weight is applied to it. A simple strut or gusset at this point is a design improvement.

Some of the worst risks occur in small undertakings or on small jobs where there is no great choice of ladders, and where use is intermittent or infrequent. There is a tendency to improvise, to keep defective ladders in service instead of throwing them away, and to repair ladders in an elementary or amateur way, such as nailing on slats of wood to prop up a cracked step.

Injuries and deaths from ladders can usually be attributed to lack of planning, which fails to ensure a safer type of access. Sometimes well-established standards for permanent ladder access are completely overlooked.

Hazard Identification and Prevention Matrix

Ladders do not provide a perfect access, but when properly designed, they provide reasonable access to infrequent locations.

	Eliminate the Hazard		Guard the Hazard		Provide a Safety Factor		Provide Redundancy		Provide Reliability
	Hazard	Safety	Hazard	Safety	Hazard	Safety	Hazard	Safety	
Natural	Gravity/ instability								
Structural/ Mechanical	Improper use (work platform)	Construct work platform		Horizontal hand grips		Built to standard	Instability due to poor mainte-nance	Anchor ladder	90%
Electrical									
Chemical									
Radiant Energy									
Biological									
Automated Systems									

Applicable Standards and Literature

OSHA

OSHA 1926.1050-1053, Subpart X Stairways and Ladders

OSHA 1926 .1053(A)(19)(i) Ladder Safety Devices

OSHA 3124 Stairways and Ladders

MSHA

CFR 30

56.11003 Construction and Maintenance of ladders

5.11004 Portable rigid ladders

56.11005 Fixed ladder anchorage and toe clearance

56.11006 Fixed ladder landings

56.11007 Wooden components of ladders

56.11011 Use of ladders

56.11017 Inclined fixed ladders

56.11025 Railed landings, backguards, and other protection for fixed ladders

56.11026 Protection for inclined fixed ladders

ANSI

A14 Safety Code for Construction, Care and Use of Ladders

A14.1 Ladders - Portable Wood Safety Requirements

A14.2 Safety Requirements for Portable Metal Ladders

A14.3 Safety Requirements for Fixed Ladders

A14.4 Job Made Wooden Ladders

A14.5 Safety Requirements for Portable Reinforced Plastic Ladders

U.S. Army Corps of Engineers

Safety and Health Requirements Manual EM 385-1-1, November 23, 2003.

Section 21 Safe Access and Fall Protection

1.D Ladders

1.D.01 Length

1.D.02 Width

1.D.03 Spacing of rungs, cleats, and steps

1.D08 Set-up

21.D.09 Use

22.A General

22.A.02 c.

22.A.04 b. & c.

22.B. Scaffolds

22.B.08 a. & b.

22.C. Metal Scaffold and Towers

22.B.03

Military Specifications

MIL-L-12197 Ladders, Fire, Extension; Ladders, Fire, Roof; and Ladders, Fire, Combination

MIL-L-13039 Ladders, Vehicle-Boarding

MIL-L-221 Ladders, Jacob's

MIL-L-52822 Ladders Vehicular Boarding

MIL-L-54002 Ladders: Aluminum, Three Combination: Step

MIL-L-6378 Ladder, Folding Refueling

MIL-T-24634 Treads, Compound-filled, for Inclined Ladders

National Fire Protection Association (NFPA)

NFPA 1931 Design and Design Verification Tests for Fire Department Ground Ladders

FPA 1932 Use, Maintenance, and Service Testing of Fire Department Ground Ladders

Underwriter's Laboratories (UL)

Underwriters Laboratories has a number of specifications and standards for portable wood and metal ladders (UL112 & UL184).

Other References

- *Accident Prevention Manual*, 11th Edition, National Safety Council (NSC), 1997.
- *Accident Facts*, 1997 Edition, NSC.
- David V. MacCollum, *Construction Safety Planning*, Van Nostrand Reinhold Co. (now John Wiley), New York, NY, 1995 pp. 100–101.
- J. Nigel Ellis, *Introduction to Fall Protection,* 3rd Edition, American Society of Safety Engineers, 2001.
- Rick Burson, "Safe at Home?" *Ohio Monitor,* October 1991.
- *Job-Made Ladders,* NSC Data Sheet 568.
- Warren Freedman, *Products Liability for Corporate Counsels, Controllers, and Product Safety Executives*, Van Nostrand Reinhold Co. (now John Wiley), New York, NY, 1984 pp. 105–106.
- *Encyclopedia of Occupational Health and Safety*, Volume 2, International Labour Office, Geneva, Switzerland, 1983.
- Wesley E. Woodson, *Human Factors Design Handbook* McGraw-Hill Book Co., 1981.
- *Human Engineering Guide to Equipment Design*, American Institute for Research, Washington, DC, 1972.
- "Bargain Ladders -Will they be safe?" *Ohio Monitor*, December 1969. Reprinted by special permission from *Consumer Bulletin*, by Consumer Research, Inc.
- "Ladders," *Hazard Information Newsletter,* Volume 3, Issue 3, June 1998.

Commentary

Portable ladders are here to stay, but safer options are available. Construction planners should examine construction tasks to avoid the use of ladders during both construction and habitation. Pole-mounted light-bulb changers and fruit pickers keep people on the ground. Pole extensions for paint rollers often eliminate the need for ladders for painting high walls and ceilings. During construction, aerial lifts or temporary scaffolding should be employed to provide access to as many elevated tasks as possible. Other alternatives, such as mobile scissor lifts, further reduce the need for portable ladders.

Since portable ladders have many uses that are inherently unsafe, design alternatives should be employed to overcome failure modes. For instance, the use of struts and gussets to reinforce the side rail beneath the bottom rung of an aluminum ladder eliminates many structural flaws and improves its safety. Architects and designers of a building should consider maintenance that does not require the use of a ladder, such as the installation of building cladding from the inside of a building instead of fasteners that must be attached from the outside of a structure.

Recommendations

If no alternative to the use of a ladder is found, the following considerations should be heeded when using or choosing a ladder:

Fixed Ladders

The considerations for design of fixed ladders define some basic concepts:

- Ladders that extend more than 20 feet require some form of protection.
- Ladders between 20 and 30 feet with a cage must have a landing and offset to reduce the fall distance within the cage.
- Continuous ladders need a safety device to secure the occupant to the ladder for ascending or descending.
- Ladders must meet specific design requirements for the size of each rung and positioning of side rails.
- Fixed ladders that provide entry to an upper level must have handholds or grab rails that extend 3 feet above the floor.
- Overhead hatches in ceilings accessed by a fixed ladder should be designed so that when the hatch is open, it locks into an upright position and provides continued handholds for gaining access to the landing.
- Fixed ladders need to be properly maintained on a regular basis to avoid rust or corrosion.

Portable Ladders

Wood and Metal Ladders

- Is it long enough? A ladder needs to extend 3 feet above the top level to provide safe access to a roof or an elevated platform. This lets you hold onto the ladder when stepping to or from the roof.
- Is the ladder secured while being transported? Metal fatigue caused by bouncing or vibration on a metal carrying rack is a problem on ladders that are customarily transported on vehicles and not properly strapped down.
- Is the wooden ladder painted? If so, this could hide cracks.
- Can the ladder be tied off to prevent sideways movement?
- Does the ladder have slip-resistant bases, shoes, spikes, or spurs to prevent slippage or sinking in soft soil?
- Is the ladder free of defects or damage such as dents in the side rails, which can cause failure of the rungs and eventual collapse?
- Is the ladder made of nonconductive plastic material if it is to be used near powerlines?
- Does the location where the ladder will be used allow sufficient space for it to be tilted at the proper pitch angle? The position of the bottom of the ladder from the vertical should be one-quarter the vertical height of the ladder's resting place at the top.

- Is the ground level where the ladder is to be placed?
- Does the ladder include identification showing it complies with pertinent standards? For instance, ladders to be used in high-voltage environments must meet industry performance standards.
- Does the ladder have a metal spreader or locking device of sufficient size and strength to hold the front and back sections securely in the open position? Make sure the spreader is locked into position.
- If a wooden ladder, does it have a water-repellent coating, varnish, or other suitable protective finish?
- If an extension ladder is raised beyond its first section, make sure the extension is latched into place.
- Never use a stepladder as a straight ladder.

ALUMINUM LADDERS

In addition to some of the many items to look for as noted above, aluminum ladders need to be inspected monthly with special attention to the following details:

- No distortion. This is usually visible by sighting along the stiles. A twisted or distorted ladder can be hazardous to use.
- No slackness of rungs in the stiles.
- Ropes, pulleys, rivets, fastenings, and other fittings should be secure and serviceable.
- No corrosion.
- No sharp edges on the stiles and rungs.
- Antislip steps or rungs. These should be in satisfactory condition and not loose.

STEP LADDERS

- Maintenance of stepladders is the key to their safe use.
- Do not stand on the top of stepladders; they have not been designed for this.

Whenever workers use ladders, the employer should give clear instructions on their safe use and ensure adequate supervision. Only one person should be on a ladder at any one time. If two persons are required to do the job, a second ladder should be provided.

Extra care should be taken when ladders are in use near entrances and gangways, particularly in places where forklift trucks or other small vehicles are operating. To avoid unexpected collision, doors should be locked and gangways blocked wherever possible; otherwise, a lookout should be posted.

In order to prevent deterioration, wooden ladders should be stored so they will not be subjected to excessive drying that will cause rungs to become loose. Ladders should be stored well away from radiators, steam pipes or other sources of heat to avoid warping. They should be protected from direct exposure to the elements in a place with free air circulation.

SECTION 5

Dangerous Access to Equipment

The Hazard Syndrome

Fixed elevated-access systems on and off construction equipment, trucks, cranes, and railroad container cars are often extremely unsafe and contribute to the hazard of falls from elevation. Fixed elevated-access systems can be considered a hazard always in the *armed* mode. Fixed ladders, steps, and handholds are often arranged in a manner that inhibits safe climbing, as they fail to provide opportunities for three-point contact (one foot and two hands) while being ascended or descended. Such ladders often present a gap as great as two to three feet between the lowest rung and the ground. Specified or horizontal handholds are often missing. These conditions fail to provide safe body movement onto an elevated platform or landing. In the far-reaching issue of rail-car use and design, when container cars move along the track, the hazard of slips and falls from dangerous access is *dormant.* When these cars are in the intermodal yard (where containers are transferred between trucks, ships, and railroad cars) the hazard is *armed*, as personnel are depending upon unsafe access methods to set the latches of the cars. The hazard is *active* when a worker slips or falls or strains a muscle due to improper use compelled by design of the accessway.

Scope

Many large construction sites still build railroad spurs to haul in equipment and materials. With the advent of shipping containers that can be easily stacked on ships, stacked two-high on railroad cars, and also be transported by truck, construction has a new efficient and popular delivery mode for supplies brought to the construction site and for trash removal.

Smaller sites use trucks with high trailer beds to transport materials. Though the cabs of these trucks are subject to National Traffic Safety and Health Administration (NTSHA) access regulations, the trailers usually are not. Frequent climbing onto and off of the trailers can result in an increased exposure to fall hazards and knee or other bodily injuries.

Standards to ensure access systems that are more ergonomic and user-friendly are often hindered by a slow process of adoption on equipment. Due to infrequent use, some aspects of equipment are overlooked and safe design can take decades to implement. For instance, a commonly overlooked access hazard is lack of safe access to the top of a crane cab to assemble or disassemble the A-Frame (which supports the pennant guy-lines to the boom) as shown in the illustration on page 245.

Peril from the Failure Mode

A prevalent example of fixed access is the design of the ladders and walkways, and the absence of handrails on intermodal railroad container cars. The design of these cars' access systems for the railroad employees or construction personnel requires disconnecting

four latches on each corner of the container that connect containers stacked two-high on a railroad car. Each latch must be latched when securing the top container to a railroad car; and each must be released when removing the top container from a railroad car. This procedure requires work crews to repeatedly get on and off the car in order to release or latch these containers as they are being transferred. Most crewmembers must walk narrow accessways to load or unload some 200-plus containers in a shift. The design and repetitive performance of this function have been the source of many falls, back injuries, and soft tissue injuries from strains. The entire process is comprised of a disregard for ergonomic sensibility and design. Unsafe and inadequate handholds make injuries even more frequent. The narrow walkway six to seven feet high without guardrails is a source of a high number of falls resulting in injury. Unsafe access presents the multiple *natural* hazards of falls due to gravity, and the *mechanical* hazards of instability and inappropriate ergonomics.

Brief Summary of the Occurrence

A railroad groundsman fell from the narrow container car walkway without handrails when the crane operator was attempting to release the upper container. Release of the container required shaking the container, and the groundsman lost his balance on the narrow walkway that had no handrails. He was thrown to the ground seven feet below and sustained crippling injuries from the fall from the railroad car.

Date of Occurrence

July 25, 2002

Source of Information

Superior court of the State of California, County of Los Angeles, Southern District Case, # NC-035309

History of the Hazard

1890s: The first Railroad Safety Appliance Act was enacted. It addressed safe access by requiring automatic couplers, and stating that if the design of the railroad car is deemed unsafe and without reasonable safeguards, and the employee is injured as a result of the hazard, the employee has the right to pursue a civil lawsuit.

1959: The U.S. Railroad Retirement Board recognized that some 3,260 persons were killed or injured while getting on or off railroad cars and locomotives.

Contributory Factors

As OSHA does not oversee railroad safety, little or no attention is given to their requirements or other supporting industry standards for safe access. The railroad companies transfer responsibility of design and assembly of container cars to car manufacturers. Cars are then leased to those transporting goods in the containers. Railroad unions have no authority to participate in the design of these cars. Railroad workers affected by the design of these cars have no way to give feedback or input on how to improve accessibility and design.

Hazard Identification and Prevention Matrix

	Eliminate the Hazard		Guard the Hazard		Provide a Safety Factor		Provide Redundancy		Provide Reliability
	Hazard	Safety	Hazard	Safety	Hazard	Safety	Hazard	Safety	
Natural	Gravity falls								
Structural/ Mechanical	Poor design: latches, access-ways	Rails, guarded accessway standard width walkway		Ergonomics	Must climb onto car to reach desired location	Mobile raised platform, mobile access steps		Pole-mounted latch release device	98%
Electrical		Wireless latching system							
Chemical									
Radiant Energy									
Biological									
Artificial Intelligence									

Applicable Standards and Literature

OSHA

56.11004 Portable rigid ladders

56.11005 Fixed ladder anchorage and toe clearance

56.11006 Fixed ladder landings

56.11007 Wooden components of ladders

56.11011 Use of ladders

56.11017 Inclined fixed ladders

56.11025 Railed landings, backguards, and other protection for fixed ladders

56.11026 Protection for inclined fixed ladders

ANSI

ANSI A14.3 Safety Requirements for Fixed Ladders

U.S. Army Corps of Engineers

EM 385-1-1 (September 1996) *Safety and Health Requirements Manual*

Section 21 Safe Access and Fall Protection.

21.D Ladders

21.D.01 Length

21.D.02 Width

21.D.03 Spacing of rungs, cleats, and steps

21.D08 Set-up

21.D.09 Use

U.S. Department of Transportation

Title 49 Code of Federal Regulations: Railroad Safety Appliance Act

231.0 Applicability and Penalties

231.1 Box and other house cars built or placed in service before October 1, 1966

231.2 Hopper cars and high-side gondolas with fixed ends

231.3 Drop-end high-side gondola cars

231.4 Fixed-end low-side gondola and low-side hopper cars

231.5 Drop-end low-side gondola cars

231.6 Flat cars

231.7 Tanks cars with side platforms

231.8 Tank cars without side sills and tank cars with short side sills and end platforms

231.9 Tanks cars without end sills

Other References

- J. Nigel Ellis, *Introduction to Fall Protection,* 3rd Edition, Chapter 4, American Society of Safety Engineers, 2001.
- "Dangerous Access and Work Platforms," *Hazard Information Newsletter*, Volume 2, Issue 1, April 1997

Commentary

Until the railroad safety laws have achieved reasonably safe access standards on and off container cars by isolating the combined hazards of poor design and natural forces working against the physical limitations of the human body, unsafe access will continue to cause injury. The railroad employee unions should have authority to retain competent engineers to review the design of cars in order to ensure that access systems are safe and ergonomic for their intended use. One way to overcome the hazards inherent in such railroad-car access systems is to develop a system wherein the railroad grounds-people either use a special tool developed to trip the latches from the ground, or a mobile platform to provide access to eliminate the need to get on and off each car. For instance, railroad workers developed a tripping and release device mounted on a pole, allowing unlatching and relatching to be accomplished from the ground. Another railroad worker developed moveable self-powered step platforms that can be moved along the side of the rail car to provide safe access to the level of the latches. In this electronic age, wireless latches that can be activated from the ground would nearly eliminate the need to climb up on the cars. The problem of unsafe access can be easily combatted by technological innovations that can provide a variety of solutions.

However, in some cases, redesign of the duty and configuration of the work crew is enough to reduce a hazard. A system safety evaluation of the task of loading and unloading the containers in order to determine crew size should be performed. Placement of a groundsman on both sides of the container car would eliminate the continual crossover action currently performed by one person.

Recommendations

The unions have the authority under the Department of Transportation to review and approve the design of these cars before they are manufactured and placed in service. Design engineers of mobile equipment should examine fixed-access systems to take a more permanent and ergonomic approach. Tank truckers could be made safe by adding collapsible guardrails in a way similar to that which is shown in the following series of illustrations.

HANS KLAESER
HERTEN
TIR
33
1203
RE KH 157

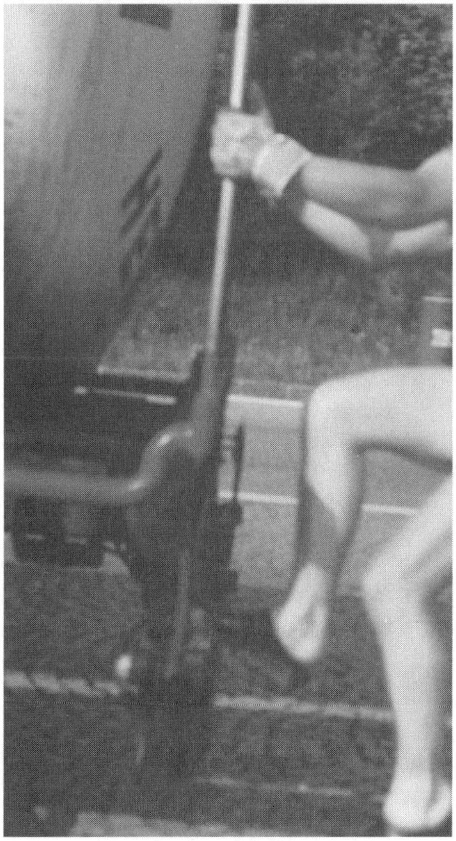

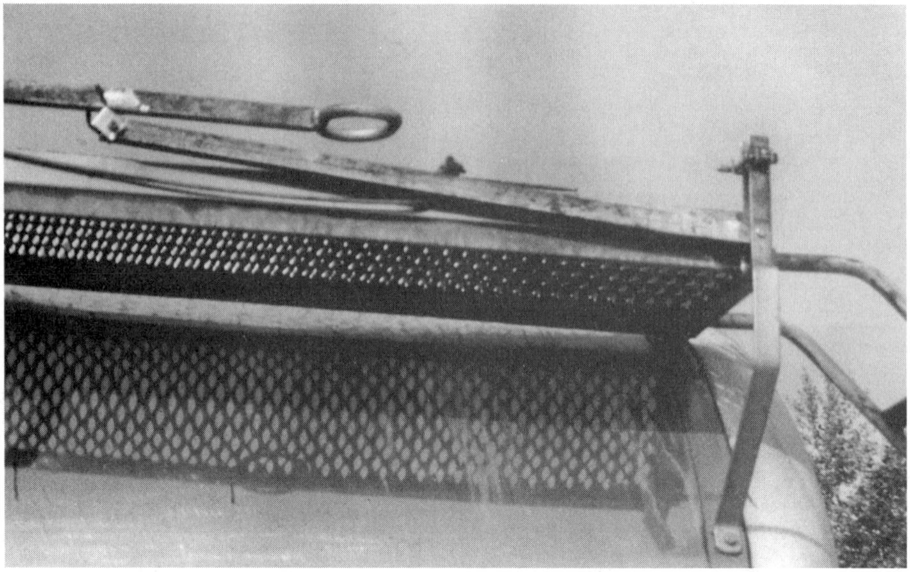

SECTION 6

Traffic Control

The Hazard Syndrome

Aging city streets, roadways, and freeways are *dormant* hazards and must be maintained, repaired, rebuilt, or improved to meet public traffic needs. Mixing public traffic with road construction or repair can be an *armed* hazard that becomes *active* when automobiles drive too close to construction workers and equipment, causing harm.

Scope

Maintaining existing traffic through the construction area is often the most hazardous traffic activity presented in construction when compared with the alternative of planning a route that detours the traffic around the construction area. Many injuries and fatalities that occur involve the vehicles and equipment used in the construction itself. This is due primarily to the small, confined work area, restricted in size to accommodate street traffic.

While construction or repairs are being made on highways, many hazards will be created, especially where high-speed traffic is involved. Traffic control is important in any activity, but especially crucial in the activities of installation or repair of underground drainage systems, installation or repair of utilities and tunnels under the roadway, installation and maintenance of overhead electrical powerlines, and trimming or removal of trees along roads or streets.

As America's population of cars and trucks continues to grow, it is obvious that the old highway infrastructure has to be updated, as it cannot handle this ever-increasing traffic. Some drivers can become outraged when traffic is delayed and they are then slowly escorted through the road-construction area. Their frustration tends to inhibit their driving abilities as they pass through the work area and beyond. However, new highway construction and the upgrading of the old routes have become priorities, especially on the state and local level. It sometimes seems that we encounter roadway construction on a daily basis as we travel from our homes to our places of business. Each construction site on our route increases the possibility that we will encounter hazards and expose ourselves or someone else to serious injury.

Peril from the Failure Mode

Traffic in a construction site presents the *natural hazards* of both impact and instability and the *mechanical* hazard of uncontrolled, unguarded machines (vehicles). Construction workers are often seriously injured or killed as a result of being struck by a vehicle passing through the work area. However, when regular traffic and construction work take place in close proximity, the reverse can also occur: a passenger vehicle can be struck by heavy construction equipment.

The State of Michigan alone has compiled five years of statistics. Michigan's statistics show that approximately 600 motorists and some 200 workers were killed in areas through which traffic had to pass while highway or roadway maintenance or construction was going on. Expanding these figures to include the other fifty states, one can estimate that several thousand people have died in the past five years due to unsafe road construction areas.

Brief Summary of the Occurrence

The operator of a road-roller pulled off to the shoulder of the road to let a passenger car by. The roller upset, resulting in the operator's death. The slope on the shoulder of the road, hidden by grass, was sufficient to upset the unguarded road-roller.

Date of Occurrence

June 8, 1992

Source of Information

Case filed in Binghamton, New York

History of the Hazard

Since the beginning of highway development in the 1920s there has always been a conflict between construction and traffic.

Contributory Factors

The activities of road construction and repair are especially dangerous, as the hazards of general construction are combined with the danger of high-speed traffic. Injuries and death will persist as long as motorists have to deal with inadequate warning signs and barriers, unsafe detours, uncontrolled traffic, mixing of off-road construction equipment with traffic, glaring lights for night construction, and confusing directions. Lack of thorough safety planning and alternate access routes for this specialized work is usually the root cause of collision, damage, serious injury, or death in and around these work areas.

Working in these locations can also be stressful because of the limited work area and the constantly moving traffic or construction equipment. When work crews and equipment have to work in very confining cuts or excavations, or on narrow fills, they can be as close as one foot away from passing traffic.

For the motorist, failure to heed warnings and to slow down while traveling on roads under repair is the main cause of many injuries and fatalities. The effectiveness of traffic control devices in construction zones depends upon their ability to satisfy the driver's need for information. Past experience indicates that the most serious failures to meet motorists needs result from the resons listed next.

- Contradictory information
- Misleading information
- Messages with incorrect distances
- Nonstandard traffic control devices
- Incorrect signs
- Transitions that are too short or are curved too sharply
- Rough and unimproved detours
- Congestion of mixing motorists with moving construction equipment
- No pilot cars to guide traffic through long areas of construction or repair

In most construction or repair sites, numerous tractors, graders, road-rollers, cranes, excavators, and big trucks are involved in the work at hand and are considered a major factor in creating hazardous conditions for the work crews and passing traffic. Failure to have vehicle backup alarms, high noise levels, materials falling from bridges or elevated roadways, flying or other falling objects, powerline contacts, and equipment upsets are but a few of the many hazards that continue to befall workers and passing motorists in the work zone.

Hazard Identification and Prevention Matrix

	Eliminate the Hazard		Guard the Hazard		Provide a Safety Factor		Provide Redundancy		Provide Reliability
	Hazard	Safety	Hazard	Safety	Hazard	Safety	Hazard	Safety	
Natural	Impact, slope								
Structural/ Mechanical	Traffic, speeding objects	Detour		ROPS		Level surfaces		Enforce low speeds, traffic control	95% 100%*
Electrical									
Chemical									
Radiant Energy									
Biological									
Automated Systems									

* Detours are 100% effective. Other controls are 95% effective.

Applicable Standards and Literature

OSHA

29 CFR Part 1926, Construction Standards

Subpart G, Signs, Signals, and Barricades

1926.200, Accident prevention signs and tags

1926.201, Signaling

1926.202, Barricades

Subpart U, Blasting and the Use of Explosives

1926.909, Firing the blast

MSHA

S 56.9313 Roadway maintenance

S 57.9300 Berms or Guardrails

S 71.301 Respirable Dust Control Plan; Filing requirements

S 75.1403-10 Criteria-Haulage; general

Subpart D. Respirable Dust Control Plans

S 77.1605 Loading and Haulage Equipment; Installations

S 90.301(a) Respirable Dust Control Plan; Approval By District Manager

ANSI

ANSI-D6.1-1988, Rev. 12/94, Manual on Uniform Traffic Control Devices for Streets and Highways, U.S. Dept. of Transportation.

U.S. Army Corps of Engineers

Safety and Health Requirements Manual (EM 385-1-1, 3 November 2003):

Section 5

(05.A.11) Personal Protective and Safety Equipment

(08.B) Signal Systems, Personnel, and Procedures

Section 21.I Access and Haul Roads

Federal Highway Administration (FHA)

Part 6-Standards and Guides for Traffic Controls for Street and Highway Construction, Maintenance, Utility and Incident Management Operations, Revision 3 of the 1988 Edition of the Manual on Uniform Traffic Control Devices, Sept. 3, 1993.

ANSI - D6.1-1988, Rev. 12/94, Manual on Uniform Traffic Control Devices for Streets and Highways, U.S. Dept. of Transportation.

Other References

- C16, "Contributing Circumstances, Road," and C19, "Work Zone Related," National Highway Transportation Safety Association (NHTSA), see information on the Search '97 Information Server, (*http://www.nhtsa.dot.gov/people/ncsa/codes/MinData/Crash.html*).

- Bruce A. Suprenant and Kim D. Basham, *Excavation Safety: Understanding and Complying with OSHA Standards,* The Aberdeen Group, Addison IL, 1993.
- Annmarie L. Geddes, "Working in the Danger Zone," *Ohio Monitor,* July 1991.
- Ruth Stidger, "The 11 Best Ways to Improve Work Zone Safety," *Better Roads*, July 1990.
- "Road Construction Traffic Control," Chapter 23, *Construction Safety*, Alliance of American Insurers, 1983.
- *A Study Concerning Driver's Attitudes Toward Construction Zones* prepared for Deere & Co., Moline, IL by Marketing Consultants, Inc., Elkhart, Indiana.
- "Traffic Control for Road Construction and Maintenance" *Hazard Information Newsletter,* Volume 3, Issue 7, October, 1998.

Commentary

Planning ahead to reroute traffic around and away from road construction benefits not only contractors but highway users. Recently, Arizona's governor has asked traffic control departments to see if alternate routes can be devised more quickly so backed-up traffic behind a collision site can be diverted. This initiative would avoid the miles of stopped vehicles and hours of waiting, not to mention the numerous rear-end collisions that result from such backups. Whenever possible, a detour of public traffic outside the work area is the best inherently safe process for traffic hazard prevention. Physical barriers separating traffic routes from construction space are also key to ensuring traffic safety.

Recommendations

OSHA requirements should address the need to detour public traffic around the construction zone rather than through it.

SECTION 7

Haul Roads

The Hazard Syndrome

Improperly designed features of haul roads such as lack of berms are *dormant* hazards because they foster the possibility of trucks leaving the roadway. With trucks on them, the hazard of these haul roads becomes *armed,* and then *active* in the event of injury or damage by any number of circumstances.

Scope

Haul roads present a hazard by

- The Lack of berms on the outside edge of the road along an embankment
- Mixing small vehicles and public traffic with construction hauling equipment

- Allowing pedestrians on haul roads
- Designing haul roads so loaded trucks ride on fill material
- Not providing blocking systems when dumping loads over an embankment
- Designing routes to have unnecessarily sharp curves

Peril from the Failure Mode

The above six hazards often result in fatalities to drivers. On occasion, pedestrians are the victims of circumstances caused by these design flaws. These hazards range from the *natural* feature of rough landscape and a lack of road designation, to the *structural/mechanical* hazard of using inadequate (fill) material that leads to an unstable surface and the lack of berms to act as barriers for trucks.

Brief Summary of the Occurrence

A driver was killed when a truck did not negotiate a curve that lacked a berm.

Date of the Occurrence

1999

Source of the Occurrence

U.S. Bureau of Mines Fatal Reports

History of the Hazard

1930s: Open-pit mines and the construction of large dams had to be accessed over previously inaccessible terrain. Haul roads were formed as a means for the travel of specialized equipment. Today, rough haul roads crisscross the landscape, many used by mining and logging trucks.

Contributory Factors

Designing roads with precarious short cuts over natural features of the landscape is the first way to create a hazardous haul road. Using the wrong materials and methodology to construct haul roads creates the hazards of potentially unstable surfaces or roads that will easily deteriorate. Failure to build stable embankments or berms is another way to invite catastrophe. Sadly, the presence of these features and the quality of the roads is determined largely by cost. Many designers believe that haul roads and temporary roads are disposable and do not ensure their safe construction or use.

Hazard Identification and Prevention Matrix

	Eliminate the Hazard		Guard the Hazard		Provide a Safety Factor		Provide Redundancy		Provide Reliability
	Hazard	Safety	Hazard	Safety	Hazard	Safety	Hazard	Safety	
Natural	Landscape	No sharp curves							
Structural/ Mechanical	Instability/ precipice			Embank-ments, berms		Easy grade slopes		No cars or pedes-trians	
Electrical									
Chemical									
Radiant Energy									
Biological									
Automated Systems									

Applicable Standards

MSHA

MSHA requires earthen protective berms on haul roads.

U.S. Army Corps of Engineers

08.D Haul Roads

08.D.01—Access/haul roads shall be designed in accordance with current engineering criteria. Prior to construction, the Contractor shall provide the GDA with a copy of the plan for review and acceptance. Work on the haul road shall not commence until the GDA has accepted the plan. The plan shall address the following items:

(a) Equipment usage, traffic density, and hours of operation;
(b) Road layout and widths, horizontal and vertical curve data, and sight distances;
(c) Sign and signalperson requirements, road markings, and traffic control devices;

(d) Drainage controls
(e) Points of contact between vehicles and the public, and safety controls at these points of contact
(f) Maintenance requirements, including roadway hardness and smoothness and dust control; and
(g) Hazards adjacent to the road such as bodies of water, steep embankments, etc.

08.D.09 Curves

(a) All curves shall have open sight lines and as great a radius as practical.
(b) Vehicle speed shall be limited on curves so that vehicles can be stopped within one-half the visible distance of the roadway.
(c) The design of horizontal curves shall consider vehicle speed, roadway width and surfacing, and super elevation.

08.D.10 Grades

(a) When necessary, based on grade and machine and load weight, machines shall be equipped with retarders to assist in controlling downgrade descent.
(b) Truck haul roads should be kept to less than a 10% grade. There should be no more than 400 ft (121.9 m) of grade exceeding 10%
(c) The maximum allowable grade shall not exceed 12%.

Commentary

Haul roads can be made safe when matched to the capabilities of the haul vehicles. Haul road grades should be consistent with the braking system of haul vehicles. Haul roads need berms to prevent driving off the edge of roadways that are constructed on a hillside. Road conditions should be a factor in determining road grades and vehicle use.

Recommendations

When applicable, haul road design should be incorporated into every pre-construction plan. To assist the formation of safe and sturdy haul roads, OSHA should provide design criteria to address construction of the road, materials used, and slope grade.

11 Types of Construction Hazards

INTRODUCTION

The following nine sections provide an overview of the basic hazards specific to particular types of construction. This topic addresses methods and practices used in the construction of houses, office buildings (small and large), warehouses, apartments, industrial developments, highways, and waterways.

1. Wood-frame construction
2. Concrete form work
3. Masonry failures
4. Tilt-up
5. Lift slabs
6. Steel erection
7. Trenching
8. Tunneling
9. Diving/Underwater construction

In such broad classes of construction, it is important to remember that many crafts and trades are involved. Construction involves people who may specialize in a craft or trade, but work in a common work area. They are all entitled to a safe workplace specific to their trade and their tools—a workplace that does not endanger the worker or workers in other crafts with the equipment and processes they use.

SECTION 1

Wood-Frame Construction

The Hazard Syndrome

House building is often performed by a variety of craftsmen who are usually employed under minimal supervision by the general contractor or landowner. Without specifying access for erecting wood-stud walls and roof trusses, many workers find it difficult to brace the trusses and secure them to the wall properly, which presents a *dormant* hazard. Workmen assembling a stud-wall house while climbing about the top of the walls to place the roof trusses is an *armed* hazard, ready to become *active* when someone falls. The following image shows how the use of wood framing as de facto guardrails is never a good idea! A worker fell multiple stories off the building shown here.

Scope

Many hazards arise from this type of construction and involve self-employed craftspeople in disciplines that are common and necessary to the homebuilding process. Trades exposed to hazards include carpenters, carpet installers, concrete and masonry workers, asphalt workers, electricians, excavators, fencing installers, heating and air-conditioning installers, insulation installers, glazers, land-clearing workers, landscapers, masons, painters, plasterboard installers, plumbers, roofers, siding installers, and stucco plasterers. The mushrooming development of tract homes erected by a variety of subcontractors has reawakened the hazards associated with multiple crews on a common work site.

Peril from the Failure Mode

The most common cause of injury in this type of construction is a fall from heights of 8 to 10 feet by workers assembling the wood framework of a structure. However, other common perils associated with wood-frame construction include

- Slippery walking surfaces (*natural, structural/mechanical*)
- Falling materials (*natural*)
- Being struck by moving vehicles or objects (*structural/mechanical*)
- Dust (nontoxic)(*natural*)
- Low-voltage electrical shock (*electrical*)
- Absence of safe access (*structural/mechanical*)

Brief Summary of the Occurrence

Three construction workers were injured when some 50 foot long, 300-pound wooden roof trusses being set were improperly braced. The trusses toppled like dominoes. Two men setting the trusses fell to the ground; one sustained a fractured skull and a broken back, the other had chest trauma. A third man was operating a fork lift when a falling truss struck him.

Date of Occurrence

August 21, 1998

Source of Information

Arizona Daily Star, August 22, 1998

History of the Hazard

Individual contractors have traditionally built wood-frame houses. This method is changing with the popularity of suburban cluster housing. Now developers who subcontract different phases of construction are erecting most wood-frame houses in huge tract developments. However, individual contractors still make up a significant portion of wood-frame constructors. Many stud-wall framers remain self-employed and are placed outside of the requirements of OSHA, so their injuries are not recordable.

1970s: The nationwide hiring of undocumented workers is increasingly prevalent. This practice creates a work force totally unprotected from dangerous workplaces. Circumstances as described above can become a source of liability to the developer or general contractor in the event of a serious injury. Usually these individuals have their own coverage and hold harmless contracts with the developer that relieve the developer of administration oversight for worker safety.

Contributory Factors

A subcontractor who has the lowest bid, based upon the agility of their employees, normally erects the stud walls, floors, and roof trusses, as well as the sheathing. A loosely-coordinated, self-employed workforce cannot participate in the coordination of a master safety plan. Neither can it typically afford equipment such as aerial lifts to provide safe, mobile work platforms. Typically, supervision is limited to the completion of the work. Safety Without a Master Plan (SWAMP) is unable to address potential hazards in the preplanning stage, and therefore is unable to implement safety precautions before workers arrive for a job.

Hazard Identification and Prevention Matrix

Wood frame construction can become inherently safe with the use of safe access techniques (see Chapter 10, Section 5).

	Eliminate the Hazard		**Guard the Hazard**		**Provide a Safety Factor**		**Provide Redundancy**		**Provide Reliability**
	Hazard	Safety	Hazard	Safety	Hazard	Safety	Hazard	Safety	
Natural	Gravity (fall)								
Structural/ Mechanical		Staging		Fall protection systems: harness or lanyard		Construction safety planning/ aerial lifts		Job site supervision	95%
Electrical	Shock			Grounded tools		Battery-powered tools		Double insulation	100%
Chemical									
Radiant Energy									
Biological									
Automated Systems									

Applicable Standards and Literature

Standards in wood frame construction are understandably varied, as the process of framing involves so many diverse tasks. Standards that must be considered in wood frame construction are listed below.

OSHA

Particular emphasis should be given to the fall protection requirements in **Subpart M 1926.500.**

Due to the great diversity of hazards that can be encountered in the construction of wood-frame structures, nearly all subparts may be applicable except,

Subpart R Steel Erection

Subpart S Underground Construction, Caissons, Cofferdams and Compressed Air

Subpart V Power Transmissions and Distribution

Subpart Y Diving

ANSI

In some circumstances, the following ANSI standards will provide pertinent information:

A10.3-1995 Operating Safety Requirements for Power Actuated Fastening Systems

A10.8-1988 Safety Requirements for Scaffolding

A10.10-1990 Temporary and Portable Space Heating Devices and Equipment Used in the Construction Industry

A14.2-1990 Safety Requirements for Portable Metal Ladders (**Note**: This states what should be provided for intended safe use.)

National Fire Protection Association (NFPA)

NFPA 251-1995 Methods of Test of Fire Endurance Building Construction and Materials

NFPA WCD Series Wood Construction Data Series 1 through 6

American Society for Testing Materials (ASTM)

ASTM 04.09-1997 Wood Construction

ASTM D 5055-1997 Standard Specification for Establishing and Monitoring Structural Capacities of Prefabricated Wood I-Joists

AFPA National Design-1997 National Design Specification for Wood Construction

AWPA

AWPA C15-1996 Wood for Commercial-Residential Construction—Preservative Treatment by Pressure Processes

German Specifications

DIN 18367-1992 Construction Contract Procedures (VOB)-Part C: General Technical Specifications in Construction Contracts (ATV) - Laying of Wood

Other References

- "Wood Frame Construction," *Hazard Information Newsletter*, Volume 3, Issue 11, February 1999.

Commentary

The engineer overseeing wood-frame-construction safety will not find answers in the thick rulebook of OSHA requirements. Safety will be achieved by requiring both management leadership and craftspeople to complete ongoing safety training in order to be eligible for their practice licenses and renewal of these licenses. State licensing programs for engineers, architects, contractors, and other specialists should include requirements for *specific safety training* that have to be met *before* a license is granted. Electricians, plumbers, and other licensed crafts should also have to meet *specific safety training* in order to qualify for a license. This training needs to be directed towards the recognition of the hazards inherent in their practice or trade and how to control them.

Architects and professional engineers especially should include this safety training in their continuing educational programs for license renewal to ensure that design and specifications address any specific hazards that may be encountered. Safety training as a requirement for the issuance of a contractor's license should be made mandatory for all contractors, including those who are self-employed. Building inspectors who come on-site to verify compliance with building permits and codes also need safety training so they are able to identify hazardous job-site conditions.

The inspector has two crucial job duties. First, the inspector should identify all potentially life-threatening conditions, such as unshored trenches or boomed equipment working next to powerlines. Second, the inspector should notify the appropriate parties of all unsafe conditions in a written report so they can take appropriate action to control or eliminate the unsafe condition.

Inspectors should not approve whatever they are inspecting at a work project until all hazardous conditions are corrected. This may seem to be an unusually antibusiness approach to safety, but it really isn't when you consider the cost of crippling injuries and wrongful deaths not only for the builder, but for the community. The implementation of such requirements, responsibilities, and practices is really cost-effective; especially when work stoppages, downtime, and retraining are factored into the equation if an injury occurs.

Having all those who are involved in any phase of the project trained to be watchdogs for safety by making them aware of inherent hazards would soon eliminate most of the life-threatening hazards in the workplace. Within this framework, both the user's and management's methods of erecting, assembling, constructing, and operating a product or procedure have to be accomplished safely, without injury to *anyone.* This is not as awesome as it seems, because with planning that includes safety, most hazards can be controlled or eliminated prior to the arrival of the work crews on the construction site.

At this point it is necessary to reiterate the need for the certification and licensing of services, crafts, and professions. It is a voluntary process that identifies that the individual has had special experience and special education.

The Certified Safety Professional (CSP) has the background to evaluate all conditions or circumstances, identify hazards, and prescribe the measures that will minimize or eliminate the opportunity for the hazard to become armed. Licensing of certain professions carries with it a legal obligation to protect the public. In this country, individual states establish boards that issue licenses that permit the holder to practice a certain profession. These boards also control license renewal. Licenses are only issued to those who can demonstrate certain skills or abilities established by the profession.

Licensing of the medical profession first began in the 1700s and then was dropped in the early 1800s. In 1889 New Hampshire led the way in regulating physicians under the state's policing powers in order to prevent the fraudulent or incompetent practice of medicine. In about 1910 licensing of engineers in the state of Wyoming was initiated because some irrigation canals were designed so improperly that water ran away from instead of to the desired fields. Licensing of engineers in the major disciplines soon became a practice in most states. Safety engineering is now licensed in both Massachusetts and California. During the twenty-first century, it is hoped that safety engineering and system safety will become licensed engineering professions.

Recommendations

The engineer has the authority to control wood-frame construction and ensure a safe work site. Creating a master safety plan and coordinating the myriad tasks that must be accomplished during such construction with any site inspectors and potential subcontractors is the right way to begin. Once safety responsibility is designated, this individual should have the authority to coordinate and enforce safety equal to other priorities of scheduling, compliance with specifications, and project completion.

SECTION 2

Concrete Form Work

The Hazard Syndrome

Today most reusable form panels are built largely with staging, except for the columns addressed in Chapter 10, Section 2. Slip forms are a common way of supporting elevated concrete pours. Slip forms are repeatedly raised from one level to the next level with a crane. Due to the adhesive nature of concrete, the cured concrete on these panels often sticks together. The common method of freeing a stuck form panel is to attempt to jerk it free with the crane that is used to lift the panel to the next level. A stuck panel is an *armed* hazard, as *it is not a freely suspended load for a crane to lift*. Overloading the crane with a stuck panel is a hazard in the *active* mode, as the crane can upset or otherwise fail.

Scope

Slip forms are used in many locations, including dams, buildings, and specially developed structures such as elevator shafts. Sticky forms are a persistent hazard, even though lubricant is used to separate them. Improved design of a form that would enable it to mechanically free itself would reduce this hazard.

Peril from the Failure Mode

Form panels that stick to concrete are not a freely suspended load and can cause the crane to overload, causing death from the collapsing boom or crane upset. The form panels that stack can easily overload the crane to cause a variety of *mechanical* and *natural* hazards.

Brief Summary of the Occurrences

(A) A crane operator sustained the loss of a foot when his cab was crushed by the crane boom, which had collapsed due to an overload caused while attempting to free a slip form from an elevator shaft. Two riggers were also struck by the falling headache ball.

(B) The deceased was decapitated by the falling boom of a crane being used to lift form panels on a concrete dam under construction. The crane's long boom was extended almost horizontally, creating maximum tension on the pin-up assembly that lowers and raises the boom. To dislodge the form panels from the concrete, the hoist line was jerked or reefed, causing the long boom to flex and the pennant-cable-guys system to go slack momentarily. The hoist line may have two-blocked and aggravated this slacking, which caused the pinup guylines to bind in the small sheaves and shear the cable, dropping the unsupported boom and jib.

Date of Occurrences

(A) June 17, 2004

(B) January 26, 1987

Source of Information

(A) U.S. District Court, Northern District of Florida, Pensacola Division

(B) 4th Judicial district, Ada County, ID, # 90527

History of the Hazard

September 1957: As already mentioned, the Portland District, U.S. Army Corps of Engineers' study showed the cost-effectiveness of including safe access in the design of concrete formwork.[1]

May 1904: The Panama Canal was built by the Americans and completed January 1, 1915—in 11½ years! Some 5,000,000 cubic yards of concrete were placed. At that time, this was the largest concrete project ever built.

[1] This study was published in an issue of *Western Construction* in the article, "How Proper Scaffolding Cuts Costs," written by the author of this book.

To build the Panama Canal, a work force of 38,000 people was needed. Approximately 900 people were killed each year. Perhaps half of this was from disease such as malaria, yellow fever, and other plagues. No records were yet developed as to the cause of workplace injuries.

Contributory Factors

Slip-form panels are always vulnerable to sticking or adhering to the concrete. Even though the forms are treated with a lubricant prior to the pouring, they persist in sticking. Weather conditions and length of curing the concrete have a minor effect on whether the forms will stick to concrete. An alternate method of mechanically freeing the panel is needed.

Hazard Identification and Prevention Matrix

	Eliminate the Hazard		Guard the Hazard		Provide a Safety Factor		Provide Redundancy		Provide Reliability
	Hazard	Safety	Hazard	Safety	Hazard	Safety	Hazard	Safety	
Natural	Falling object								
Structural/ Mechanical	Sticking panel	Self-freeing panel	Compromised load/ overload/ upset	Hydraulic jacks					1*
Electrical						LMI on crane			2*
Chemical	Adhesive properties	Lubricant to prevent sticking							3*
Radiant Energy									
Biological									
Automated Systems									

* Controls from 1,2, and 3 together would result in 100% reliability.

Applicable Standards and Literature

Nothing has been written about the hazard of slip form sticking, and no requirement for ensuring a freely suspended load before lifting with a crane currently exists. However, the following standards are applicable to other scaffolds.

OSHA

1926.451 Subpart L contains requirements for scaffolds.

1926.500 Subpart M lists requirements for fall protection.
Comment: This sets forth very detailed requirements for body harnesses, positioning belts, buckles, connectors, lifelines, and so on, which are costly but necessary requirements for preventing injuries. To prevent falls, the use of staging, step rungs, and anchors, in almost all instances, is a profit-oriented solution.

1926.703 Requirements for cast-in-place concrete

(a)(2) Drawings or plans, including all revisions, for the jack layout, formwork (including shoring equipment), working decks, and scaffolds, shall be available at the job site.

(c)(3) vertical slip forms shall be provided with scaffolds or work platforms where employees are required to work or pass.

1926.1050-1053 Subpart X, Stairways and Ladders, also contains requirements for safe access.

ANSI

A10.9-1997 Safety Requirements for Concrete Construction Masonry Work

Paragraph 7.3 Formwork Design
Comment: Any reference to a provision for safe access for workers has been omitted in 7.3.1 & 7.3.2. If fall prevention is to be achieved on concrete formwork, then design requirements must include provisions for safe access.

Paragraph 7.4.4 states: "When forming and stripping operations of formwork cannot be done from the floor, ground or other solid construction, ladders, scaffolds or other approved means of access shall be used."
Comment: This reference should also include guidelines recommending that safe access and staging that meet the ANSI requirements for scaffolding and fixed ladders should be an integral part of the design.

A48.1-1986 Forms for One-Way Concrete Joist Construction

A48.2-1986 Forms, Two-Way Concrete Joist Construction

A145.1 Recommended Practice for Concrete Formwork

American Concrete Institute (ACI)

ACI 347-1978 Recommended Practice for Concrete Formwork

ACI 347R-1994 Guide for Formwork for Concrete. Paragraph 3.1.e, "Incorporation of scaffolds, working platforms and guardrails into formwork design."

ACI SP 4, 6th Rev. *Formwork for Concrete,* M.K. Hurd, 1998

U.S. Army Corps of Engineers

EM 385-1-1 (3 November 2003)

Section 21, Safe Access and Fall Protection:

Paragraph 27.B, Formwork and Shoring, states that planning and design shall meet American Concrete Institute's publication ACI 347R-1994, *Guide to Formwork for Concrete.*

Section 22, Work Platforms, page 429.

Section 27, Section A, Concrete and Masonry Construction, page 479.

Other References

- *Hazard Information Newsletter*, Volume Two, Issue One, *Dangerous Access and Work Platforms.*
- S.W. Nunnally, *Construction Methods and Management,* Prentice-Hall. Chapters 10 and 11 both apply to concrete formwork.
- S269.3 Concrete Formwork, Canadian Standards Association, 1992.
- German Standard 4235 P4, Compacting of Concrete by Vibrating; Compacting of In-Situ by Formwork Vibrators, 1978.
- *Handbook of Heavy Construction*, Section 21, "Conventional Concrete," Part D, "Stationary Forms," 1971.
- WCD 3 Design of Wood Formwork for Concrete Structures, National Forest Products Association, 1961.
- 1995 Formwork for Concrete, Standards Association of Australia.
- "Concrete Formwork" *Hazard Information Newsletter* Volume 3, Issue 8, November 1998.

Commentary

In most instances, a set of hydraulic jacks could be incorporated in the design of the form panel and the structure being erected in order to free the form panel from the concrete wall to which it is sticking. The design of a form panel could include a small portion that could be jacked away from the concrete structure. Then this panel section could be retracted as the entire form panel becomes disconnected from the concrete surface. This action would ensure a freely suspended load for the crane and prevent overload. Engineering design to incorporate hand-operated hydraulic jacks (rams) would provide a failure-free method of releasing form panels prior to lifting.

Recommendations

This is a hazard that could be eliminated if engineering research developed a design for self-freeing slip forms to make them inherently safe. Use of a more effective chemical lubricant or mechanical devices, such as the aforementioned hydraulic ram, to free the form panels before they are lifted would both be effective means of reducing this hazard.

SECTION 3

Masonry Failures

The Hazard Syndrome

Erecting masonry walls requires staging to support the masons and the materials being used. These requirements are meant primarily to prevent the hazards of falls and falling objects (covered in Chapter 10, Sections 1, 2, and 3). The hazard of the failure of unsupported walls from wind during the period of time required for the mortar to set is often overlooked. Mortar that has been compromised during the curing process creates an unsafe structure. Unbraced, uninsulated brick or block walls with uncured mortar present an *armed* hazard that can become *active* during freezing weather, heavy wind, or any other element that causes walls to fail.

Falls from an elevation are the principal source of injuries to block layers, mainly due to poorly constructed scaffolding that has no railings, stairways, ladders, or other means of safe access. Scaffolding often has improper planking that does not provide a safe walking surface or is unsecured and can slip out of place. In some instances, the scaffolding or staging collapses or upsets due to overload or unbalance. The causes for such imbalance are usually either too many concrete blocks or bricks have been placed on it, or the blocks have been stacked improperly.

Power saws used to cut concrete, cinder blocks, and tile can produce so much dust or debris that workers often sustain skin and eye injuries or experience breathing problems if not protected by proper clothing, eye protection, and respirators.

Old masonry walls, with lime mortar joints weakened by either weathering or by the continuous vibration of traffic, can lose part of their facings or collapse entirely. Even buildings under construction using newer masonry blocks or bricks can fail if they are not braced or secured properly. These structures will always pose a threat to workers and the public alike.

Scope

Masonry construction composed of stone, brick, concrete block, hollow clay tile, and even unfired adobe set in mortar and laid up by unit is one of the oldest forms of construction. Specialists in the trade, including masonry engineers, designers, and construction managers, must be familiar with the properties of bricks, mortars, clays, and other structural materials. Some mortars can withstand higher temperatures; others hold up in humid climates. Specific knowledge of materials and processes used in masonry construction ensures both the safety of the construction process and the prolonged life of the structure.

Specific statistics are difficult to find, because injury-reporting data addresses only the type of injury sustained and almost universally fails to address the hazard. Until reporting is improved, we will not have as much information as we need to target hazards, so little progress can be anticipated in reducing injuries in masonry construction.

Peril from the Failure Mode

Walls fail during erection for a number of reasons. *Natural* elements such as wind and freezing weather can easily cause collapse, sometimes without warning. This creates the hazard of being struck by the falling wall.

Brief Summary of the Occurrence

A falling concrete-block wall, erected the day before, struck a worker on an Air Force construction project. During the night the temperatures dipped to the low twenties, and the mortar froze. In the morning, a section of the wall collapsed as the worker was passing, killing him. The work specifications did not require insulating materials or supportive bracing during the curing process.

Date of Occurrence

Late 1990s

Source of Information

Litigation under protective order

History of the Hazard

Ever since people began building adobe masonry walls of unrefined earthen materials (sometimes waterproofed with the dung of domesticated animals), the purpose has been to keep the weather out of the enclosure. In many instances, brick masonry has stood the test of centuries. However, some masonry units fail, and continue to fail, due to aging, construction errors, water, and a host of other problems. Listed below are just a few of the successes and failures in older masonry buildings:

~700 BC: The Romans developed masonry into a fine building art, with the firing of easily portable bricks and mortar, which were made with fine volcanic dust. The Romans were excellent engineers and designed temporary forms to support brick arches. Some of the high brick aqueducts made nearly 2000 years ago are still standing, and some still continue to be functional.

1922: The Majestic Theater in Pittsburgh, Pennsylvania, originally built as a church, lost part of its roof when the end wall, made of 13-inch-thick brick and rising 50 feet high, bowed and pulled over a 60 foot timber roof truss.

1947: A brick pier on the Empire Apartments in Washington, DC collapsed after 54 years of satisfactory support. The structure was seven stories high with wall thicknesses varying from 13 inches to 27 inches.

Over time, the weathering of the mortar joints caused both of the aforementioned failures.

1930: A 70-story warehouse collapsed in New York when the wall moved outward and allowed the floor beams to drop. The cause of this collapse was due to the accumulated effect of traffic vibration on the walls from an adjoining rough-surfaced street.

1942: A 26-unit two-story apartment dwelling constructed with face brick backed with load-bearing structural clay developed cracks that broke up the face brick.

Incompatibility of materials used in the construction was determined to be the cause, as the red-clay-tile backing had absorbed much moisture and expanded.

Even today, unfired, sun-dried adobe-brick structures are still functional. One of the few two-story adobe buildings is the Schefflin Hall in Tombstone, Arizona, which is 120 years old. To this day the King Solomon Masonic Lodge #5 meets on the second floor. These adobe walls rise over 35 feet above the ground level and are structurally sound, as they are protected and weatherproofed by a thick plaster outercoating.

Contributory Factors

Injuries in masonry construction are primarily the result of a failure to include safety planning before the work commences. Aside from the aging process, masonry structures fail for many reasons:

- Improperly shored and braced structures can collapse.
- The hanging of work platforms on the side of a wall being built invites disaster, but the practice is common.
- When concrete floor slabs must be tied into masonry walls, the four-inch brackets on the joists that are embedded in the walls can pull out and collapse the wall if an overload occurs and the joists become stressed. (To prevent this type of failure, joists should have additional support in the middle while supporting the extra weight of the wet concrete.)
- Fancy brick decorative panels used on the walls of shopping centers, schools, and industrial buildings are sometimes placed in metal frames that allow twisting and eventual failure.
- Steel spandrel beams used as wide doorway or window supports can become overloaded and can twist, resulting in either failure or bulging of the masonry wall above the beams.
- When steel beams supporting masonry walls above them are not tightly bolted to their support columns, they can rotate just a mere fraction of an inch and cause the masonry wall to crack or bulge dangerously.
- Temperature fluctuations, rain exposure, moisture absorption, chemical alteration of the masonry units and mortar, all acting in many combinations, can develop elastic and plastic strains and alter the strength of masonry.
- All masonry units, from sponge-like porous cinder block to almost impervious glazed brick, will absorb water in varying degrees. Even with nonporous masonry units, water can still saturate the mortar if it hasn't been sealed. Saturation brings possible frost damage and, where conditions permit, sulfate reaction.
- Presacked mortar that uses limestone rather than lime easily absorbs water and can cause a brick or block wall to become waterlogged, causing the mortar to fail. Architects and designers usually specify lime mortar, but some bags are not labeled with their content. Since it is assumed the presacked bags of mortar contain lime, buyers are unaware of its hazardous failure mode.
- When masonry walls enclose roofs, the expansion or retraction of roof areas that are exposed to high sun temperatures or surfaced with materials having high expansion

coefficients may crack and bulge the walls or make the corners unstable. Roofs should be on top, or detached from the masonry wall to give them freedom of movement. For instance, a metal roof has to be anchored so that any expansion will not exert any force upon the masonry walls.

- Thermal change upon masonry materials is not completely reversible. When expansion causes the formation of minute cracks or separation between brick and mortar, grit produced at the cracks from airborne dust can partly fill the gap. On cooling, the wall cannot come back to its original dimension. Every succeeding cycle then adds to the total distortion. Not many repetitions are needed to cause stress.
- When both brick and block units are used in a wall, their incompatibility sometimes makes the wall susceptible to cracking along both horizontal and vertical lines.
- Sometimes the use of substitute materials when a low budget must be met results in serious failure. The tendency toward denser masonry units and stronger mortars results in minute separations along all brick or stone edges. A square foot of brick wall has about 6.5 linear feet of mortar joints. The walls of a building 100 × 200 feet in area and 20-feet high have more than 15 miles of joints. If only 15% of those joints are slightly open, imagine the amount of water that can enter the masonry.

Hazard Identification and Prevention Matrix

	Eliminate the Hazard		Guard the Hazard		Provide a Safety Factor		Provide Redundancy		Provide Reliability
	Hazard	Safety	Hazard	Safety	Hazard	Safety	Hazard	Safety	
Natural	Wind/ freezing								
Structural/ Mechanical		Bracing		Insulation					100%
Electrical									
Chemical	Improper chemical content for intended use	Famili-arity with properties of mortar		Specified safe use		Instructions on mortar bags		Water/ weather-proofing	100%
Radiant Energy									
Biological									
Automated Systems									

Applicable Standards and Literature

OSHA

1910 Subpart I, App. B, Non Mandatory Compliance Guidelines for Hazard Assessment & Personal Protective Equipment Selection

1926 Subpart H, General Requirements for Storage Materials Handling, Storage, Use & Disposal

1926.250(b)(7) Masonry

1926 Subpart L, Scaffolds

1926.450(b) Scope, application, and definitions

1926.451(d)(3)(iii,) Counterweights

1926 Subpart M, Fall Protection

1926.500(b) Scope, application, and definitions

1926.502(j)(6) Fall protection systems criteria and practices

1926 Subpart M, App E, Sample Fall Protection Plan—Non mandatory guidelines for complying with W/1926.502(k)

1926 Subpart Q, Concrete and Masonry Construction

1926.700 General Requirements

1926.701 Requirements for Equipment and Tools

1926.706 Requirements of Masonry Construction

1926 Subpart T, Demolition

1926.854 Removal of walls, masonry sections, and chimneys

ANSI

A1.3 Masonry Cement

A10.9 Safety Requirements for Concrete Construction Masonry Work, 1983

A103.1 Chemical-Resistant Masonry Units

A41.1 Building Code Requirements for Masonry

A41.2 Building Code Requirements for Reinforced Masonry, 1970

A62.2 Coordination of Masonry

A62.3 Clay Concrete Modular Masonry Units

B94.12 Carbon-Tipped Masonry Drills & Blanks for Carbide, 1977

B212.15 Cutting Tools—Carbon-Tipped Masonry Drills and Blanks for Carbon-Tipped Masonry Drills, 1994.

U.S. Army Corps of Engineers

EM 385-1-1 (September 1996) *Safety and Health Requirements Manual*

Section 27 Concrete and Masonry Construction and Steel Erection

27.A, Concrete and Masonry Construction—General

27.G, Masonry Construction

The following are some important safety tips and hazard-prevention methods outlined in this manual. These should be included in management's construction-safety plans whenever concrete, steel, and masonry construction is to be done:

27.A Concrete and Masonry Construction—General

27.A.01 Construction loads shall not be placed on a concrete or masonry structure or portion of a concrete or masonry structure unless the employer determines, based on information from a person who is qualified in structural design, that the structure or portion of the structure is capable of supporting the loads.

27.A.02 Employees shall not be permitted to work above or in positions exposed to protruding reinforcing steel or other impalement hazards unless provisions have been made to control the hazard.

27.G Masonry Construction

27G.01 A limited access zone shall be established whenever a masonry wall is being constructed. The limited access zone shall conform to the following:

(a) The limited access zone shall be established prior to the start of construction on the wall.

(b) The limited access zone shall be equal to the height of the wall to be constructed plus 1.2 m (4 ft), and shall run the entire length of the wall.

(c) The limited access zone shall be established on the side of the wall that will be un-scaffolded.

(d) The limited access zone shall be restricted to entry by employees actively engaged in constructing the wall; no other employees shall be permitted to enter the zone.

(e) The limited access zone shall remain in place until the wall is adequately supported to prevent overturning and to prevent collapse unless the height of the wall is over 2.4 m (8 ft), in which case the limited access zone shall remain in place until the requirements of 27.G.02 have been met.

27.G.02 All masonry walls over 2.4 m (8 ft) in height shall be adequately braced to prevent overturning and to prevent collapse unless the wall is adequately supported so that it will not overturn or collapse. The bracing shall remain in place until permanent supporting elements of the structure are in place.

27.G.03 Scaffolds for masonry construction workers shall not be used to provide temporary lateral support of masonry walls.

27.G.04 Cleanouts shall be on the side of the masonry wall opposite the scaffolding.

27.G.05 Fall protection shall be provided to masonry workers exposed to falls of six feet or more.

Other References

- Jacob Field, *Construction Failure,* John Wiley & Sons, 1968.
- *Uniform Building Code, Volume I,* (Intl. Conference of Building Officials, Pasadena, CA).
- *Manual of Accident Prevention in Construction,* Association General Contractors of America, Inc., Washington, DC, 1971.
- *Engineering News Record*
  "Training," July 20 1998, Page 20.
  "Accidents," September 7, 1998, Page 22.

- "Masonry Failure Hazards," *Hazard Information Newsletter,* Volume 3, Issue 10, January 1999.

Commentary

The vulnerability of masonry walls is not usually addressed in construction plans. It is difficult to find literature explaining how to provide insulation to prevent freezing of the mortar. There should be specific special conditions in any construction or masonry contract to take measures that prevent failure from wind, freezing, and other natural conditions. However, reinforcement and compatibility with other materials and the environment are two important steps toward sound masonry.

Masonry must be properly reinforced with steel or built in separate units with isolation joints to give freedom of expansion so that cracking will not occur. Masonry, concrete, steel, or glass surfaces must be chemically cleaned, allowed to dry and maintained at proper temperature if a good joint is to be expected. Joint sealants must be compatible with materials next to the joint.

Masonry enclosures of roof areas that are exposed to high temperatures or surfaced with materials having high expansion coefficients must be detailed to give the roof freedom of movement. Compatibility—almost equality—is needed in the moisture and thermal changes of the various masonry units making up a wall. Provision for differential dimension change, in addition to lateral thermal length change, must be made for brick facing on tall concrete frames.

Recommendations

The following is a list of prevention methods that should be included in masonry-construction specifications to prevent injury to workers:

- Insulation should be provided to prevent "green mortar" from freezing.
- Scaffolds for brick or stonework should conform to standard specifications.
- Basement walls should be braced sufficiently to withstand any load that may be placed upon them during construction.
- Backfilling should not be done against "green" walls.
- Supports that cause vibration should not be placed on walls until they have properly set.
- On high walls, especially where openings are not plentiful, good bracing is necessary during construction to withstand wind and other pressure.
- Guys or other stays should not be put through brickwork until it has set firmly, and should be anchored only to places of sufficient stability to withstand stress.
- Mortar tubs should be kept free from ragged edges, which may cut the hands and arms of the bricklayer dipping mortar from them.
- Power saws used for cutting brick or stone should be equipped with dust collectors. Men using saws should wear eye protection; if dust collectors are not used, respirators are necessary. Power switches should be so placed so that the operator does not have to reach across the blade to start or stop the motor.
- Appropriate protective cream or lanolin ointment should be available for use on hands to prevent lime burns.

- A bottle of glycerin or olive oil should be included in the first-aid kit where men are handling lime. These should be applied when lime gets into open cuts.

SECTION 4

Tilt-Up

The Hazard Syndrome

The "tilt-up" procedure is the method of casting reinforced concrete walls on the floor of a structure, curing the concrete, and raising the walls to a vertical position with a crane. The walls are then anchored to the footing and braced in place. When the slab is first cast, the hazard of wall collapse or crushing by falling object is *dormant.* The hazard is *armed* while the slab is being lifted, and while it is supported with temporary bracing. When the slab is finally anchored within the completed roof substructure, the hazard becomes dormant again. This hazard does not become *active* until the building wall is demolished.

Scope

The tilt-up of precast concrete slabs has become a well-recognized method of erecting building sidewalls. The hazard is armed for the entire cycle of lifting the slab into place, as there are many possible failure modes in lifting the slab and while it is being secured by temporary bracing. Failures during the lifting process happen when the lifting anchor pulls free, or the anchor does not freely rotate to allow a direct pull on the lifting cables, resulting in a falling slab.

Peril from the Failure Mode

A falling slab is a dangerous situation. A faulty lifting release device will cause the slab to fall before the lift is completed, becoming a dangerous source of energy. A falling slab can crush people and equipment. Though a falling slab of this type presents the *natural* hazard of gravity and the *structural/mechanical* hazards of structural failure and improper bracing, the hazard of tilt-up is different than many typical falling-object hazards. Tilt-up is a procedural hazard, when the possibility of a worker being crushed by a falling slab is created by faulty lifting processes, improper or poorly used equipment, and inadequate bracing methods while the wall is being placed.

Brief Summary of the Occurrence

A slab was being lifted, and at the point it was raised approximately 20 degrees from the ground, the release line appeared to become looped and entangled. The owner and superintendent of the operation stepped up onto the lifted slab and attempted to untangle the release line. The hoist line attachment to the slab did not rotate for a direct pull and then snapped into place. This caused a "bull whip" effect on the release line used to disconnect the hoist cable from the anchor. The release line struck the superintendent's right shoulder, instantly amputating his arm.

Date of Occurrence

2000

Source of Information

Under protective order

History of the Hazard

Mid 1950s: Tilt-up technology was developed and employed. On the completion of the Dulles Dam, the U.S. Army Corps of Engineers, Portland District modified the design of a warehouse building to allow for the use of tilt-up. Cost savings were found in reduced erection time and lower labor costs. In addition, structural strength and the building's fire resistance rating were increased. These advantages have caused tilt-up to become a recognized construction method.

Contributory Factors

A new product that had not been tested or analyzed before being marketed caused the aforementioned injury. Absence of product testing and evaluation before marketing presents a serious hazard in concrete-form development, because it transfers product testing to the workers. The act of lifting a heavy concrete slab, or performing any task where equipment is necessary, is totally reliant upon the quality of the equipment used. When faulty, inappropriate, or untested equipment is used for a lifting task, catastrophe can occur.

Hazard Identification and Prevention Matrix

	Eliminate the Hazard		Guard the Hazard		Provide a Safety Factor		Provide Redundancy		Provide Reliability
	Hazard	Safety	Hazard	Safety	Hazard	Safety	Hazard	Safety	
Natural	Gravity								
Structural/ Mechanical	Faulty equipment	Stable connections						Pre-testing past rated capacity	100%
Electrical									
Chemical									
Radiant Energy									
Biological									
Automated Systems									

Applicable Standards

OSHA
See **1926**

U.S. Army Corps of Engineers
Page 544- 27.6

27.C Precast Concrete Operations:

27.C.01—Precast concrete members shall be adequately supported to prevent overturning or collapse until permanent connections are complete.

27.C.02 Lifting inserts and hardware:

(a) Lifting inserts which are embedded or otherwise attached to tilt-up precast concrete members shall be capable of supporting at least two times the maximum intended load applied or transmitted to them.

(b) Lifting inserts which are embedded or otherwise attached to tilt-up precast concrete members, other than tilt-up members, shall be capable of supporting at least four times the maximum intended load applied or transmitted to them.

(c) Lifting hardware shall be capable or supporting at least five times the maximum intended load applied or transmitted to the lifting device.

27.C.03—No employee shall be permitted under precast concrete members being lifted or tilted into position except employees required for the erection of those members.

Commentary

Anchor products used in lifting processes should require certification that they are safe for their intended use. Construction sites should not be the testing ground for anchor systems.

Recommendations

The European common market (CE) program should be adopted for application in the United States. This program is outlined in the International Standards Organization (ISO). ISO series 9000 and 14000 provide detailed implementation of the CE certification process. Design engineers should specify the Z10 ANSI/AIHA standards when applicable.

SECTION 5

Lift Slabs

The Hazard Syndrome

Lift slabs are concrete slabs cast one on top of another and raised into place to serve as the deck of each floor. The slabs are raised onto columns, but because more must be lifted, they cannot be braced laterally until the end of the application. A multistoried building constructed with lift slabs can collapse if each slab is not anchored properly at its designated floor level. The primary danger of lift slabs comes from the slabs not being properly anchored until each one is put into place, and the instability of the building until the slabs are anchored. If the slabs are in the process of being pulled apart and set, and one collapses, there is no chance of escape for anything or anyone underneath. The hazards are *dormant* when the slabs are cast in place. The hazard is *armed* when the jacking (lifting process) begins. The hazard is *active* when the anchor system connecting the slab (deck) to the building column fails. When this occurs, the slabs (deck) can fall all the way to the ground floor. In some instances, several slabs (decks) for upper floors drop in succession, breaking loose previous deck slabs that have been raised and anchored in place.

Scope

Lift slabs are an efficient form of construction. Shorter buildings (usually 1–5 stories) are most likely to utilize this process. When lift-slab work is completed successfully, it is very economically sound; when it is not, it leads to catastrophe. Design failure of anchors of the slab (deck) to the building columns can be a "self-destruct" hazard, bringing most of the structure down with it.

Peril from the Failure Mode

The falling slabs can trap anyone on any of the floors and crush them. If an anchor fails, collapse happens so quickly that no one can escape the building. Danger in the lift-slab process is due to the *natural* hazard of gravity and the *structural/mechanical* hazard of anchor-point failure.

Brief Summary of the Occurrence

In 1987, the collapse of several floors of the L'Ambiance Plaza killed 28 workers. This tragedy initiated investigations concerning the design of the jacking system and the connectors of the slab to the steel pillars.

Date of Occurrence

1987

Source of Information

Bridgeport, CT newspaper

History of the Hazard

1970s: Lift-slab methodology was developed and still remains a viable construction procedure.

Contributory Factors

The engineering design did not provide a fail-safe connection of the slab to each of the building columns.

Hazard Identification and Prevention Matrix

	Eliminate the Hazard		Guard the Hazard		Provide a Safety Factor		Provide Redundancy		Provide Reliability
	Hazard	Safety	Hazard	Safety	Hazard	Safety	Hazard	Safety	
Natural									
Structural/ Mechanical	Structural failure, inadequate supports			Im-proved connec-tors		Fail-safe jacking system		Brackets in addition to connectors	100%
Electrical									
Chemical									
Radiant Energy									
Biological									
Automated Systems									

Applicable Standards

OSHA

http://www.osha.gov/Publications/3106.html

U.S. Army Corps of Engineers

27D Lift Slab Operations

27.D.01—Lift-slab operations shall be planned and designed by registered engineers or architects. Such plans and design shall include detailed instructions and sketches indicating the prescribed method of erection and shall be submitted to the GDA for review.

27.D.02 Jacking Equipment

(a) The manufacturer's rated capacity shall be legibly marked on all jacks and shall not be exceeded.

(b) Threaded rods and other members that transmit loads to the jacks shall have a minimum safety factor of 2.5.

(c) Jacks shall be designed and installed so that that they will not continue to lift when overloaded.
(d) All jacks shall have a positive stop to prevent overtravel.
(e) Hydraulic jacks used in lift slab construction shall have a safety device that will cause the jacks to support the load in position if the jack malfunctions.

27.D.03 Jacking Operations

(a) When it is necessary to provide a firm foundation, the back of the jack will be blocked or cribbed. Where there is a possibility of slippage of the metal cap of the jack, a wood block shall be placed between the cap and the load.
(b) The maximum number of manually controlled jacks or slabs shall be limited to 14, and in no event shall the number be too great to permit the operator to maintain the slab level without specific tolerances.
(c) Jacking operations shall be synchronized to ensure even and uniform lifting of the slab.
(d) During lifting, all points of the slab support shall be kept within $^1/_2$ inch (1.2 cm) of that needed to maintain the slab in a level position.
    (i) If leveling is automatically controlled, a device shall be installed which will stop the operation when the $^1/_2$ inch (1.2 cm) leveling tolerance is exceeded.
    (ii) If leveling is manually controlled, such controls shall be located in a central location and attended by a trained operator while lifting is in progress.
(f) No one shall be permitted under the slab during jacking operations.

Commentary

No industry design standard has been developed for locking and securing the lifted slab at the floor level where it is to be permanently attached. There is a need for coordinating lift-slab design with construction engineers to ensure inherently safe connection of the lifted floor slabs to the supporting columns.

Recommendations

The design of a lift-slab connection to the supporting pillars involves the expertise of the structural engineer, construction-management engineers, and safety engineers. In this manner, the design of the connection of the lift slab to the supporting pillars is suitable for field/site conditions and circumstances. The connectors used to support each floor level must be strong enough to support the lift slabs for the floors above, as well as the stresses encountered when jacking the next level. The support-column connecting device must be simple and not error provocative, as human error cannot be tolerated in this instance.

SECTION 6

Steel Erection

The Hazard Syndrome

Approximately 250,000 people are employed in steel erection. In many cases, the erection of steel framework includes the installation of steel decking, roofing, and sheathing. All these processes require work in elevated areas, which is a *dormant* hazard. Loss of foot traction by workers without fall protection *arms* the hazard for falls on the same level and falls from an elevation.

Additional major hazards include the collapse of a beam or joist (most often at the point of connection) of the structure being assembled, which can *arm* the hazard for the falling or buckling of structural members, or the falling of other building materials temporarily placed on the structural members. Inadequate or improper lifting or hoisting equipment and rigging is another prevalent hazard in steel erection, as machines and bracing methods must be properly suited to a task.

Scope

Fatal falls by assemblers are the most prevalent hazard associated with steel erection. The ironworkers' union (to which about 44% of all ironworkers belong) records approximately 6200 injuries each year. Twenty-four percent of these (1500) are from falls. Nearly half (600) of these falls were from an elevation. More than 11% of all injuries (700) were due to crane failure or misuse. Union records show about 46 fatalities a year, with 11 of these concluded to be from slipping.

Looking at the big picture of *all* steel erection (by union ironworkers and erectors of pre-engineered steel buildings who may work in other crafts), it can be estimated that probably four times more injuries occur than the union's figures indicate. Predesigned packaged-steel buildings are assumed to be easy to erect and are often assembled by untrained personnel without proper equipment. This practice leads to exacerbation of a host of hazards inherent to the tasks of steel erection by a worker's unfamiliarity with procedures or lack of equipment. This practice, too, leads to misreporting or underreporting. Because such construction is often a one time activity, injury reports do not always cite steel erection as the source of the hazard. Statistical estimates of all injuries and deaths sustained by workers in steel erection could be as high as 25,000 injuries and deaths a year, with approximately 6000 due to falls, of which 2500 would probably be from an elevation. As shown on the opposite page, proper latching procedures are crucial to successfully preventing falls.

Additional injuries attributed to getting the steel into place using lifting equipment can be estimated at almost 3000 per year. A significant portion of steel-erection injuries is due to crane misuse.

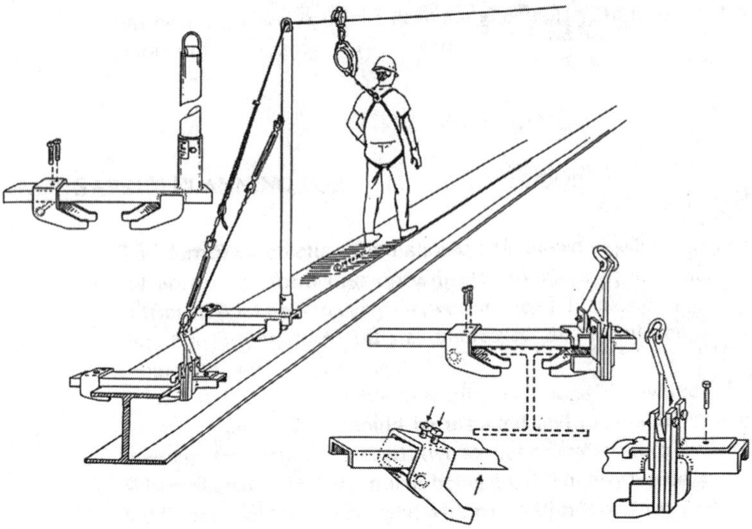

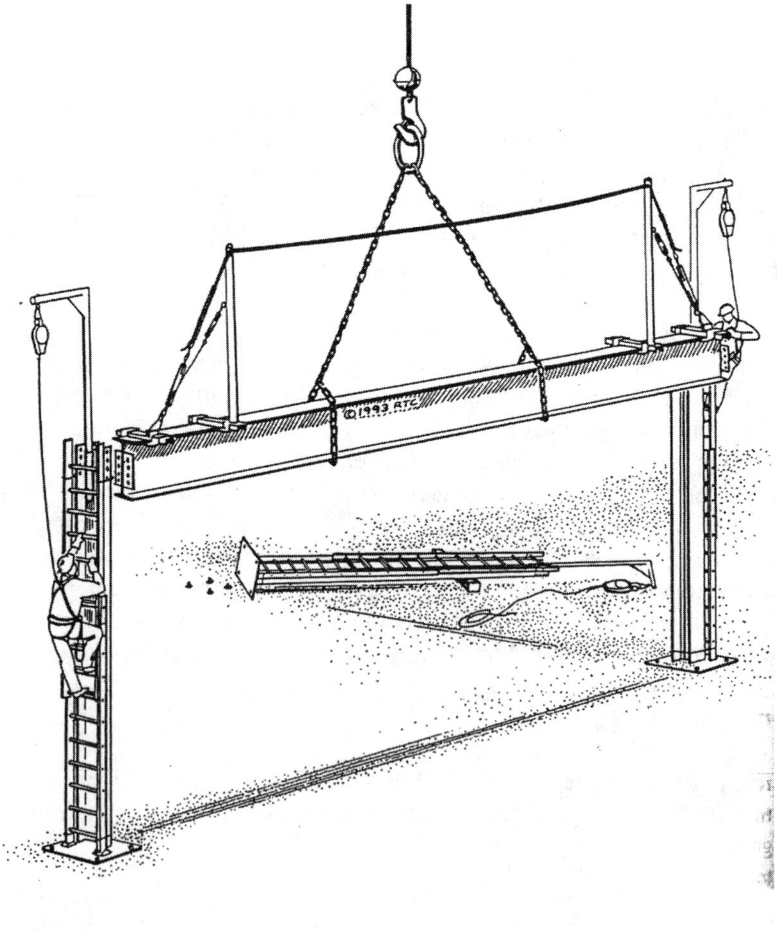
©1993 RTC

Peril from the Failure Mode

Today's construction methods include two types of steel erection. The traditional method of steel erection is the assembly of steel framework piece-by-piece, story-by-story, for skyscrapers, large commercial/industrial buildings, and bridges. However, pre-engineered buildings, with their promise of efficient use of material and assembly instructions, are gaining popularity. A pre-engineered building is built from kits or packages containing assembly drawings and all structural members, fasteners, roofing, and sheathing materials.

Both methods present the *natural* hazards of falls from elevation and the *structural/mechanical* hazards associated with lifting equipment and connectors. Another hazard presented by steel erection is the secondary-voltage *electrical* hazard associated with welding.

Brief Summary of the Occurrence

One man was killed when a roofing panel collapsed during installation, causing him to fall from an elevation of 60 feet. The panel was being used as a temporary walkway for installing insulation. These panels are approximately 2.5 feet wide with a 4-inch curb on either side to enhance structural strength. Panels can withstand the weight of a man in the center, but any pressure on the curbs of the panel will collapse it. The deceased was kneeling on the panel when his knee inadvertently pressed on the curb, causing the curb to buckle and collapse the panel.

Date of Occurrence

1980

Source of Information

Federal Court of Oklahoma City, OK

History of the Hazard

As early stone and brick buildings gained in height, they required stronger and stronger walls. A building ten stories high needed ground floor walls at least four feet or more thick. This limited the height of buildings until the mid-1800s.

1884: The Home Insurance Company erected a ten-story-high steel-framed building in Chicago. This was the birthplace of the steel skyscraper.

1935: Steel-erection safety came to prominence with the building of the Golden Gate Bridge in San Francisco. Safety nets were introduced on this project, and nineteen steelworkers were saved from a plunge into the unforgiving straits.

World War II: Henry Kaiser's shipyards built liberty ships on assembly ways (an inclined structure on which ships are built and launched). Large preassembled sections of the ship's hull and superstructure were put together in a work area behind the ways, and huge track-mounted cranes picked up the sections, carried them to the ways, and lifted them into place. This method reduced two hazards: falls from an elevation, and welding in enclosed compartments. Incidentally, travel alarms were also used on these moving tower cranes, which became the forerunner of the back-up alarms that were first required by the Corps of Engineers on its large dam-construction projects after World War II.

Post–World War II: Prengineered, packaged off-the-shelf steel buildings began to dominate the marketplace. As the demand grew for quickly erected steel warehouses, business centers, and schools, many new workers who had little previous experience or training in this field were given one time employment.

Contributory Factors

There are two predominant reasons why structural-steel erectors are extremely vulnerable to crippling injuries and death. The first reason is that *safety standards fail to address the root cause of the life-threatening and often life-taking hazards inherent in steel erection.* For example, there are no governmental or industry requirements or standards that establish a specific safe coefficient of slip resistance on the steel on which workers must walk during the erection or installation of steel beams, flooring, roofing, or other components.

The second reason is *an absence of an industry-wide practice to require a written steel-erection safety plan prepared with the collaboration of the structural engineer and the construction manager to ensure its adequacy and compliance.* When safety planning is incorporated as part of operating practices, injury rates are dramatically lowered. The lack of OSHA requirements for a written construction safety plan presents an anomaly to the reality of steel erection. The January 1982 Business Roundtable's A-3 report, *Improving Construction Safety Performance,* recommends that construction safety be successfully influenced by owners who require a formal site-safety program that defines the performance required. The absence of such a safety plan allows equipment that is not safe or adequate to be brought onto the job, and methods or procedures that are dangerous to be overlooked by management or supervision. These omissions plague the steel-erection process with unnecessary hazards that cause injury, damage, and downtime that soon run costs substantially above the bid.

Because of the availability of pre-engineered, packaged steel buildings, which *appear* to be simple to assemble, erection is sometimes attempted using managers and workers untrained in this type of construction. Operation of rental cranes and other equipment by these workers, which is typically outside their usual work assignment, can easily result in injury.

Hazard Identification and Prevention Matrix

	Eliminate the Hazard		Guard the Hazard		Provide a Safety Factor		Provide Redundancy		Provide Reliability
	Hazard	Safety	Hazard	Safety	Hazard	Safety	Hazard	Safety	
Natural	Gravity (fall)								
Structural/ Mechanical	Structural failure	Stronger curb		Self-supporting insulation panels		Warning on roof panels with load limits		Fall protection	100%
Electrical									
Chemical									
Radiant Energy									
Biological									
Automated Systems									

Applicable Standards and Literature

OSHA

On July 10, 1995, James Stanley, the Deputy Assistant Secretary of Labor for OSHA issued an Interpretation of Subpart M for Fall Protection in Steel Erection, in which he cited the recent OSHA decision to delay application of Subpart M to steel-erection activities on nonbuilding structures (it already did not apply to steel-erection activities on skeleton steel buildings), and made a restatement of the interim enforcement policy to be applied to fall hazards in steel erection activities:

"Effective immediately and until further notice:

1. New Subpart M does not apply to steel erection activities. The term "steel erection activities" means the movement and erection of skeleton steel members (structural steel) in or on buildings and non-building structures. It includes initial connecting, moving point-to-point, installing metal floor or roof decking, welding, bolting, and similar activities. It does not mean the erection of steel members such as lintels, stairs,

railings, curtain walls, windows, architectural metalwork, column covers, catwalks, and similar non-skeletal items, nor does it mean the placement of reinforcing rods in concrete structures.

"For the purposes of the stay action and this policy, the term "buildings" includes tiered and non-tiered single-story and multistory buildings, warehouses, gymnasiums, stadiums, power plants, theaters, mill buildings, and similar structures. The term "tiered" means the skeleton steel framework is erected in vertically stacked columns; it is not limited to multi-floored structures. The term "non-building structures" includes bridges, viaducts, overpasses, towers, tanks, billboards, antennas, and similar structures.

2. Fall protection for steel erection activities shall be provided consistent with the following:
   (a) Tiered buildings:
      (i) Exterior fall hazards of 25 feet or more are covered by 1926.105(a). Exterior fall protection is not required for fall hazards of less than 25 feet.
      (ii) Interior fall hazards of 30 feet or more on buildings which have floors or are adaptable to temporary floors are covered by 1926.750(b)(2)(i). Fall protection is not required for fall hazards of less than 30 feet. Temporary floors will generally be practicable in the construction of a typical multi-floored building.
      (iii) Interior fall hazards of 25 feet or more on tiered buildings that are not adaptable to temporary floors are covered by 1926.750(b)(1)(ii). Fall protection is not required for fall hazards of less than 25 feet.
   (b) Non-tiered buildings. Exterior and interior fall hazards of 25 feet or more are covered by 1926.105(a). Fall protection is not required for fall hazards of less than 25 feet.
   (c) Non-building structures. Exterior and interior fall hazards of 25 feet of more are covered by 1926.105(a). Fall protection is not required for fall hazards of less than 25 feet.
3. Former (i.e., prior to February 6, 1995) paragraphs 29 CFR 1926.104, 1926.105, and CFR1926.107(b)(c) and (f) will continue to apply to steel erection activities.

CFR 1926.500(a)(2)(iii), Subpart M, Fall Protection

CFR 1926 Subpart R, Steel Erection (ongoing negotiated rulemaking activity)

CFR 1926.750 Flooring Requirements (summarized and not complete)

(a) Permanent flooring-skeleton steel construction in tiered buildings
   1. The permanent floors shall be installed as the erection of structural members progresses, and there shall be not more than eight stories between the erection floor and the uppermost permanent floor, except where the structural integrity is maintained as a result of the design.
   2. At no time shall there be more than four floors or 48 feet of unfinished bolting or welding above the foundation or uppermost permanently secured floor.

(b) Temporary flooring-skeleton steel construction in tiered buildings
   1. (i) The derrick or erection floor shall be solidly planked or decked over the entire surface except for access openings. Planking or decking of equivalent strength shall be of proper thickness to carry the working load. Planking shall be not less than 2 inches thick full size undressed, and shall be laid tight and secured to prevent movement.

(ii) On buildings or structures not adaptable to temporary floors, and where scaffolds are not used, safety nets shall be installed and maintained wherever the potential fall distance exceeds two stories or 25 feet. The nets shall be hung with sufficient clearance to prevent contact with the surface of structures below.

(iii) Floor periphery-safety railing. A safety railing of $^1/_2$-inch wire rope or equal shall be installed, approximately 42 inches high, around the periphery of all temporary-planked or temporary metal-decked floors of tiered buildings and other multi-floored structures during structural steel assembly.

2. (i) Where skeleton steel erection is being done, a tightly planked and substantial floor shall be maintained within two stories to 30 feet, whichever is less, below and directly under that portion of each tier of beams on which any work is being performed, except when gathering and stacking temporary floor planks on a lower floor, in preparation for transferring such planks for use on an upper floor. Where such a floor is not practicable, paragraph (b)(1)(ii) applies.

(ii) When gathering and stacking temporary floor planks, the planks shall be removed successively, working toward the last panel of the temporary floor so that the work is always done from the planked floor.

(iii) When gathering and stacking temporary floor planks from the last panel, the employees assigned to such work shall be protected by safety belts with safety lines attached to a catenary line or other substantial anchorage.

(c) Flooring—other construction

1. In the erection of a building having double wood floor construction, the rough flooring shall be completed as the building progresses, including the tier below the one on which floor joists are being installed.
2. For single wood floor or other flooring systems, the floor immediately below the story where the floor joists are being installed shall be kept planked or decked over.

CFR 1926.751 Structural Steel Assembly

(a) During the final placing of solid web structural members, the load shall not be released from the hoisting line until the members are secured with not less than two bolts, or the equivalent at each connection and drawn up wrench tight.

(b) Open web steel joints shall not be placed until such framework is safely bolted or welded.

(i) In steel framing, where bar joists are utilized, and columns are not framed in at least two directions with structural steel members, a bar joist shall be field-bolted at columns to provide lateral stability during construction.

(ii) Where long-span joists or trusses, 40 feet or longer, are used, a center row of bolted bridging shall be installed to provide lateral stability during construction prior to slacking or hoisting line.

(iii) No load shall be placed on open web steel joists until these security requirements are met.

(iv) Tag lines shall be used for controlling loads.

CFR 1926.752 Bolting, Riveting, Fitting-up, and Plumbing-up

(a) General Requirements
  (i) Containers shall be provided for storing or carrying rivets, bolts, and drift pins, and secured against accidental displacement when aloft.
  (ii) Pneumatic hand tools shall be disconnected from the power source, and pressure in hose lines shall be released, before any adjustments or repairs are made.
  (iii) Air line hose sections shall be tied together except when quick disconnect couplers are used to join sections.
  (iv) Eye protection shall be provided in accordance with Subpart E of this part.

(b) Bolting
  (i) When bolts or drift pins are being knocked out, means shall be provided to keep them from falling.
  (ii) Impact wrenches shall be provided with a locking device for retaining the socket.

(c) Riveting
  (i) Riveting shall not be done in the vicinity of combustible material unless precautions are taken to prevent fire.
  (ii) When rivet heads are knocked off, or backed out, means shall be provided to keep them from falling.
  (iii) A safety wire shall be properly installed on the snap and on the handle of the pneumatic riveting hammer and shall be used at all times. The wire size shall be not less than No. 9 (B&S gauge), leaving the handle and annealed No. 14 on the snap, or equivalent.

(d) Plumbing-up
  (i) Connections of the equipment used in plumbing-up shall be properly secured.
  (ii) The turnbuckles shall be secured to prevent unwinding while under stress.
  (iii) Plumbing-up guys and related equipment shall be placed so employees can get at the connection points.
  (iv) Plumbing-up guys shall be removed only under the supervision of a competent person.
  (v) Wood planking shall be of proper thickness to carry the working load, but shall be not less than 2 inches thick full size undressed, exterior grade plywood, at least $^3/_4$-inch thick, or equivalent material.
  (vi) Metal decking of sufficient strength shall be laid tight and secured to prevent movement.
  (vii) Planks shall overlap the bearing on each end by a minimum of 12 inches.
  (viii) Wire mesh, exterior plywood, or equivalent, shall be used around columns where planks do not fit tightly.
  (ix) Provisions shall be made to secure temporary flooring against displacement.
  (x) All unused openings in floors, temporary or permanent, shall be completely planked over or guarded in accordance with Subpart M of this part.
  (xi) Employees shall be provided with safety belts in accordance with 1926.104 when they are working on float scaffolds.

CFR 1926.753 Safety Nets, Subpart R Steel Erection—Safety Nets are required if there is a "fall potential distance that exceeds 25-feet or two stories, and shall be hung with sufficient clearance to prevent contact with the surface of the structure below."

Comment: To date, the requirements of Subpart M Fall Protection are still in limbo for steel erection.

ANSI

A10.13-1989 Steel Erection—Safety Requirements

U.S. Army Corps of Engineers

Structural Steel Assembly, November 3, 2003, summarized as follows:

Material and Equipment

- Impact wrenches shall have a locking device for retaining the socket.
  - Containers shall be provided for storing or carrying rivets, bolts, and drift pins, and secured against accidental displacement when aloft.
  - A safety wire shall be properly installed on the snap and on the handle of the pneumatic riveting hammer and shall be used at all times: the wire size shall be not less than No. 9 (B&S gauge), leaving the handle and annealed No. 14 on the snap or equivalent.
  - Structural and reinforcing steel for walls, piers, columns, and similar vertical structures shall be guyed and supported to prevent collapse.
- During the final placing of solid web structural members, the load shall not be released from the hoisting line until the members are secured with not less than two bolts or 10% of the bolts, whichever is greater, at each connection, drawn up wrench tight.
- Open web steel joists shall not be placed on any structural steel framework until the framework is permanently bolted or welded.
  - No loading shall be placed upon steel joists until all bridging is completely and permanently installed.
  - For individual joists, the hoisting cables shall not be removed until sufficient top and bottom chord bridging is attached to provide lateral restraint.

Lateral bracing

- In steel framing, where bar joists are used and columns are not framed in at least two directions with structural steel members, a bar joist shall be field-bolted at columns to provide lateral stability during construction.
- Where longspan joists or trusses, 12 m (40 ft.) or longer, are used, a center row of bolted bridging shall be installed to provide lateral stability during construction prior to slacking or hoisting line.
- No load shall be placed on open web steel joints until the stability requirements of 27.E.03 are met.
- Workers shall be provided fall protection whenever they are exposed to falls of 1.8 m (6 ft) or more.
- Riveting shall not be done near combustible materials unless precautions are taken to prevent fire.
- When bolts, drift pins, or rivets are being knocked out, they shall be kept from falling.

Flooring

- Temporary flooring-skeleton steel erection.
- Derrick or erection floors shall be planked or decked over their entire surface except for access openings.
- On structures not adaptable to temporary floors, and where scaffolds or personnel lifting platforms are not used, safety nets shall be installed and maintained whenever the potential fall distance exceeds two stories or 7.5 m (25 ft)
- Temporary flooring shall be maintained within two stories or 9 m (30 ft), whichever is less, below and directly under that portion of each tier of beams on which any work is being performed, except:
  - when removing temporary flooring on a lower floor in preparation for transfer to an upper floor, or
  - where such a floor is not practical, in which case safety nets shall be used.
- The design of temporary flooring shall be approved by a registered engineer.
- Planking or metal decking shall be of sufficient strength and thickness or carry the working load: planking shall be not less than 5 cm (2 in) thick (full size undressed).
- Planking and metal decking shall be laid tight and secured to prevent displacement.
- A fall protection system, delineated in an activity hazard analysis and accepted by the designated authority, shall be provided for employees placing and removing temporary flooring.

MIl Specs

IL-E-3677 Erection Equipment for Bridge, Fixed, Steel, Panel, Bailey Type

IL-H-52802 Hook, Steel, Bolted Tank Erection

American Institute of Steel Construction (AISC)

S310 Design, Fabrication and Erection of Structural Steel

S326-1978 Design, Fabrication and Erection, Structural Steel

S32701994 Specification for the Design, Fabrication, and Erection of Steel Safety-Related Structures for Nuclear Facilities

Other

Steel Joint Institute

JI TD9-1987 Handling and Erection of Steel Joists and Joist Girders.

Steel Deck Institute

SDI 105-1992 Erection Instructions for Steel Frames.

Other References

- *Engineering News Record*, "Oregon OSHA hits Midwest Steel with Proposed $1.5-million fine," February 2, 1998.
- David V. MacCollum, *Construction Safety Planning*, Chapter 25, Phase Planning for Steel Erection, John Wiley, 1995.
- J. Nigel Ellis, PhD, CSP, PE, *Introduction to Fall Protection,* 3rd Edition, ASSE, 2003. Mr. Ellis also conducts training in fall protection and can be reached at (302) 571-0767.

- Lindley R. Higgins, PE, *Handbook of Construction Equipment Maintenance*, McGraw-Hill, New York, 1979 Chapter Seven.
- Merrit and Ricketts, *Building Design and Construction Handbook*, McGraw-Hill, 1994. Section Seven, "Structural Steel Construction" and Section Eight, "Cold-Formed Steel Construction."
- *Handbook of Heavy Construction*, edited by Havers and Stubbs, McGraw-Hill. Section 23, "Steel Construction."
- "Steel Erection," *Hazard Information Newsletter*, Volume 3, Issue 2 May 1998.

Commentary

Many design engineers are not aware of the multifaceted hazards of steel erection and how design can minimize hazards by looking beyond the structural integrity of the completed building. The erection *process* must be examined in order to meet the needs of the variety of erectors. Erection planning should include the design of mobile scaffolds to support workers while the structure is being built. Inherent structural features should include permanent anchor points for lanyards and harnesses, and reinforced or redundant bolts, pins, and connectors. Specific instructions should detail how the structures should be anchored and fastened at the time of erection to prevent collapse due to wind, vibration, or other variables that may arise in the erection sequence.

Recommendations

The hazards of steel erection are multifaceted in nature and present specific problems that must be addressed by the architect, design engineer, and construction manager. Workshop study sessions by designers and construction managers should be developed to examine actual erection methods on typical projects. Such programs would identify how the steel components and the design of connecting systems could be improved to aid the workers who assemble structures.

SECTION 7

Trenching

The Hazard Syndrome

The number of utilities placed underground continues to grow. Water, sewage, and gas lines have long been placed underground, but now powerlines as well as telephone and cable systems are also buried. Placing these systems underground normally requires digging a trench. Any trench deeper than five feet should be considered a *dormant* hazard. The unguarded trenching site shown on the next page exposes workers to the hazard of a fall from elevation. Someone in an unsupported trench *arms* the hazard, and soil failures causing a landslide *activate* the hazard. A common source of disastrous soil failure is the operation of equipment above a bank that is being excavated from below.

Scope

Installation or repair of underground utilities is usually subcontracted to a small contractor on a low-bid basis with little safety planning or job-safety overview. Unfortunately,

those generally responsible for design, installation, or repair are not concerned with specific requirements for shoring, or aware of other hazardous conditions such as buried utilities or the danger to excavation equipment from overhead powerlines. The use of equipment or access granted to heavy machines or vehicles next to an excavation presents a serious hazard with many legal ramifications. Too often designers or architects assume that the contractor will follow up on all these issues. When the work is inspected for compliance with design specifications, some inspectors are not aware of special soil properties or other conditions that could be considered unsafe. Few specific requirements for trenching safety are spelled out in many contracts. Even the overbroad contract provisions (28 pages) set forth in OSHA Subpart P 1926.650 that "shall be followed" are hollow admonitions that are not able to account for site-specific details that may affect working conditions.

The excavation flow chart on the following page determines the selection of the protective measure.

Peril Presented by the Hazard

Depending on the type of soil, trenching can also be a *chemical* hazard requiring specific equipment and methodology. However, injuries in trenching are caused primarily by *natural* hazards such as gravity and soil instability. Any trench deeper than five feet should be considered hazardous, because it can cave in and suffocate or immobilize

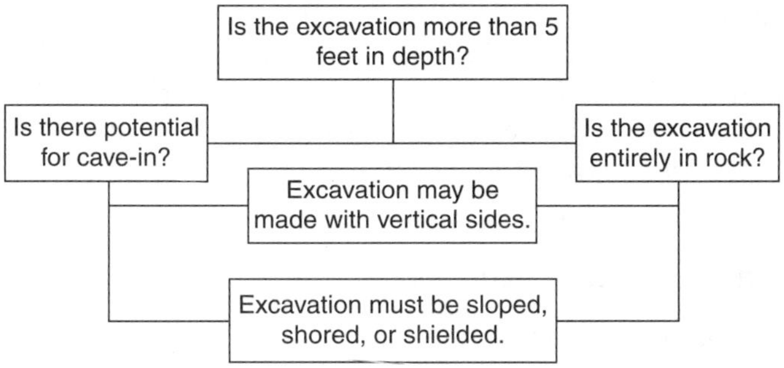

workers in the trench. Unstable soil can cause collapse without warning, and collapse can be prompted by any number of external factors, including weather conditions and proximity of other types of equipment. Flooding due to seepage from the water table, broken pipelines, or surface drainage into the trench could overload the shoring and is a common cause of collapse. Sides of the trench can collapse when people are going down into it or climbing up from it.

Besides trench collapse, here are other hazards inherent to trenching:

- Oxygen might be deficient or dangerous vapors present.
- Toxic material may be present in the soil.
- Flooding can cause damage to equipment and workers may be unable to escape the water.
- No ladders to provide safe access.
- Underground utilities are often damaged because their presence was unknown to those doing the excavation
- No barricades or covers to prevent unintended falling or driving into the open excavation. (People and vehicles often fall victim to an unguarded hole in the ground.)
- No curbs around an open trench. (Excavated soil or other materials piled next to a trench often slip and roll into it, striking workers below. Workers in the bottom are vulnerable to falling objects from above.)
- Powerline contact that occurs when lowering pipe or other materials by crane or excavator into a trench when the boom or hoist cable strikes an overhead powerline.
- Upset of heavy equipment from the unstable surface caused by a vertical embankment being excavated from below.

Brief Summary of the Occurrences

(A) One man was excavating a trench for a water line leading to a Connecticut wholesale shopping club when the unsupported trench caved in. Two other men jumped in to save their friend, and the trench caved in a second time. All three men were rescued, but the two men who attempted to rescue the first victim suffered permanent injuries and were successful in their claim against the general contractor, as he had failed to ensure a properly shored trench.

(B) The operator of a front-end loader was required to remove topsoil from the top of a glacial gravel till. The 15-foot high vertical bank was being excavated from below. The unstable vertical bank caved in and the loader upset and slid to the bottom. The loader cab was crushed. The operator could not be rescued and died of suffocation from the weight of gravel.

Date of Occurrences

(A) October 30, 1998

(B) April 25, 2003

Source of Information

(A) *The Jere Beasley Report,* June 2005

(B) Case currently being filed in New Hampshire

History of the Hazard

Starting around 100 BC, Romans trenching during the construction of aqueducts experienced the same hazards that apply to construction today.

Contributory Factors

When a vertical incision is made into soil to make a trench, the weight of the surrounding soil becomes unsupported and can easily collapse. Sometimes the soil from the trench itself is thrown out on either side of the trench, and this additional weight creates even more pressure on the unsupported walls. The weight and vibration of equipment in use next to the trench can play a major role in trench collapse by destabilizing the surrounding soil. The greatest unknown is the moisture content of the soil, as well as the strata of sand or mud, which can be transformed into a slip plane.

The reason for most trench failures is unqualified or incompetent persons who did not recognize the hazards inherent to this practice or chose to ignore them! Often laborers at the work site do not recognize what appears to be a stable vertical earth bank as an armed hazard waiting to collapse into the action mode. Any trench over five-feet deep has no real standing time and needs to be controlled by immediate placement of a trench box or shoring *before* entry. For this reason, proper soil support should be included as a design-planning requirement.

Hazard Identification and Prevention Matrix

	Eliminate the Hazard		Guard the Hazard		Provide a Safety Factor		Provide Redundancy		Provide Reliability
	Hazard	Safety	Hazard	Safety	Hazard	Safety	Hazard	Safety	
Natural	Gravity/ unstable soil				Unknown soil conditions				
Structural/ Mechanical	Structural failure of trench	Shoring/ trenchless machines		Steel trench boxes		Soil testing		On-site review	100%
Electrical									
Chemical									
Radiant Energy									
Biological									
Automated Systems									

Applicable Standards and Literature

OSHA

Subpart P, Excavations, 1926.650 through 1926.652, and appendixes A through F have 28 pages of Definitions (650), General Requirements (651), and Requirements for Protection Systems (652), which cover in great detail the points outlined in the section "How Trenching Injuries and Death can be Prevented."

ANSI

In 1944, the A10.2 American Standard Safety Code for Building Construction, Part II, "Excavation Work," Section 6 divided its discussion of trenching into its "General Requirements," "Trenches of Varying Widths and Depths," and "Trenches with Hydrostatic Pressure."

In 1973 the National Safety Council sponsored a draft for an A10.12 "Safety Requirement for Excavation." To date, there is still no standard that addresses trenches. **Comment:** Considering that for over a quarter of a century ANSI has not developed requirements for trenching, when it is estimated that in the same period of time perhaps 1,000 lives have been lost, isn't it time to address this hazard?

U.S. Army Corps of Engineers

EM385-1-1 (November 3, 2003) *Safety and Health Requirements Manual*

The U.S. Army Corps of Engineers Safety and Health Requirements Manual has had a long and continuing emphasis on safe trenching since 1958. For nearly 40 years it has developed easy-to-follow, detailed instructions on how trench excavation can be made safe. The current update of this manual, on November 3 2003, released, contains 12 pages on this subject in Section 25, Excavations. The most significant requirements are

25.A.01(a)(b) Planning

(a) Prior to opening an excavation, underground installations (e.g., sewer, telephone, water, fuel, electric lines) shall be located and protected from damage or displacement: utility companies and other responsible authorities shall be contacted to locate and mark the locations and, if they so desire, direct or assist with protecting the underground installations.

(b) Where excavations are to be performed in areas known or suspected to be contaminated with explosives, unexploded munitions, or military ordnance, surface and subsurface clearance by qualified explosive ordnance disposal (EOD) personnel shall be accomplished prior to excavation work.

25.A.02 (a)(b)(c) Excavation inspection and testing

(a) When persons will be in or around an excavation, the excavation, the adjacent areas, and protective systems shall be inspected daily, as needed throughout the work shifts, and after every rainstorm or other hazard-increasing occurrence by a competent person.

(b) If evidence of a situation which could result in possible cave-ins, slides, failure of protective systems, hazardous atmospheres, or other hazardous condition is identified, exposed workers shall be removed from the hazard and all work in the excavation stopped until all necessary safety precautions have been implemented.

(c) In locations where oxygen deficiency or gaseous conditions are known or suspected, air in the excavation shall be tested prior to the start of each shift or more often if directed by the designated authority: a log of all test results shall be maintained at the work site.

25.A.03(a)(b)(d)(e) Protective systems

(a) The sides of all excavations in which employees are exposed to danger from moving ground shall be guarded by a support system, sloping or benching of the ground, or other equivalent means.

(b) Excavations less than 1.5 m (5 ft) in depth and in which a competent person examines and determines there to be no potential for cave-in do not require protective systems.

(c) Support systems shall be in accordance with Section 25.D

(d) Protective systems shall have the capacity to resist without failure all loads that are intended or could reasonably be expected to be applied to the system.

25.A.04(a)(1)(2)(b)(c) Stability of adjacent structures

(a) Except in stable rock, excavations below the level or the base of footing of any foundation or retaining wall shall not be permitted, except in the following conditions.

(i) a support system, such as underpinning, is provided to ensure the stability of the structure and to protect employees involved in the excavation work or in the vicinity thereof; or

(ii) a registered professional engineer has approved the determination that the structure is sufficiently removed from the excavation so as to be unaffected by the excavation and that the excavation will not pose a hazard to employees.

(b) If the stability of adjoining buildings or walls is endangered by excavations, shoring, bracing, or underpinning designed by a qualified person shall be provided to ensure the stability of the structure and to protect employees.

(c) Sidewalks, pavements, and related structures shall not be undermined unless a support system is provided to protect employees and the sidewalk, pavement, or related structure.

25.A.05 Where it is necessary to undercut the side of an excavation, overhanging material shall be safely supported.

25.A.06(a)(b)(1)(2) Protection from water

(a) Diversion ditches, dikes or other means shall be used to prevent surface water entering an excavation and to provide good drainage of the area adjacent to the excavation.

(b) Employees shall not work in excavations in which there is accumulated water or in which water is accumulating unless the water hazard posed by accumulation is controlled.

(i) Freezing, pumping, drainage, and similar control measures shall be planned and directed by a registered engineer. Consideration shall be given to the existing moisture balances in surrounding soils and the effects on foundations and structures if it is disturbed.

(ii) When continuous operation of ground water control equipment is necessary, an emergency power source shall be provided. Water control equipment and operations shall be monitored by a competent person to ensure proper operation.

25.A.07(a)(b)(c) Protection from falling material

(a) Employees shall be protected (by scaling, ice removal, benching, barricading, rock bolting, wire mesh, or other means) from loose rock or soil which could create a hazard by falling from the excavation wall. Special attention shall be given to slopes which may be adversely affected by weather, moisture content or vibration.

(b) Materials such as boulders or stumps, that may slide or roll into the excavation shall be removed or made safe.

(c) Excavated material shall be placed at least 0.6 m (2 ft) from the edge of an excavation or shall be retained by devices which are sufficient to prevent the materials from falling into the excavation. In any case, material shall be placed at a distance to prevent excessive loading on the face of the excavation.

25.A.08(a)(b)(c) Mobile equipment and motor vehicle precautions

(a) When vehicles or mobile equipment are utilized or allowed adjacent to an excavation, substantial stop logs or barricades shall be installed: the use of a ground guide is recommended.

(b) Workers shall stand away from vehicles being loaded or unloaded to avoid being struck by spillage or falling materials.

(c) Excavating or hoisting equipment shall not be allowed to raise, lower, or swing loads over personnel in the excavation without substantial overhead protection.

25.A.11 Employees entering excavations classified as confined spaces or which otherwise present the potential for emergency rescue shall wear a harness with a lifeline securely attached to it.

25.B Safe Access

25.D. Support Systems

25.D.01 Support systems shall be in accordance with one of the systems outlined in (a) through (c) below:

(a) Designs drawn from manufacturer's tabulated data shall be in accordance with all specifications, limitation, and recommendations issued or made by the manufacturer.
   (i) Deviation from the specifications, recommendations, and limitations are only allowed after the manufacturer issues specific written approval.
   (ii) A copy of the manufacturer's specifications, recommendations, and limitations—and the manufacturer's approval to deviate from these, if required—shall be in written form and maintained at the job site during excavation.

(b) Designers shall be selected from and be in accordance with tabulated data (such as tables and charts). At least one copy of the tabulated data shall be maintained at the job site during excavation. The tabulated data shall include:
   (i) identification of the parameters that affect the selection of the protective system drawn from such data,
   (ii) identification of the limits of use of the data, and
   (iii) explanatory information as may be necessary to aid the user in correctly selecting a protective system from the data, and
   (iv) the identity of the registered professional engineer who approved the data.

(c) Designed by a registered engineer. At least one copy of the design shall be maintained at the job site during excavation. Designs shall be in writing and include:
   (i) a plan indicating the sizes, types, and configurations of the materials to be used in the protective systems, and
   (ii) the identity of the registered engineer who approved the design.

25.D.02 Materials and equipment used for protective systems

(a) Materials and equipment shall be free from damage or defects that might impair their proper function.

(b) Manufactured materials and equipment shall be used and maintained in a manner consistent with the recommendations of the manufacturer and in a manner that will prevent employee exposure to hazards.

(c) When material or equipment is damaged, a competent person shall examine the material or equipment and evaluate its suitability for continued use.

25.D.03 Installation and removal of support systems

(a) Members of support systems shall be securely connected together to prevent sliding, falling, kickouts, or other predictable failure.

(b) Support systems shall be installed and removed in manners that protect employees from cave-ins, structural collapses, or from being struck by members of the support system.

(c) Individual members of a support system shall not be subjected to loads exceeding those which they were designed to withstand.

(d) Before temporary removal of individual members, additional precautions shall be taken to ensure the safety of employees, such as installing other structural members to carry the loads imposed on the support system.
(e) Removal shall begin at and progress from the bottom of the excavation; members shall be released slowly as to note any indication of possible failure of the remaining members or possible cave-in on the sides of the excavation.
(f) Backfilling shall progress together with the removal of support systems from excavations.

25.D.04(a)(b)(c) Shield systems
(a) Shields shall be installed in a manner to restrict lateral or other hazardous movement of the shield in the event of the application or sudden lateral loads.
(b) Employees shall be protected from the hazard of cave-ins when entering or exiting the area protected by shields.
(c) Employees shall not be allowed in shields when shields are being installed, removed, or moved vertically.

25.D.05 (a)(b)(c)(d) Additional requirements for trenching
(a) Installation of support systems shall be closely coordinated with excavation of trenches.
(b) Bracing or shoring of trenches shall be carried along with the excavation.
(c) Backfilling and removal of trench supports should progress together from the bottom of the trench: jacks or braces shall be released slowly and, in unstable soil, ropes shall be used to pull out the jacks or braces from above after personnel have cleared the trench.
(d) Excavation of material to a level no greater than 0.6 m (2 ft.) below the bottom of the members or a trench support system (including a shield) shall be permitted, but only if the system is designed to resist the forces calculated for the full depth of the trench and there are no indications while the trench is open of a possible loss of soil from behind or below the bottom of the support system.

Other References

- Bruce A. Suprenant and Kim D. Basham, *Excavation Safety*, The Aberdeen Group, 1993.
- North American Society for Trenchless Technology, 435 North Michigan Avenue, Suite 1717, Chicago, IL, e-mail 75147.2600@compuserve.com; Web page http://www.bc.irap.nrc.ca/nodig/.
- W. Martin "Red" Cass, PE, *Common Sense in the Common Trench*, Equipment Guidebook Company, Palo Alto, CA. date
- David V. MacCollum, Chapter 23, "Phase Planning for Excavating." *Construction Safety Planning*, Van Nostrand Reinhold, 1995.
- Gary Owensby, *OHIO Monitor*, "Down Below," March/April 1994.
- "Down Under," *OHIO Monitor*, November/December 1993 issue.
- "Trenching," *Hazard Information Newsletter*, Volume 2, Issue 2, May 1997.

Commentary

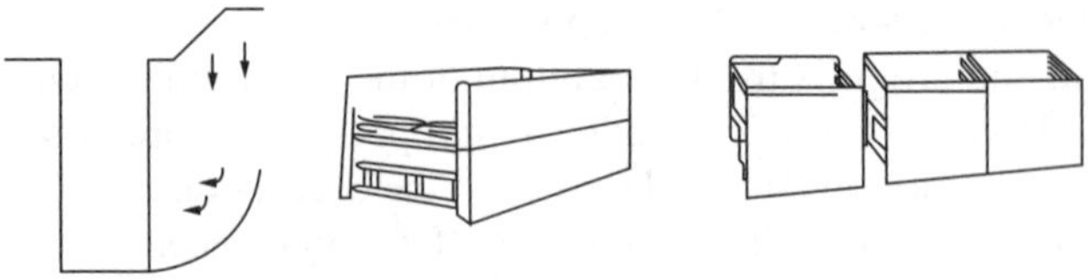

Fortunately, the new approach of using horizontal-direction trenchless excavators to automatically create trenches can be utilized in most cases.

The use of horizontal-direction excavators has a lot of advantages over conventional trench digging. When a designer specifies the trenchless excavation, workplace safety is enhanced and the long-term advantage of not disturbing the existing soil allows for the replacement of conductors or tubing without digging. The image below shows such as trench being excavated without workers. (Although it should be gaurded with barricades to prevent workers or pedestrians from falling into it.)

When confronted with no other alternative, the work should be overseen by designating a *qualified person*, preferably a registered, licensed professional engineer who is experienced and knowledgeable in soil mechanics and in determining soil classifications. This individual should be familiar with OSHA 1926.652, be able to determine whether the excavation will endanger employees or adjacent structures, and be able to design or recommend an appropriate trench or excavation support system and ensure it will be used.

Site assessment is imperative to trenching. Learning the characteristics of the site helps the qualified person determine the appropriate type of excavation, either vertical or sloped, that presents the fewest hazards. There are many types of soils and rock formations, each of which have their own individual failure modes. Inherent properties of the ground in excavation must play a prominent role in determining safety plans for trenching. Water seepage can cause a change of conditions, just as drying will induce failure in some cases. Because of these complexities, a qualified engineer needs to visually examine the soil/rock and determine whether manual tests need to be made so that proper trench or bank support can be prescribed. Previously excavated ground is even more unpredictable, as the material used as fill usually has entirely different characteristics than virgin soil. Addressing the issue of *how* and *where* a utility will be installed underground ensures soil stability and prevents ground failures. The *site specificity* assures knowledge and documentation for an exact location if needed for future excavations.

Next, a *competent person* should be assigned to oversee the work, identify existing and predictable hazards that are dangerous, and take prompt corrective measures. The qualified person should then identify all other hazards and define appropriate safeguards during the design phase of trench excavation.

Recommendations

At the time of design, management needs to become more closely involved in the design of excavations, which require methodology inclusive of either shoring or other safe underground systems suitable for the existing soil conditions. When trenches are required, the type of shoring system should be included as a special condition, with specifications suited to the site and possible equipment hazards designed out.

The good news is that with modern trenchless horizontal-direction boring equipment, many trenches do not need to be dug. Trenchless or "no-dig" technology includes various methods and machines used to install, replace, rehabilitate, or inspect underground utility and piping systems. The goal is to reduce to a minimum the need to excavate in order to avoid disrupting existing roadways and other structures, thereby minimizing costs and eliminating the hazards inherent in trenching. To better understand what trenching equipment is available and how it functions, it is best to request information from manufacturers and suppliers, and ask for on-site demonstrations to ensure getting the safest and most productive equipment for your applications. Some "no-dig" systems include:

- Slurry and auger systems for microtunneling that have very close tolerances for both line and grade
- Directional boring machines
- Pipe jacking or ramming—a nonsteerable system

Often a cost analysis could support trenchless excavation for installing utilities underground.

SECTION 8

Tunneling

The Hazard Syndrome

Tunnel construction has many inherent hazards. Besides the hazards of collapse, rock falls, and flooding, the most devastating can be the chance of explosion from methane gas when tunneling in ground rich in carbonous material, such as narrow seams of coal or oil-soaked sandstone. The hazard is *dormant* as long as it is not excavated with a tunnel or shaft. The hazard is *armed* when the tunnel creates an air space for explosive fumes to collect. When ignited, the hazard is in the *active* mode, causing an explosion that is a potentially massive source of destruction.

Scope

Tunnels are constructed or excavated for highways, subways, railroad routes, and to provide fresh-water system irrigation and hydroelectric power. Ubiquitous mine tunnels provide access to both coal and mineral deposits. Because of their potential for catastrophic failure modes that claim numerous lives, tunnel and underground failures are always newsworthy, whether in the United States or foreign countries. The January 2006 tunnel explosion in a West Virginia mine served as a wake-up call to many industry officials: that such incidents can and should be prevented by the application of technological advances in tunnel construction processes and equipment.

Peril from theFailure Mode

Tunneling encompasses the hazards of gas and flammable dusts, both of which can explode. The same hazards are also present in the mining of coal. The *chemical* hazard that causes an explosion of gas or dust can turn a tunnel into a gun barrel that propels any equipment down the tunnel like a missile and in some cases can cause the collapse of the tunnel. When oxygen is insufficient for complete combustion, carbon monoxide gas is generated. This gas suffocates those entering the tunnel, after the explosion but before it is ventilated especially workers with inadequate or faulty self-breathing apparatus.

Brief Summary of the Occurrence

A three-mile-long water-supply tunnel at Port Huron became contaminated with methane gas during the drilling of a vent shaft on the other end of the tunnel and exploded, killing 21 workers. The force of the blast propelled a Pumpcrete machine down the three-mile tunnel like a bullet in the barrel of a gun, killing 26 additional workers. Testimony revealed that the construction company had placed insufficient concrete behind the tunnel form panels, leaving voids that allowed methane gas to enter the tunnel.

Date of the Occurrence

December 11, 1971

Source of Information

OSHA investigative hearings in early 1972

History of the Hazard

Coal mine explosions from methane gas are an ongoing hazard that has existed since Europe began coal mining in the 1600s and continues today. Methane gas makes no distinction between tunneling and coal mines.

1800s: Railways were becoming an increasingly popular form of transport. Tunnels were carved into the landscape for railroad lines.

1825: Construction began on the tunnel under the Thames river in England. Though it took 40 years to complete successfully, it was the first soft-soil project ever started, and it revolutionized tunneling.

1900s: The development of the pneumatic drill increased dust levels in tunnels and mines. With the dust came increased risk of explosion.

March 25, 1985: Unvented methane gas leaking from the soil below it accumulated in the basement of a Los Angeles Ross clothing store and exploded. The blaze caused by the explosion went on for days, with flames shooting out from fissures in the earth nearby. The undisputed presence of methane-rich soil derailed plans for an L.A. subway line in that area.

June 1997: An article describing the autopsies of four men who died after an explosion in a tunnel in an urban area in Japan notes that inhaled methane levels were not high enough to have killed these men, but were high enough to show that methane had been present in low levels at the construction site throughout the length of the project where the explosion occurred. Methane gas is present in many sites, and soil must be assessed.

Contributory Factors

The design of ventilating systems for the elimination of methane gas from underground excavations is well known, but is often ignored in order to save on immediate costs. Coal mines and tunnels can extend for miles underground without alternate shafts for ventilation and worker access. The absence of a short-distance air-moving system creates a dangerous mix of dead air and possibly poisonous or explosive gas.

Hazard Identification and Prevention Matrix

	Eliminate the Hazard		Guard the Hazard		Provide a Safety Factor		Provide Redundancy		Provide Reliability
	Hazard	Safety	Hazard	Safety	Hazard	Safety	Hazard	Safety	
Natural									
Structural/ Mechanical	Sealed tunnel wall			Provide a safe haven with air and food supplies for emergencies		Dust collection system		Breathing apparatus/ communication devices for workers	
Electrical									
Chemical	Explosive/ combustible gas			Methane gas extraction system					
Radiant Energy									
Biological									
Automated Systems									

Applicable Standards

OSHA

See **1926**

U.S. Army Corps of Engineers

EM 385-1-1 (November 3, 2003) *Safety and Health Requirements Manual*

Section 26 Underground Construction (Tunnels, Shafts, and Caissons)

26.C Air Monitoring, Air Quality Standards, and Ventilation

General Requirements for Fire Prevention and Protection 09.A

Other References

Harold W. Richardson, Robert S. Mayo, *Practical Tunnel Driving,* McGraw Hill, New York, 1941.

David V. MacCollum, PE, CSP, *Construction Safety Planning,* Wiley & Sons, 1993.

Commentary

Tunneling requires extensive planning at the time of design to ensure that the control of hazards is defined as specific cost- and management-oversight requirements. Several redundant ways of preventing the accumulation of methane gas in a tunnel include:

- Ventilation
- Sealing the tunnel walls
- Piping installed into the seams of a tunnel to collect the methane gas. This is often a safe and profitable process.
- A redundant system that applies two or more of the above

Recommendations

Tunnel designers and construction engineers need to develop a working liaison to overcome changing circumstances so engineering controls can be immediately installed. The design engineer and the construction manager must work with the geologist or sales specialists to develop the most effective excavation and support system.

Australian coal mines require a steel compartment (safe haven) for miners, with 72 hours of oxygen, water, food, and alternate communication lines. U.S. mines should provide this type of safe haven in mines.

SECTION 9

Diving/Underwater Construction

The Hazard Syndrome

The principle hazard of diving or working underwater is the onset of a condition called dysbarism, which results from the body's nitrogen absorption at a greater rate than normal due to the pressure increases on the body. As a diver descends or a worker enters a caisson, pressure in the body's air cavities changes and dissolves extra air into the blood. Rapid decompression allows the dissolved air to form nitrogen bubbles that course through the bloodstream. This condition is called decompression sickness, known as "the bends." The moment that the diver descends into the water, the hazard is *armed.* Rapid ascent makes the hazard *active.*

Scope

The hazard of dysbarism ("the bends") is ever-present for diving, underwater construction, and the use of caissons in which compressed air is used in an underwater working chamber for excavation. Caissons are pre-engineered structures built onto a foundation in a waterway for the purpose of providing a dry work area for construction workers. Caissons are commonly used in the construction of bridges and piers. To create a caisson, divers excavate an underwater foundation, which is usually the bedrock beneath the waterway, though sometimes it is hard mud. A concrete structure is then prefabricated and anchored onto the foundation. Air inside the structure must be pressurized from within to keep the force of the water from collapsing the walls. A rough diagram of a caisson is depicted below.

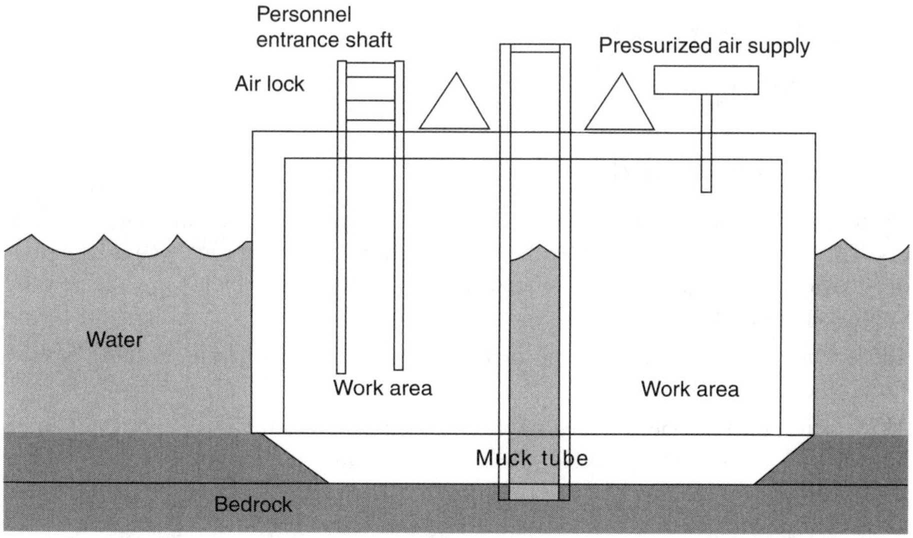

Though not technically "diving," workers in a caisson are subjected to pressure changes almost as extreme as divers and can suffer from the same afflictions. In fact, Caisson's Disease is another term for decompression sickness. The same hazard exists in tunneling when an airlock is used to provide back pressure to prevent tunnel collapse and flooding.

Peril from the Failure Mode

Serious and disabling injury from decompression sickness can lead to a variety of afflictions. Nitrogen bubbles can disrupt tissues in virtually any organ, including the brain, lungs, and spinal cord. Bubbles may deposit in joints, causing soreness or pain. Fatigue is the most widely experienced symptom of decompression sickness, but paralysis or death from conditions left untreated is not uncommon. Arterial gas embolisms can form when the alveoli in the lungs rupture from rapid depressurization, allowing gas bubbles to escape into the arteries. These may lead to strokes, paralysis, or numbness of the body. The bends can be classified as a *natural* hazard because it is caused by the human body's reaction to pressure change. However, such pressure changes could not be brought about without *structural/mechanical* innovations, such as an underwater breathing apparatus and caisson design, so this natural condition is a *structural/ mechanical* hazard as well.

Brief Summary of the Occurrence

A young naval diver was ordered to surface without sufficient time to decompress. The boat that he returned to did not have a decompression chamber. It took several hours via air-evacuation to reach a hospital where a decompression chamber was available. The diver suffered serious injuries and brain damage.

Date of Occurrence

1998

Source of Information

Case under protective order

History of the Hazard

1845: Symptoms consistent with decompression sickness were reported in coal miners. This condition was caused by the use of compressed air to force water from tunnels. After miners left the pressurized tunnels of the mine, they were plagued by cramps and muscular pain.

1854: Pol and Watelle discovered that recompression would relieve symptoms of decompression sickness.

1878: Paul Bert connected decompression sickness with the phenomenon then known as Caisson's Disease. He named the condition "the bends" because of the bent over posture assumed by most of those afflicted. He recommended recompression, then gradual decompression as a treatment, achieving stunning success.

1970: The Jones Act, 46 U.S.C. 688, was enacted to protect seamen and workers on offshore vessels against occupational injuries. However, protection under this act does not cover any worker who is not permanently employed on an offshore vessel.

1983: The death rate of men who were working under the Hudson River in a tunnel into New York City dropped dramatically from almost one worker in four to about one in sixty after a recompression chamber was used.

Contributory Factors

Due to Henry's Law, which states that "the concentration of a solute gas in a solution is directly proportional to the partial pressure of that gas above the solution," the frequency and depth of dives and exposure to work in caissons plays an active role in determining the amount of gas in a worker's blood. Frequent or intense exposure to pressurization can leave increased nitrogen and other gases in a workers blood, even if the worker has decompressed properly. For this reason, workers should not be asked or allowed to make frequent transitions between pressures. Improper decompression methods and inadequate recompression measures for workers afflicted with decompression sickness is the reason for this occupational hazard.

Hazard Identification and Prevention Matrix

	Eliminate the Hazard		Guard the Hazard		Provide a Safety Factor		Provide Redundancy		Provide Reliability
	Hazard	Safety	Hazard	Safety	Hazard	Safety	Hazard	Safety	
Natural	Nitrogen bubbles in blood								
Structural/ Mechanical		Job-site decom-pression chamber				Limited time and frequency of exposure to pressuri-zation		Limit air travel	
Electrical									
Chemical									
Radiant Energy									
Biological									
Automated Systems									

Note: Navy diving manuals provide one of the most respected standards for diving.

Applicable Standards

OSHA

OSHA 29 CFR 1910.401-441

U.S. Army Corps of Engineers

EM 385-1-1 (November 3, 2003) *Safety and Health Requirements Manual*

Section 30 (pp. 611–630)

Commentary

Diving operations and work using compressed air require compliance with specific standards and the provision for an on-site recompression chamber. Additionally, those engaged as divers should not participate in air travel for at least 12 hours after any dive. This international standard should be extended to 24 hours following multiple days of repetitive dives.

Recommendations

All diving needs to be conducted consistent with current U.S. Navy requirements.

12 Operation and Maintenance Hazards

INTRODUCTION

Construction projects present ongoing hazards during the erection phase that extend to the operation of the completed facility. Architects and design engineers who address solutions to foreseeable operational hazards in the design phase of a project will save future time, trouble, and lives. These potential hazard areas include:

1. Fire protection
2. Confined spaces
3. Light
4. Toxic fumes and gases
5. Wind and flooding
6. Sanitation

These subjects must be incorporated into the facility design as well as the master construction plan.

SECTION 1

Fire Prevention

The Hazard Syndrome

Flammable building materials and refuse create an *armed* hazard. When flammable materials are used in components of a facility, the hazard of fire extends into the life of that facility, waiting for the right conditions to become *active* and cause damage. Fire prevention during and after construction includes more than just removing the flammable trash from the work site. Materials must be carefully balanced; amounts of gases and all sources of ignition must be monitored. When considering fire prevention, one must remember that materials can become more flammable in the presence of other substances. Some combinations of materials may lead to spontaneous combustion. The following illustration reminds the user that combustion is dependant upon the following fire source triangle:

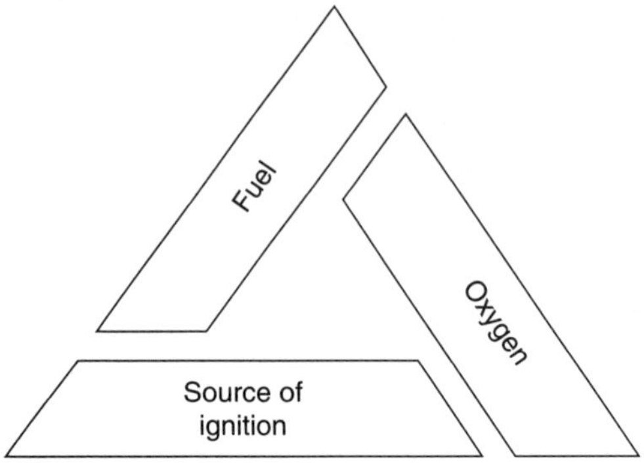

Scope

Fires during construction, maintenance, alteration, or demolition operations are an ever-present threat. The fire potential is inherently greater during short-term operations due to the presence of large quantities of combustible materials and debris, together with such ignition sources as temporary heating devices, operations with cutting/welding/plumber's torches, open fires, and close-proximity smoking. There are many highly flammable materials used as part of the structure being erected, as well as exposed sources of ignition during the construction process. Burning insulation with no viable exit route can incinerate construction workers and, later, inhabitants of the facility.

Peril from the Failure Mode

Fire hazards can arise from *natural, mechanical, electrical, chemical*, and *radiant* sources. The "Contributory Factors" part of this section offers a comprehensive layout. Fires and burns are the fifth leading cause of unintentional injury or death for all ages. Unintentional fires and burns cause between about 3000 and 5000 civilian deaths, and close to 20,000 injuries per year. Fires also present major costs in materials and damages. Property loss in fires was estimated to be $8.2 billion in 1995 according to the National Safety Council (Accident Facts, 1995). On an average, damage from industrial and commercial fires cost about $4 billion per year.

Fires arise from sources other than exposed flames. For instance, more than 100 fires and 10 deaths have been attributed to halogen light bulbs. The 300-watt halogen bulb can get as hot as 970 degrees Fahrenheit, compared with a 150-watt incandescent bulb, which reaches approximately 340 degrees. Another danger of halogen bulbs is from shattering caused by hot spots attributed to moisture, dirt, or skin oil on the bulb surface.

In construction activities and finished buildings, all sources of ignition must be closely monitored. Building materials often contain toxic or flammable components. Polyurethane materials include a broad family of plastics ranging from the styrofoam-type of insulation or packing materials, to the more solid materials such as sheeting, panels, and structural items, which, when ignited, burn rapidly and produce intense heat with dense smoke and toxic fumes. Rags soaked with linseed oil or paint products can create a source of spontaneous combustion.

Brief Summary of the Occurrence

Two welders working on a suspended scaffold below the twelfth floor of a county jail under construction were gassed and incinerated when hot slag set fire to flammable materials installed beneath them. During the five days prior to this particular fire, three trash fires had occured at the base of the building had occurred, presumably started by hot slag falling from the welding being done above. However, the presence of fires did not prevent styrofoam blueboard (a highly flammable product) from being installed on brackets from the first floor to the tenth floor, when in fact, it should not have been installed until all welding operations were completed. Stand-alone scaffolding was installed up to the fifth floor, on which rolls of petroleum-based felt paper were stored, ready for installation over the blueboard. Hot slag from the welding equipment fell on the felt paper, and ignited. The fire quickly spread and engulfed the blueboard, causing a conflagration within seconds.

No escape routes were available because the jail's windows were barred, and there was no escape ladder or access to the roof, which was one floor up. The welders were unable to raise the scaffolding system to escape and quickly asphyxiated from the toxic fumes of phosgene gas (a byproduct of burning styrofoam). They never had a chance.

No fire safety plan was in effect during construction.

Date of Occurrence

April 29, 1993

Source of Information

General Court of Justice, Superior Court Division, Forsyth County, NC, # 93-CvS-7156

History of the Hazard

1667: Following the Great Fire of London, the first fire insurance office was established by Dr. Nicholas Barbon.

1736: Benjamin Franklin established the first American volunteer fire department in Philadelphia.

1792: The Insurance Company of North America was formed. It was the first American fire and marine insurance company.

1808: The first leather fire hose seamed by copper wire rivets was produced. The same year, the first working fire hydrant was installed in New York City, attached to an underground wooden water main.

1840: The first steam fire engine in the United States was designed and constructed by P.R. Hodge, CE of New York City at the instigation of insurance companies.

1852: A patent was issued for the first perforated-pipe sprinkler system designed by James Pichen Francis, proprietor of Locks and Canals at the Merrimac River in Lowell, MA. This probably was the first recognized installation of fixed fire-protection equipment, but the full, expansive development of sprinkler systems did not really get underway for another twenty years.

1874: Henry S. Parmelee invented and patented a sprinkler that would discharge water automatically when fire occurred.

1872–1914: Some 450 patents were granted for components of sprinkler systems. From this time on, inventions and other fire-protection developments occurred at an increasing pace. Sprinkler systems were delayed until metal piping and substantial pumps were developed. The fusible sprinkler head was still a few years away; alarms, post indicator valves, Siamese connections, and accurate hydraulic data would not be available until the twentieth century.

1911: The Asch Building (Triangle Shirtwaist Company) fire, which killed approximately 150 people in New York, inspired some of the early laws.

1936: Nine persons lost their lives in a New York City restaurant when fire spread as a result of combustible silk cloth ceiling decorations. This incident brought about fire safety codes prohibiting furnishings and decorations of an explosive or highly flammable character from being used in any place of assembly or other occupancy.

1942: In Boston, 492 persons were killed in the Coconut Grove nightclub fire. The lessons learned in this fire made way for improvements in fire laws and ordinances throughout the United States. Essentially, they were:

- No place of public assembly should be filled beyond authorized seating capacity.
- No combustible material should be used for decorations in places of public assembly.
- Every building used for public assembly, and every room and section thereof, considered separately, must have at least two means of exit located as remote from each other as practical.
- If a building or room accommodates more than 600 persons, at least three exits must be provided; for more than 1000 persons at least four exits.
- Exits must be maintained free from obstructions, clearly marked, and lighted.
- Exits must provide a clear path of travel to the street or open yard. Exit doors must swing with direction of exit travel.

1952: The spray sprinkler head was developed, which provided more efficient distribution of fine water spray with less water flow.

1968: Congress established the Fire Research and Safety Act.

1990: The Hotel and Motel Fire Safety Act of 1990 was passed by Congress after a series of hotel fires killed more than 400 people in the preceding five years. It required government employees to stay only at hotels and motels with installed hardwired smoke detectors and automatic sprinkler systems in buildings of three or more stories. Retrofitting was on a "voluntary basis" under the threat of losing government business (17,300 hotels and motels appear on the Federal Emergency Management Agency list). The American Hotel & Motel Association claims the industry spent more than $1 billion to comply with the act in order to qualify for the government's lodging business, which tops $1.5 billion annually. However, without debate or public comment, a tiny rider was buried in a massive Defense Department authorization bill that knocked out compliance requirements and eliminated the General Accounting Office's enforcement role.

Contributory Factors

A study of these fires shows that the following mistakes have frequently set the stage for destruction:

- Dangerous accumulations of flammable trash and materials
- Absence of portable fire hoses with adequate water and sprinkler systems
- Delay and confusion in calling the fire department

Few construction fires result from natural causes such as lightning. A study of more than 25,000 fires reported over a ten-year period reveals the causes shown in Table 12-1.

According to Factory Mutual, electricity is the most common and costly source of fire or explosion on its insured properties. Next to faulty wiring, motors are the most frequent source of electrical ignition. Other causes of workplace fires include hot surfaces, overheated materials, cigarette smoking, sparks, friction, and lightning. It should be noted that arson is fast becoming a leading cause of construction and demolition fires.

Hazard Identification and Prevention Matrix

	Eliminate the Hazard		Guard the Hazard		Provide a Safety Factor		Provide Redundancy		Provide Reliability
	Hazard	Safety	Hazard	Safety	Hazard	Safety	Hazard	Safety	
Natural	Friction/ sparks			Flammable trash containment					1*
Structural/ Mechanical	Hot surfaces			Fire hydrants		Sprinkler system		Safe exit	2*
Electrical	Sparks/ wiring			Overloaded circuit breaker		Ground-fault system			3*
Chemical	Flamma-bility, sponta-neous combustion			Airtight/ controlled atmosphere container		Non-flammable insulation			4*
Radiant Energy	Overheat			Temperature control. Automatic warning or shutoff					
Biological									
Automated Systems									

*1,2,3, and 4 combined create a reliability of 99%.

Table 12–1

Cause	Percentage*	Notes
Electrical	23%	The leading cause of industrial fires. Most start in wiring and motors.
Smoking	18%	A potential cause of fire almost everywhere.
Friction	10%	Hot bearings, misaligned or broken machine parts, choking or jamming of material, poor adjustment of power drives on conveyers.
Overheated materials	8%	Abnormal process temperatures, especially those involving heated flammable liquids and materials in dryers.
Hot Surfaces	7%	Heat from boilers, furnaces, hot ducts and flues, electric lamps, irons, and hot-process-metal igniting flammable liquids and ordinary combustibles.
Burner flames	7%	Improper use of portable torches, boilers, dryers, ovens, furnaces, portable heating units, and gas or oil-burner flames.
		Note: A majority of fires in which cutting and welding is a factor have been caused by sparks. These globules of molten metal have scattered horizontally as far as 35 feet, setting fire to all kinds of combustible material. They also fall through cracks, pipe holes, or other small openings in floors and partitions, starting fires that reach serious proportions before being noticed.
		Heat from metal being welded or cut has caused fires when the hot pieces were permitted to rest or fall upon combustible materials. Fires and explosions have also been caused where this heat was transmitted, as in the case of a container, through the metal to a flammable atmosphere or to combustibles within the container.
Combustion sparks	5%	Sparks and embers released from incinerators, foundry cupolas, furnaces, fireboxes, various processing equipment, and industrial trucks.
Spontaneous ignition	4%	In oil waste and rubbish; deposits in dryers, ducts and flues; materials susceptible to heating; and industrial wastes.
Cutting and welding	4%	Sparks, arcs, and hot metal from cutting and welding operations.
Exposure, Incendiarism, Mechanical sparks, Molten substances, Chemical action,		(Each of these is <4% for a total of <24%)
Static sparks, Lightning, Miscellaneous		

Applicable Standards and Literature

Industrial workers in furnace operations, welding and cutting, and other hot environments face a wide range of fire hazards. OSHA's personal protective equipment rule, 29 CFR 1910.132(d)(1) General Requirements (revised April 6, 1994) states that "the employer shall assess the workplace to determine if hazards are present, or are likely to be present, which necessitate the use of personal protective equipment (PPE)." In the hot environment, the assessment should consider hazards presented by sparks, hot flying particles, and spray or splatter from molten metal.

OSHA

OSHA 1910.156 Subpart L Fire Protection, "Fire brigades and portable fire suppression equipment," September 1996

OSHA 1910.164 Fire Detection Systems, September 1996

OSHA 1910.252, **Subpart Q** General Requirements, Welding, Cutting and Brazing (a), September 1996

OSHA 1926.24, **Subpart C** Fire Protection and Prevention, September 1996

OSHA 1926, **Subpart F** Fire Protection and Prevention, September 1996

OSHA 1926.352, **Subpart J** Fire Prevention, Welding and Cutting, September 1996

OSHA 1960.2 Definitions, **Subpart A** General, September 1996

MSHA

Code of Federal Regulations 30 CFR, Chapter 1, Parts 18, 26, 31, 32, 35 & 36 have substantial information on testing and design of equipment and materials used in mining to reduce fire hazards.

Part 56, Subchapter N Metal and Nonmetal Mine Safety and Health; Subpart C - Fire Prevention and Control 56.4000–56.4604

Part 57, Subpart C Safety and Health Standards—Underground Metal and Nonmetal Mines 57,4000–57.4761

ANSI

T1.307 (Rev.1/1/90) Fire Resistance Criteria-Part I: Ignitability Requirements for Equipment Assemblies, and Fire Spread Requirements for Interconnection Wire and Cable

U.S. Army Corps of Engineers

EM 385-1-1 (3 November, 2003) Section 9 Fire Prevention and Protection

National Fire Prevention Association (NFPA)

NFPA-502—Fire Protection for Limited Access to Highways, Tunnels, Bridges, Elevated Roadways, and Air Tight Structures

NFPA BR 12 SPANI (Rev. 1/1/95) 10 Tips for Fire Safety

NFPA 51B Cutting and Welding Processes, 1989 Edition, published by the National Fire protection Association (NFPA)

NFPA 241 Safeguarding Construction, Alteration, and Demolition Operations

NFPA 1500HB (Rev. 1/1/92) Fire Department Occupational Safety and Health Program Handbook

NFPA 101-81 (Rev. 1/1/81) Code for Safety to Life From Fire in Structures

Note: NFPA publishes some 15 volumes of National Fire Codes.

American Society of Testing Materials (ASTM)

ASTM E 1740 (Rev. 11/1/95) Determining the Heat Release Rate and other Fire-Test-Response Characteristics of Wallcovering Composites using a Cone Calorimeter

ASTM D 5664 (Rev. 3/15/95) Evaluating the Effects of Fire-Retardant Treatments and Elevated Temperatures on Strength Properties of Fire-Retardant Treated Lumber

ASTM STP 1150 (Rev. 1/1/92) Fire Hazard and Fire Risk Assessment

ASTM D 5516 (Rev. 4/10/96) Standard Method for Evaluating the Flexural Properties of Fire-Retardant Treated Softwood Plywood Exposed to Elevated Temperatures

ASTM E 1546 (Rev. 8/15/93) Guide for Development of Fire-Hazard-Assessment Standards

ASTM E 1513 (Rev. 3/15/93) Application of Sprayed Fire-Resistive Materials (SFRMs)

ASTM F 1312 (Rev. 8/31/90) Brick, Insulating, High Temperature, Fire Clay

ASTM STP 826 (Rev. 1/1/83) Fire Resistive Coating

ASTM E 69 (Rev. 9/10/95) Combustible Properties of Treated Wood/Fire Tube Apparatus

ASTM E 176 (Rev. 11/10/95) Terminology of Fire Standards

ASTM E 163 (Rev. 6/1/93) Fire Tests of Window Assemblies

ASTM E 152 (Rev. 6/1/93) Fire Tests of Door Assemblies

ASTM E 119 (Rev. 4/15/95) Fire Test of Building Construction and Materials

ASTM E 108 (Rev. 9/10/95) Test Method for Fire Tests of Roof Coverings

ASTM D 1360 (Rev. 10/1/95) Fire Retardancy of Paints (Cabinet Method)

ASTM O 4.07 (Rev. 12/1/96) Building Seals and Sealants; Fire Standards; Building Constructions

Underwriters Laboratories Inc. (UL) was established in 1893 and has a number of fire references, including a fire resistance index published by its fire prevention department. Its main office is located at

207 East Ohio Street
Chicago, IL 60611

Other References

- John M. Sadler, "How to Prevent Construction Fires," *Occupational Health & Safety*, July 1995.
- *Fire Protection for Facilities Engineering, Design, and Construction.* MIL-HDBK-1008 (Rev. 1/15/94).
- Kenneth A. Kander, "Avoiding Costly Construction Fires," *Constructor*, July 1992.
- Fire Protection: Design Criteria, Options, Selection. Means Fire Protection (Rev. 1/1/91).
- J. Davitt McAteer "Fire Safety," *Miner's Manual* Third edition Chapter 5, Crossroads Press, published by the Occupational Safety & Health Law Center, January, 1985.
- Factory Mutual Engineering Corp., *Handbook of Industrial Loss Prevention*, McGraw-Hill Book Company.
- Fire Protection. Department of Energy (DOE) Order 5480 (Rev. 12/18/80).
- Paul R. Lyons, *Fire in America,* National Fire Protection Association, Boston, MA., 1976.
- "America Burning," The Report of the National Commission on Fire Prevention and Control, U.S. Government Printing Office, Washington, DC, 1973.
- "How Safe is the Channel Tunnel?" *Condé Nast Traveler*, February 1997.
- Robert Laford, "The Missing Cog in Fire Safety," *Occupational Health & Safety*, February 1997.
- *Travel Weekly*, February 17, 1997.
- *Safety and Health*, National Safety Council's (NSC), September 1996.
- *Accident Facts*, National Safety Council, 1995.
- *Industrial Fire World*, December 1991.
- David Von Drehle, *Triangle: The Fire that Changed America*, Grove Press, 2003.
- "Fire Prevention," *Hazard Information Newsletter,* Volume 2, Issue 3, June 1997.

Commentary

The construction manager should review all contracts to ensure an effective safety plan that involves close communication with fire prevention agencies, material manufacturers and handlers, and site supervisors. Architects and designers must gain knowledge of the flammability of construction materials, including interactive properties, and make an informed decision concerning the optimal materials to use for specific situations. Use of some materials, such as various styrofoams, even in the event of containment by the structure, creates a dangerous condition during construction. Design engineers should find a less flammable substitute and more highly controlled sources of ignition during construction.

Recommendations

A fire-safety plan should be included in all construction, alteration, or demolition contracts, and the owner's involvement in fire safety should be established, even though the work may be entirely under the jurisdiction of the contractor. Large-scale construction sites change rapidly as construction progresses. The fire-safety plan should be flexible throughout the different stages of construction. Essential items to be included in the plan (discussed in detail below) are:

- A fire-safety plan developed with the local fire department
- Rapid communication system
- Installation of fire-protection systems as construction progresses
- Preservation of existing sprinkler and fire-prevention systems during demolition
- The organization and training of an on-site fire brigade
- Consideration of special hazards resulting from previous occupancies
- No accumulation of trash or combustible materials should be allowed where they are vulnerable to ignition and could endanger buildings or facilities
- On-site security to keep trespassers out
- Smoking only in designated areas

The plan should include on-site inspections by the fire agency and a drill to determine if public or private fire equipment actually has unobstructed and fast access. There should also be a readily available public fire alarm box near the premises and telephone service to the fire department with the fire department's telephone number and site address posted near each telephone. Access for heavy fire-fighting equipment should be provided to the immediate job site at the start of the project and maintained until completion.

A suitable location at the site should be designated as a command post and provided with plans, emergency information, keys, and communication or other equipment as needed. The person in charge of fire protection should go immediately to the command post whenever a fire occurs.

Accumulations of combustible waste material, dust, and debris should be removed from the structure and from the immediate vicinity at the end of each work shift, or more frequently as necessary for safe operations. Rubbish should not be burned on the premises without first obtaining a permit. Many communities require permits for open fires for trash and warming. These fires can present a serious threat due to windblown embers.

All materials that are susceptible to spontaneous ignition, such as oily rags, should be stored in a proper disposal container. Smoking should not be permitted in and around areas that have combustible material or liquids. These areas should have appropriate "No Smoking" signs posted.

In all buildings more than one story in height, *at least one stairway in usable condition at all times should be provided* and should meet the requirements of 5-2.2 of NFPA 101 Life Safety Code. This stairway should also be properly lighted. At least one approved fire extinguisher should be provided in plain sight on each floor, at each usable stairway, and wherever there is an accumulation of combustible material. Large warehousing facilities should have fire doors with approved closing devices and appropriate hardware.

Equipment to be installed or combustible construction materials should not be stored any closer than 30 feet from the structure under construction or alteration.

Water supply for fire protection, either temporary or permanent, should be an integral part of the safety plan. Ground water mains and hydrants are to be provided, and they should be *installed, completed, and in service prior to construction work.*

As 23% of the industrial fires arise from faulty electrical equipment or wiring, it is recommended that only equipment that meets the Underwriters Laboratories (UL) or the National Fire Protection Association (NFPA) requirements should be used.

A permit system should be used for all hot work operations (welding or cutting with flame or electrodes), or other flame or spark-producing activities on the job site, and should be under the supervision of the fire-prevention manager. A permit should not be issued until

- it has been determined that hot work can be safely conducted at the desired location;
- combustibles have been moved away or covered by an approved fireproof tarpaulin;
- the atmosphere is nonflammable; and
- a fire watch (with a dedicated extinguisher) is posted for the duration of the work and for 30 minutes thereafter (60 minutes for torch-applied roofing operations) to see that sparks or drops of hot metal do not start fires. All cracks and openings in floors should be covered or closed.

Although a cutter or welder has the best opportunity to avoid fire or injury by proper control of the equipment he is using, there are many circumstances where fires, explosions, or severe injuries are inevitable if an oxy-fuel gas torch or an electrode is to be used. Such circumstances can arise if the cutter or welder is not aware of (1) the proximity or flammable nature of nearby combustible solids, liquids, or dusts; (2) the presence of explosive mixtures of flammable gases or vapors and air; or (3) the presence of an oxygen-enriched atmosphere. Precautions taken by a cutter or welder will often be governed by the desire of others for speed or economy in the work, or by the failure of management to emphasize the possible extent or seriousness of a fire in the work area. All three—the cutter or welder, the supervisor, and management—share full responsibility for the safe use of cutting or welding equipment.

Safe cutting and welding requires approved areas and established procedures for approving cutting and welding. This step cannot be implemented without an individual responsible for authorizing cutting and welding operations. This individual should define the fire hazards and be familiar with fire prevention requirements.

Internal combustion engines and associated equipment, such as air compressors, hoists, derricks, pumps, and similar devices, should be located so that the exhaust points away from combustible materials. Where the exhausts are piped outside the structure under construction, alteration, or demolition, a clearance of at least nine inches should be maintained between the piping and combustible material.

Asphalt and tar kettles for roofing operations should be located in a safe place outside the building, or on a noncombustible roof, so as to avoid igniting combustible material below. Used roofing mops and rags should be cleaned or discarded and placed away from the building and combustible materials.

SECTION 2

Confined Spaces

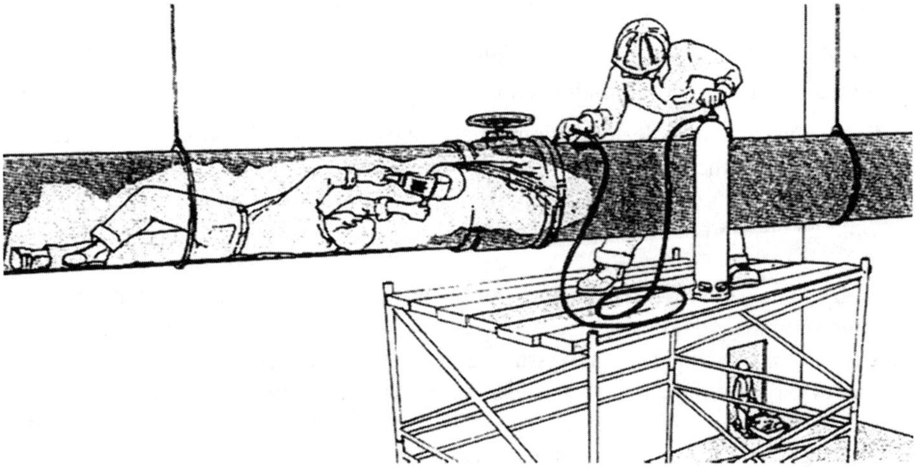

The Hazard Syndrome

The American Society of Safety Engineers' (ASSE) *The Dictionary of Terms used in the Safety Profession*, Third Edition, defines a confined space as "a compartment or area having limited access and (usually) no alternate escape route, having severely limited natural ventilation or an atmosphere containing less than 20 percent oxygen, and having the capability of accumulating a toxic, flammable, or explosive atmosphere, or of being flooded. Examples include tanks, vaults, ducts, conduits, wells, stacks, tunnels, silos, pits or trenches more than 4 feet (1.22 meters) deep, ship holds, pipelines, manholes, and sewers." Confined spaces are found everywhere and require special equipment and other precautions to work in safely. The picture in the previous page shows a typical confined space.

Unless properly planned and used as a jail cell, a confined space is a *dormant* hazard for at least one of three reasons:

- Lack of oxygen
- Dangerous gases
- Entrapment

Entry into a confined space without the necessary safeguards is an *armed* hazard. It is *active* when the occupant is unable to exit and asphyxiates.

Scope

Confined spaces include structures designed to store products; to enclose materials, liquids or gases; and to transport products. These structures are very common on construction sites, and they are usually very dangerous to enter for inspection, maintenance, repair, or cleanup because many of them have some type of mechanical entrapment or environmental hazard. Confined spaces containing hazardous fumes are especially dangerous and should be entered with extreme precaution. Holding tanks for mining or industrial processing materials are likely to contain caustic materials such as cyanide or other toxic or corrosive substances.

Peril from the Failure Mode

Confined spaces present the *natural* and *chemical* hazards of possibly toxic gases that can lead to asphyxiation. Conflicting statistical data on injury and death due to confined spaces abounds. According to a NIOSH study, 60% of confined-space fatalities occur when the initial victim enters the confined space and is then followed by untrained and unprotected standby personnel or would-be rescuers. It is common for two or three rescuers to become victims, too. *Mechanical* factors of faulty or inadequate accessways compound this hazard and are a significant contributor to injuries and death from confined spaces.

The statistics shown on the next page were compiled by the National Traumatic Occupational Fatalities' (NTOF) surveillance system maintained by NIOSH's Division of Safety Research. Based on a sampling of a few key hospitals nationwide that covers only a small percentage of national unintended deaths, statistics show that during a ten-year period, 670 people fell victim to confined spaces in 585 separate incidents, which indicates that approximately twelve percent of the incidents involved multiple victims. The following chart depicts confined space deaths by industry, according to the article "Worker Deaths in Confined Spaces," *Professional Safety*, November 1996.

When those 670 victims are added to the 606 victims of trench cave-ins and 207 who were engulfed in loose materials, a total of 1483 victims emerges. That's an average of 140 people killed each year in some sort of a confined space.

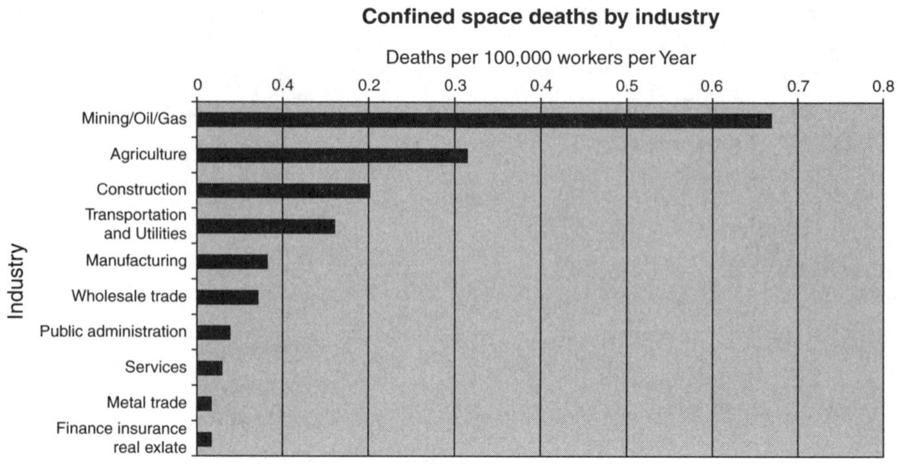

A twenty-year study conducted by the University of Massachusetts at Lowell on the health, attitudes, and culture of construction workers found that those who work in confined spaces suffer higher-than-average occurrences of pulmonary, cardiovascular, and muscular disorders, as well as hearing loss. Their life spans are ten years shorter than the national average.

Brief Summary of the Occurrence

Scott Dominguez was ordered to enter a 25,000-gallon storage tank to remove sludge from the bottom and clean it out. The reaction of water from a fire hose with the sludge released poisonous hydrogen cyanide gas. Prior to this incident, OSHA had provided guidance on how the tank could be safely cleaned. Scott Dominguez was issued no protective clothing or safety gear of any kind. When he collapsed due to toxic fumes, an entry hole had to be burned into the side of the tank with an acetylene cutting torch. However, due to the lack of a confined space permit, the contents of the tank were unknown, and it took 45 minutes to an hour to cut the tank open. Exposed to poison for so long, Scott Dominguez remains to this day severely and permanently brain damaged.

Date of Occurrence

August 27, 1996

Source of Information

U.S. District Court, Sixth District, State of Idaho, County of Caribou, # CV 98-02936. Prior to this case, the EPS had successfully conducted an investigation and gathered data from many sources. Documentation is listed in the book *The Cyanide Canary,* listed in the "Literature" section.

History of the Hazard

The dangers associated with working in confined spaces were first recorded when Emperor Trojan reportedly had criminals clean sewers in ancient Rome. This form of punishment often had fatal results, which the emperor considered to be appropriate treatment of criminals, as it was a way to condemn them to death.

1556: It was recognized that stagnant air in mines was the cause of many breathing difficulties and that there were too many fires and explosions in mines from methane gas.

1925: Alice Hamilton observed that decomposing organic matter in vats, tanks, and manholes emitted hydrogen sulfide, and that this caused death by asphyxiation.

Contributory Factors

Confined-space deaths occur in all kinds of ways. This hazard is often overlooked because those involved don't recognize it as a hazard until someone is trapped. For instance, on July 10, 1998 in Manhattan, New York City, a laborer was inspecting a 6-foot-deep by 6-foot-long by 3-foot-wide pit to see if gas and electric lines were present as marked on a utility map. Unaware that someone was below, a backhoe operator began to fill the pit with construction debris and concrete slabs. The victim was buried for eight hours before he was noticed to be missing. Most often confined spaces contain:

Oxygen deficiency—the most prevalent factor. It is often caused by rotting materials, corrosion, and so forth that deplete or displace the oxygen.

Carbon monoxide—a frequent offender.

Carbon dioxide—the next most frequent killer. It is 1.5% denser than air and sinks to a lower level in basements, pits, and shafts, displacing oxygen. Improper ventilation systems are often responsible for the "dead air" composed of carbon monoxide in such spaces.

Other gases found in confined spaces include:

Nitrous oxide—a product of diesel engine combustion

Hydrogen sulfide—a naturally occurring gas

Sulfur dioxide—a naturally occurring gas

Methane—a naturally occurring gas

Natural gas—a naturally occurring gas that includes methane, propane, butane, and other trace gases

Nitrogen—used as a fire suppressant

Storage bins that contain grain, mineral ores, aggregates, fertilizers, or other materials that are loaded into the top and processed by extracting them from the bottom using continuous conveyors, screw augers, or gravity are very hazardous. Materials often adhere on top and form an arched cavity as they are pulled out from underneath. Workers can be entrapped when they step onto this crust. This situation is an ongoing invitation to enter and die, as workers can be either engulfed or sucked under the material and quickly suffocated.

A fire in Haysville, Kansas, in which workers were entombed in a tunnel beneath grain silos, was caused by exploding dust that ruptured the silos and caused mountains of grain to flow like water and flood the tunnel, making rescue very difficult.

Open wells, holes, accessible mine shafts, frac tanks used in oil- or gas-well-drilling operations, farm silos or manure storage pits, and other entrapments sometimes do not have safe access or egress from which to escape dangerous fumes or gases. Some confined spaces may have good ventilation but are so cramped that avoidance of uninsulated electrical conductors or moving parts is next to impossible.

Supervisors must always remember that a narrow aisle in a storeroom in which unstable boxes, sacks, or pallets are stored well above someone's head is really a confined space, as it is almost identical to an unshored trench that could be affected by vibration or surface instability.

Hazard Identification and Prevention Matrix

	Eliminate the Hazard		Guard the Hazard		Provide a Safety Factor		Provide Redundancy		Provide Reliability
	Hazard	Safety	Hazard	Safety	Hazard	Safety	Hazard	Safety	
Natural	Lack of oxygen								
Structural/ Mechanical	Limited access			Ventilation system/ safe access		Safety harness			100%
Electrical	Source of ignition	Insulation		Dust detectors		Commu-nication system		Safe zone or room used in mines to wait until help comes	
Chemical	Explosive or toxic gases/ unknown substance	Material flush before entrance		Chemical respirator					
Radiant Energy									
Biological									
Automated Systems									

Applicable Standards and Literature

OSHA

29 CFR 1910.5(c)(1) states that when permit-required confined space hazards are identified through compliance with paragraph (c)(1) of this section and they are also covered in **58 CFR 4469**, the telecommunications standard, not the permit-space standard, applies.

29 CFR 1910.119 Process Safety Management

29 CFR 1910.120 Hazardous Waste and Emergency Responses

29 CFR 1910.132 Personal Protective Equipment

29 CFR 1910.134 Respiratory Protection

29 CFR 1910.146 Confined Space Entry Standard, Section (c)(1) requires all employers, including telecommunications companies, to determine if their workplaces contain any confined spaces that meet the definition of a "permit-required confined space." If a telecommunications employer has permit-required spaces that pose hazards that are not addressed by 1910.268(o), the employer should protect any employees assigned to enter those spaces through compliance with the pertinent provisions of 58 CFR 4469, Footnote 4.

29 CFR 1910.146 Appendix C, Examples of Permit-required Confined Space Programs

29 CFR 1910.146 Appendix D, Confined Space Pre-Entry Check List

29 CFR 1910.147 Lock Out/Tag Out

29 CFR 1910.252(b)(4)(iv) requires that an attendant be stationed outside a confined space, such as a manhole or other small opening through which a welder must enter, "to observe" the welder at all times. This requirement is based upon a similar requirement in ANSI Z49.1-1967. However, Appendix C of 1910.146 lists "tapping or rapping codes on tank walls" as an example of an adequate means of communication and observation between an attendant and a welder working in a permit-required confined space.

29 CFR 1910.268(o) Underground lines—The provisions of this paragraph apply to the guarding of manholes and street opening and to the ventilation and testing for gas in manholes and unvented vaults where telecommunications field work is performed on underground lines.

29 CFR 1910.272 Grain Handling Facilities

29 CFR 1910.1000 Air Contaminants

29 CFR 1910.1200 Hazard Communication

29 CFR 1926.21 Safety Training and Education

29 CFR 1928 General Duty Clause

OSHA Fact Sheet 93-28 highlights the Grain-Handling Final Standard.

MSHA

CFR Title 30:

S 48.5 Training of new miners; minimum courses of instruction; hours of instruction

S 48.6 Training of newly employed experienced miners; minimum courses of instruction

S 48.6 Training of miners assigned to a task in which they have had no previous experience; minimum courses of instruction

S 48.8 Annual refresher training of miners; minimum courses of instruction; hours of instruction

S 50.20-3 Criteria—Differences between medical treatment and first aid

56.4400–56.4430 Flammable and Combustible Liquids and Gases

56.4600–56.4604 Welding/Cutting/Compressed Gases

56.16012 Storage of Incompatible Substances

57.4400–57.4463 Flammable and Combustible Liquids and Gases

57.4600–57.4660 Welding/Cutting/Compressed Gases

MSHA does not have its own respirator requirements for confined spaces but uses one consistent with the requirements of ANSI Z88.2-1969. OSHA's permit-required confined-spaces standard does not cover the 102,000 coal miners and approximately 97,000 workers in metal and nonmetal mining who often use respirators.

NIOSH

95-109 Bulletin on Safe Grain and Silage Handling

93-116 Bulletin warning farmers of Deadly Risk of Grain Suffocation

87-113 Guide to Safety in Confined Spaces

42 CFR Part 84 Respiratory Protective Devices, *Federal Register*, Part II, published June 8, 1995

NIOSH published *Worker Deaths in Confined Spaces* in January 1994 in an effort to help protect the more than 2 million workers in confined spaces. This document describes the circumstances of more than 100 workers who died in confined spaces.

ANSI

Z49.1-1988 Safety in Welding and Cutting states that the attendant shall "observe the workers inside or be in constant communication with them."

Z88.2-1992 Respiratory Protection

Z117.1-1995 Safety Requirements for Confined Spaces

Environmental Protection Agency (EPA)

In 1988, the EPA implemented underground-storage-tank regulations that require underground-storage-tank-systems (UST) owners to upgrade or remove tanks that do not meet EPA standards. Since implementing this regulation, many occupational injuries and fatalities have occurred as a result of cave-ins, explosions, chemical overexposure, and many other hazards encountered during removal.

Air Force Occupational Safety and Health (AFOSH)

Document 127-22 Walking surfaces, guarding floor and wall openings and holes, fixed industrial stairs, and portable and fixed ladders confined spaces, 9/28/92.

Document 127-25 Confined Spaces

New England Institute Water Pollution Control Commission

This Commission has prepared a booklet and video entitled *Tank Closure without Tears*, which can help workers to recognize, evaluate, and control potential hazards.

Compressed Gas Association (CGA)

The CGA has brochures on transferring liquid carbon dioxide, which are available to assist workers, designers, engineers, safety and training personnel, distributors, restaurant personnel, inspectors, and other users.

American Society of Safety Engineers Journal (*Professional Safety*)

Patricia T. Utley, *System Safety Hazard Analysis: Determining confined space entry requirements*, September 1994.

Jim Roughton, *Confined Space Entry: An overview*, September 1993.

Ernest B. Jorgensen Jr., *Confined Space Entry*, February 1992.

Mahendra Roychowdhury, *Confined Space Entry Program for any Industry*, February 1992.

James C. Wernicke, *Confined Spaces Can Mean Your Death*, October 1990.

Chris W. Campbell, *All about Confined Spaces*, February 1990.

Other References

- Hilldorfer, Joseph & Robert Dugoni, *The Cyanide Canary,* Free Press, (Simon & Schuster) New York, NY, 2004.
- *Guidelines for Work in Inert Confined Spaces in the Petroleum Industry*, 11/1/87.
- National Technical Information Services, *Safety Requirements for Confined Spaces,* 1/1/95.
- "Confined Spaces," *Ohio Monitor,* March 1989.
- Henry Weinstein, "U.S. Bureaucracy Blamed in Confined-Space Deaths" *Arizona Daily Star*, February 1989.
- "Confined Spaces," *Hazard Information Newsletter*, Volume 3, Issue 6, September 1998.

Commentary

Confined spaces should be avoided whenever possible. When a confined space is necessary, a host of precautions can be taken to make them inherently safe. All confined spaces should be identified at the time of design. Design specifications need "special conditions" in the contract specifications on how the confined space will be controlled. Adequate ventilation systems are the easiest way to enhance the safety of a confined space. A most casual review of detailed requirements for respirators reveals that reliance on respirators is probably going to be more costly, labor-intensive, and burdensome than installation of a ventilation system.

In grain and silage handling, the suffocation hazard can be eliminated by never allowing workers to enter a silage or grain-storage structure when it is being loaded or unloaded. Power to all automatic conveyor equipment should be shut off, locked, and tagged to prevent unexpected operation. Locking and tagging of manual equipment also needs to be done. Consider building a permanent ladder and a service platform when possible on the inside of all grain bins and silos on which workers can stand to break up an arched crust or clog and not be sucked under if the grain suddenly starts to cave in.

Recommendations

Construction management should be sure workers have been trained in confined-space entry and rescue procedures, and have a concise plan of action should an emergency occur. Procedures should be established well before any entry and should be very specific. No rescue should be attempted unless personnel are properly trained, authorized, and properly equipped. Listed below are some of the hazard-prevention measures that can be taken.

A system safety review will usually reveal the pitfalls of a confined space. Many design options to correct these can be achieved. For example:

- Equip control valves located in pits with extensions so they can be operated from above the pit instead of requiring someone to descend into it. Piping can also be configured to locate valves above ground.
- Design large boilers/furnaces, hydroelectric turbines and penstocks, storage tanks, and other confined spaces with two entry portals to give better ventilation and safer access.
- Install agitators to prevent arching of materials, or build platforms inside storage silos, bins, and tanks so workers do not have to stand on materials that have crusted over and arched or clogged.
- Provide a built-in exhaust system to evacuate toxic or dangerous air so hazardous accumulations do not occur.

The other dangers posed by confined spaces that cannot be eliminated by design should be covered in a comprehensive written program that includes:

- **A management survey** to identify all confined spaces, since many workers do not recognize unsafe places that are obvious to the trained safety professional.
- **Inspection** to determine whether the entry of personnel into a confined space is absolutely necessary, whether the work could be done from the outside, or whether contaminates could be eliminated by a forced-air ventilation system.
- **Identification of all hazards** associated with the confined space.
- **A job-hazard analysis** for each task to be performed inside the confined space that explains each step to be taken and what pre-entry procedures are needed. A review of reference materials will assist in finding out the hazards that workers are likely to find, their exposure limits, the personal protective equipment they will need, the symptoms of adverse exposure, and what first-aid might be needed.
- **Preparation of confined spaces** that includes lockout/tagout, purging, ventilation, and procurement of special equipment, tools or any other necessary safety devices. Procedures for inspection and maintenance of all piping, tubing, hoses, and fittings

by qualified personnel should be established and performed in accordance with the manufacturer's specifications.

- **Specifying the type of communication to be used** between the worker in the confined space and the standby person.
- **Specifications** for all clothing to be worn and equipment to be used.
- **Issuance of written entry permits** that comply with OSHA's *Confined Spaces Entry Permit* requirements:
  - They must specify the location, type of work, and any specific procedures.
  - They must certify that the space has been evaluated and tested by a qualified person.
  - They must verify that all necessary protective measures have been taken.
- **Appropriate rescue response modes, which include**
  - **Category I**—Rescue Available (RA). A general rule of thumb is that Category I rescues are non-life-threatening entrapments and those in which the entrant can be expected to self-rescue (exiting the space with no assistance) under normal circumstances. Category I rescues involve permit-required confined spaces that
    - do not require entrants to wear fresh-air breathing equipment,
    - do not expose the entrant to any obvious immediate hazard, and
    - do not warrant rescue personnel standing by during the entry.
  - **Category II**—Rescue Standby (RS). Atmospheric and engulfment hazards require Rescue Standby. This category covers permit-required confined spaces that warrant standby rescue personnel at the entry site and are
    - spaces in which entrants are required to use fresh-air breathing equipment,
    - spaces in which obvious extreme hazards exist, or
    - spaces in which an entrant would be expected to have difficulty with self- rescue.

When work is planned within a confined space, the entry supervisor and those who are to carry out the work should be briefed at a safety meeting so they will have an understanding as to what is to be done, the hazards that may arise, and the necessary protective measures to be taken. Reviewing written standard operating procedures and using the entry permit as a preparation checklist are recommended. Even a skilled, seasoned employee needs a reminder and review. Anyone authorized to enter a confined space should have an up-to-date medical assessment and proper vaccinations such as tetanus.

Confined spaces should be adequately ventilated to maintain a safe working environment. Ventilation systems should be designed to exhaust from the lowest level and allow make-up air to enter at a higher point. The following will ensure efficient ventilation:

- Develop and implement a procedure to monitor the atmosphere.
- Use elephant-trunk type ducts.
- Keep the exhaust vent of a gasoline-powered blower at least six feet away and downwind from the entry into a confined space and the fresh-air intake duct.
- Keep the length of air-ducts and 90 degree bends to a minimum, because they severely reduce a blower's capacity. Two 90 degree bends can cut the blower's efficiency almost in half.

Atmospheric testing and monitoring should include:

- A detection instrument with alarm points set at OSHA limits. For example, 19.5% minimum and 23.5% maximum for oxygen and 10% below the explosive level for any flammable vapors that may be anticipated, thus ensuring that the explosive limits are never reached.
- A measuring instrument with an alarm that can't be turned off until reset.
- A test for oxygen concentration before testing for flammable vapors.
- Taking at least three sets of readings—one before you ventilate, one after you ventilate, and one after entry into the confined space.

Training of personnel should include:

- Use of entry permits.
- Selection of safe entry procedures.
- Selection of proper respiratory protection.
- Selection of lifelines and retrieval systems.
- Selection of protective clothing.
- The admonition to never enter a confined space without being authorized and suitably trained and equipped.
- A review of confined-space rescue plans.
- Lockout and tagging procedures for electrical sources to be followed before entering a confined space.
- Proper cleaning, disinfection, storage, and inspection instructions.
- Surveillance of work-area conditions.
- Information on the nature of the toxic fumes that may be encountered in the confined space being entered.
- A medical review that includes where the first-aid kit is located on the job site and how to give CPR. Stress the need to drink lots of liquids in hot weather; heat stroke can be a killer. Explain the difference between heat stroke and exhaustion. If a worker suffers from heat stroke, he or she should be removed from the heat and, if in high humidity, should be placed in cool water or, if in low humidity, sprayed with cool water from a hose. Training should stress that immediate medical attention is the best choice.

Requirements for a standby person include:

- Completing confined-space training.
- Being trained in rescue procedures, including the proper use of rescue equipment such as safety lines and harnesses, which includes knowing how to sound the alarm to summon rescue personnel if someone is in need of help. Stress the importance of keeping the entrance clear so rescue crews have unobstructed access.
- Knowing what equipment is required for the job of standby person and having it available.
- If any auxiliary ventilation is used, being sure the source of air doesn't draw carbon monoxide or other contaminants into the confined space and that the exhaust is dispersed downwind away from the intake duct.

- Coordinating entry and rescue procedures with contractors on the job site.
- Being able to recognize the hazards to be encountered in the confined space.
- Knowing how to test to determine if conditions are safe for entry and how to set up rescue procedures. OSHA specifies that the entry supervisor should know first-aid and how to perform CPR.
- Ensuring that those entering a confined space always wear a harness and lifeline.
- Seeing that only qualified people are allowed into a confined space and keeping unauthorized or unprotected people out.
- Knowing who and how many qualified workers are in the confined space.
- Having telephone, radio, or other appropriate equipment available to set up in order to communicate with workers inside or to summon rescue personnel.
- Being alert for any dangerous conditions that may be developing inside the space, which includes being aware of any changes in a worker's behavior that may indicate an exposure to a harmful atmosphere. The standby person should know the symptoms of someone in distress before sounding the alarm for help. Blue fingernails and mild feelings of dizziness and confusion are the early stages of a lack of oxygen. An oxygen-deficient atmosphere can cause unconsciousness almost immediately.
- Knowing when not to enter the confined space. Generally it is not prudent to enter a confined space until additional help arrives, but this isn't always possible. Don't try it alone if you can help it.
- Notifying those inside the space when you can no longer perform your duties so they can evacuate until an alternate standby person arrives.
- Returning the entry permit to management after the job is completed. OSHA requires that companies retain permits for a year and review them annually to improve entry procedures.

Safety equipment should include but not be limited to:

- Respiratory protection. Respirators, air-purifying filters, and self-contained breathing apparatus (SCBA) are all subject to new OSHA regulations in CFR 1910.134, 1915.1052 and 1926.1152. These standards have detailed requirements for program management, evaluating exposure, medical examinations, training, and fit testing of the respirator to an individual's face. A casual review of these detailed requirements for respirators reveals that reliance upon respirators is probably going to be more costly, labor-intensive, and burdensome than the installation of an efficient ventilating system.
- Safety belts, lifelines, harnesses, static-line restraint devices in conjunction with a lifeline and harness, mechanical lift devices, and head, hearing, hand, foot, and body protection.
- A harness with a lifeline when the hazard of engulfment is present. The lifeline should be of such length that it will not allow the employee to sink any further than waist deep.
- Proper protective clothing (hard hats, steel-lined boots, protective eyewear, earmuffs, gloves).
- Proper support equipment that includes gas detectors, communications equipment, and winches for lowering and raising people in and out of confined spaces.

Required signs warn people not to enter and inform people about the toxic fumes that may be present, as illustrated below.

DANGER PERMIT REQUIRED CONFINED SPACE DO NOT ENTER

- At least three people should be present when a confined space is entered.
- Persons entering the area should wear a safety harness attached to a lifeline.
- The standby person should remain at the entrance to watch the person inside the confined space, keep tension on the lifeline at all times to prevent the worker from slipping or falling, and maintain constant communication with the person inside.
- The third person should be trained in rescue procedures and remain close by to go for help or assist in rescuing the person in the confined space if necessary.

SECTION 3

Lighting

The Hazard Syndrome

Insufficient light is an *armed* hazard, as it inhibits vision. Too much light is also an armed hazard, as it can cause glare and lead to momentary blindness, also inhibiting vision. Lighting hazards become *active* when a site mishap is the direct result of visual impairment caused by adverse lighting conditions, or when glare weakens the eye and vision is permanently lost or decreased.

Scope

Construction work is best performed in full daylight. Work at night, inside buildings, and underground all require continuous artificial lighting.

Daylight is not always an option in many construction situations. Most work on city roads is performed at night to decrease traffic hazards. In addition, work in confined spaces or inside buildings under construction by their nature deny natural light. Work in such areas must be provided with an adequate alternative lighting source.

Peril from the Failure Mode

Without light, it is difficult to see without night-vision devices, which are a costly substitute for lighting. Being unable to see is disabling and dangerous, both, for lack of vision, which inhibits vision of many surrounding areas and the task at hand, and for the potential damage caused to a worker's eyes. Darkness is a *natural* hazard. However, too much natural light can also create a hazard. Glare from the sun can inhibit vision, and exposure to harmful rays can damage the eyes if not properly protected.

Brief Summary of two Occurrences

(A) A journeyman electrician arrived at a construction site one winter morning before dawn. Due to the lack of light, he could not see adequately in front of him and fell into an unbarricaded pit.

(B) Chapter 8, Section 5 featured the example of an operator of an aerial lift who rode it into live powerlines and became disabled and disfigured for life. Revisitation of the scenario reveals that the major contributory factor was the glare from looking into the sun, which blinded him. As a result, he was unable to see the powerlines.

Date of the Occurrence

(A) February 27, 1995

(B) February 1, 1993

Source of Information

(A) Case files, Charleston, NC

(B) Case files, New Hanover, SC

History of the Hazard

Since the beginning of time, darkness and glare have imposed limitations upon everyone. Before the advent of child labor laws, children who made lace and wove rugs commonly went blind. With the advent of electric welding, many workers have sustained permanent eye damage from blinding flashes in unshielded welding operations.

Contributory Factors

Lack of light is one of the most frustrating reasons for discontinuing a task. Many people are pressed to work in conditions that provide minimal light or exposure to glare in order to increase productivity.

Hazard Identification and Prevention Matrix

	Eliminate the Hazard		Guard the Hazard		Provide a Safety Factor		Provide Redundancy		Provide Reliability
	Hazard	Safety	Hazard	Safety	Hazard	Safety	Hazard	Safety	
Natural	Night-time darkness or glare from the sun								
Structural/ Mechanical	Dark-ness due to confined/ indoor spaces	Monitor light source/ provide proper artificial light		Safe access/ remove mechan-ical hazards or provide barriers					95%
Electrical		Remove electrical hazards							100%
Chemical									
Radiant Energy		Provide light source							100%
Biological									
Automated Systems									

Applicable Standards

U.S. Army Corps of Engineers

EM 385-1-1 3 November, 2003

07.A General

07.A.01—While work is in progress, offices, facilities, accessways, working areas, construction roads, etc. shall be lighted by at least the minimum light intensities specified in table 7.1 (below).

07.A.02—Office lighting shall be in accordance with ANSI/Illuminating Engineering Society of North America (IESNA) RP-1.

07.A.03—Roadway lighting shall be in accordance with ANSI/IESNA

07.A.04—Marine lighting shall be in accordance with ANSI/IESNA 07.A.05 Means of Egress

(a) Means of egress shall be illuminated, with emergency and non-emergency lighting, to provide a minimum of 1 footcandle (fc) (lumens per square foot (lm/ft^2) [11 lux (lx), measured at the floor]. Reference NFPA 101.

Table 12–2
U.S. Army Corps of Engineers

Facility or function	Illuminance-lx (lm/ft²)
Accessways	
• General indoor	55 (5)
• General outdoor	33 (3)
• Exitways, walkways, ladders, stairs	110 (10)
Administrative areas (offices, drafting and meeting rooms, etc.)	540 (50)
Chemical laboratories	540 (50)
Construction areas	
• General indoor	55 (5)
• General outdoor	33 (3)
• Tunnels and general underground work areas (minimum 110 lx required at tunnel and shaft heading during drilling, mucking, and scaling)	55 (5)
Conveyor routes	110 (10)
Docks and loading platforms	33 (3)
Elevators, freight, and passenger	215 (20)
First-aid stations and infirmaries	325 (30)
Maintenance/operating areas/shops	
• Vehicle maintenance shop	325 (30)
• Carpentry shop	110 (10)
• Outdoors field maintenance area	55 (5)
• Refueling area, outdoors	55 (5)
• Shops, fine detail work	540 (50)
• Shops, medium detail work	325 (30)
• Welding shop	325 (30)
Mechanical/electrical equipment rooms	110 (10)
Parking areas	33 (3)
Lavatories, wash, and dressing rooms	110 (10)
Visitor areas	215 (20)
Warehouses and storage rooms/areas	
• Indoor stockroom, active bulk storage	110 (10)
• Indoor stockroom, inactive	55 (5)
• Indoor rack storage	270 (25)
• Outdoor storage	33 (3)
Work areas-general (not listed above)	325 (30)

(b) The illumination shall be arranged so that the failure of any single lighting unit, including the burning out of an electric bulb, will not leave any area in total darkness.

07.A.06 Lamps and fixtures will be guarded and secured to prevent injury to personnel. Open fluorescent fixtures will be covered with wire guards, lenses, tube guards, and locks, or safety glass that require force in the horizontal axis to remove the lamp.

07.A.07—Lamps for general illumination shall be protected from accidental contact or breakage. Protection shall be provided by elevation of at least 7 ft (2.1 m) from normal working surface or suitable fixture or lamp holder with a guard.

Commentary

The Biblical quotation "let there be light" is an easy way to overcome many hazards. It is natural for children to be afraid of the dark, but it makes no sense that grown men are afraid of the "light" regarding the need for hazard prevention with engineering controls.

Recommendations

All general work-area lights on construction projects should be raised to five candle-power. This action would raise the light level to where reasonable safe access is clearly visible. Workers exposed to glare should report the conditions immediately. Work should be recommenced only after taking precautions to minimize the glare and its effects on work areas, including access points, use of hand tools, and proximity to larger hazards such as powerlines.

Electric arc welding needs to be in a location where curtains or screens are used to confine the high-intensity light so bystanders without appropriate protection are not exposed.

SECTION 4

Toxic Fumes and Gases

The Hazard Syndrome

Leaking toxic fumes and gases from damaged hoses or piping are often a hazard that can arise on construction sites and during maintenance of a completed facility. Carbon monoxide (CO) as an exhaust gas from gasoline engines is the most prevalent toxic gas found on the construction site, but there are many other sources of toxic fumes arising from construction processes. Paint, solvents, and adhesives are a predictable source of toxic fumes during construction or maintenance. Improper ventilation can lead to a high concentration of fumes that would not otherwise be dangerous. A running gasoline engine in an enclosed shop is an *armed* hazard that can easily become *active* when its exhaust causes people to exhibit adverse bodily effects such as dizziness, shortness of breath, or headache from the buildup of toxic fumes.

Scope

The incomplete combustion of fuels from gasoline and diesel engines usually creates a source of carbon monoxide, but can also synthesize gases such as nitrous oxide. Tools or machines powered by gasoline or diesel are an unavoidable hazard on construction sites. Carbon dioxide fire extinguishers contain heavy gas that seeks low spaces, where it accumulates. Welding presents sources of dangerous fumes that can be harmful or fatal due to prolonged exposure in an unventilated space.

In addition to these common construction circumstances, carbon dioxide presents a hazard in many household situations. Its properties as a colorless, odorless gas make it especially dangerous to unsuspecting persons.

Peril from the Failure Mode

Toxic gas can cause serious and permanent lung damage that can lead to disability or death. The central nervous system is frequently affected by toxic gases at very low doses. Though naturally occurring in some situations, gasses such as carbon monoxide can synthesize during industrial processes and create a *chemical* hazard.

Brief Summary of the Occurrence

Two young servicemen were stationed in Sierra Vista, AZ in the winter of 1967. They were also musicians in a local band. They arrived back at their house after playing a gig in December, turned on the heater, and went to sleep. One of the young men was engaged to be married, and his fiancée went to wake him up the next morning. She received no answer when she tapped on the door. Feeling something was out of place, she peered through the window and discovered the bodies of the two young men. They had been asphyxiated by carbon monoxide exhaust from the unvented space heater.

Date of Occurrence

December 10, 1967

Source of Information

This incident prompted testimony from several sources presented to the National Consumer Product Safety Commission, which led an extensive investigation into the hazards associated with unvented gas heaters.

History of the Hazard

The lethal nature of high concentrations of carbon monoxide was recognized by the Greeks and Romans, who sometimes used the gas to execute criminals or to commit suicide. Even so, it was not until Haladane's experiments in 1895 that the relationship between carbon monoxide, hemoglobin, and carboxyhemoglobin was understood.

1930s: During the construction of Hoover Dam and the excavation of its two diversion tunnels, gasoline-powered dump trucks were used for hauling out the muck and debris, and little ventilation was provided. At the time, the rumor was widespread that many workers died of carbon monoxide poisoning, but were diagnosed as succumbing to tuberculosis in order to avoid being counted as workplace fatalities.

1972: In the Sunshine silver-mine fire in Idaho, carbon monoxide inhalation snuffed out the lives of 91 miners and spurred the Department of the Interior's response to the weaknesses inherent in the administrative setup of the safety and health program for miners.

1970s: EPA standards prompted automakers to improve basic engine design. The EPA motor-vehicle program has achieved considerable success in reducing carbon monoxide emissions. By 1975 most new cars were equipped with catalytic converters designed to convert carbon monoxide to carbon dioxide. Catalysts typically reduce carbon monoxide emissions upwards of eighty percent. In the early 1980s automakers introduced converters of a more sophisticated nature, plus on-board computers and oxygen sensors to help optimize the efficiency of the catalytic converter.

1992: Carbon monoxide levels exceeded the Federal air quality standard in twenty U.S. cities, home to more than 14 million people.

Until **1994**, vehicles were tested for carbon monoxide emissions only at 75°F. Recognizing the effect of cold weather, the 1990 Clean Air Act calls for 1994 and later cars and light trucks to meet a carbon monoxide standard at 20°F.

January 2, 2006: Recent mining endeavors at the Sago mine in West Virginia caused the mines to be contaminated with carbon monoxide and resulted in the death of eleven miners.

Contributory Factors

Because this silent killer is odorless, many people are totally unaware that the fumes are poisoning them.

Burning fuel will always emit certain levels of carbon monoxide until a better and cleaner source of combustion or fuel is developed and used exclusively. The risk of CO poisoning increases during winter months when windows and doors are closed and furnaces are in operation. Without proper equipment maintenance, ventilation, and CO-detection devices, closed working and living spaces can become potential death traps. This becomes even more critical at higher elevations, where persons must inhale a larger volume of air to extract sufficient amounts of oxygen than they would at lower elevations. Also, the constant depletion of oxygen by unvented heaters that emit carbon monoxide is more rapid at higher elevations.

Over a half million lift trucks (forklifts) are used in the United States, the majority powered by internal-combustion engines. Both gasoline- and propane-powered engines produce carbon monoxide. Running these vehicles inside buildings and other confined spaces creates the potential for a dangerous buildup of carbon monoxide. Too often, employees will rely on their noses and eyes to alert them to any odors or irritating properties of unburned fuel. This nonmechanical method should not be encouraged.

Some people are just plain lucky! While a family physician was flying his single engine plane, he was overcome by carbon monoxide from a faulty fuel line. The plane flew 250 miles on autopilot until it ran out of gas and glided into a snow-covered field and crashed into a row of trees. The doctor awoke with a terrific headache and a broken wrist.

The cause is still unknown: was it a design defect or poor maintenance that caused engine fumes to be captured inside the plane?

Despite numerous and widespread warnings, far too many people are still dying from carbon monoxide poisoning in trucks and cars. Typically, the hurried driver on a cold winter day starts the machine and lets it "warm up" without opening the garage door. With the engine running, a vehicle can spew forth enough carbon monoxide to kill a person. Older vehicles and even some new ones can develop exhaust-system leaks and spread sufficient carbon monoxide fumes into the driver's compartment to drug or kill you if all windows are closed.

Common sources of carbon monoxide include:

- Clogged chimneys or damper flues
- Restricted furnace intakes
- Unvented space heaters and burners
- Indoor use of charcoal blacksmith forges and barbecues
- Weather-tight and unvented buildings
- Indoor operation of gasoline engines
- Defective heat exchangers
- Heating of trailers, tents, and so on with camp stoves
- Blast furnaces and smelters
- Highway tunnels
- Defective exhaust systems in vehicles
- Compressed-air tanks that are polluted by compressor exhaust and then used by scuba divers and rescue workers
- Overheated compressors, machines, or other equipment, or partially burned lubricating oil
- Fumes from small gasoline engines powering portable electric generators, lawn movers, power saws, and other shop and garden tools
- Mine or tunnel ventilating-system intakes next to the exhaust pipe of a motor powering the fan
- The exhaust on transoms that suck the fumes into the stern of boats
- Fumes from the auxiliary power generator on recreational vehicles and houseboats
- Open windows that allow exhaust fumes to blow in
- Unvented wells, sump holes, or abandoned mines
- Furnace rooms without adequate fresh-air intakes for the combustion chamber

Hazard Identification and Prevention Matrix

	Eliminate the Hazard		Guard the Hazard		Provide a Safety Factor		Provide Redundancy		Provide Reliability
	Hazard	Safety	Hazard	Safety	Hazard	Safety	Hazard	Safety	
Natural	Fire*								
Structural/ Mechanical		Ventilation systems		Welder's air source				Under-ground chambers in coal mines.	
Electrical								CO_2 detector and alarm	
Chemical	Toxic fumes	Nontoxic welding rods							
Radiant Energy									
Biological									
Artificial Intelligence									

*As carbon monoxide is a product of incomplete combustion, a fast-burning fire may produce of carbon monoxide. An occupational reality of firefighters is the constant presence of carbon monoxide; many use a portable air system when entering a burning building.

Applicable Standards and Literature

OSHA

OSHA's permissible exposure limit (PEL) of 50 ppm is based on an eight-hour time weighted average (TWA) exposure.

1910.94(c)(7)(iv)(f)(2) Ventilation, Subpart G

1910.134(d)(2)(ii) Respiratory protection, Subpart I

1910.146 Appendix E Sewer system entry

1910.178(i)(1) Powered industrial trucks, Subpart N

1910.261(m)(2) Pulp, paper and paperboard mills, Subpart R

1910.430(b)(3)(i) Equipment, Subpart T

1910.1000 Table Z-1 Limits for air contaminants

1915.7(c)(4) Competent person, general provisions

1915.136(c) Internal combustion engines other than ships equipment, Subpart H

1917.23 Hazardous atmospheres and substances, Subpart B

1917.25 Carbon Monoxide, Subpart B

1918.93 Ventilation and atmospheric conditions, Subpart I

1918.102(b)(1) Respiratory protection, Subpart J

1926.57(h)(7)(iv)(f)(2) Ventilation, Subpart D

1926.103(f)(2)(ii) Respiratory protection, Subpart E

1926.800(j)(1)(iii)(A) & 1926.800(k)(10)(ii) Underground construction, Subpart S

Carbon Monoxide Poisoning, OSHA Fact Sheet No. 92-11, U.S. Department of Labor Program Highlights, U.S. Government Printing Office, 1992

MSHA

Part 31 Diesel mine locomotives
- **31.4** General requirements
- **31.5** Diesel mine locomotives - Methods of testing
- **31.9** Recommendations on the use of diesel locomotives underground.

Part 32 Mobil diesel-powered equipment for non-coal mines
- **32.4** General requirements
- **32.5** Inspection and tests
- **32.9** Recommendations on the use of mobile diesel-powered equipment in non-coal mines.

Part 36 Mobil diesel-powered transportation equipment for gassy non-coal mines and tunnels
- **36.35** Engine exhaust systems
- **36.26** Composition of exhaust gas
- **36.43** Determination of exhaust-gas composition.

Part 75 Mandatory Safety Standards—Underground coal mines

NIOSH

Recommended Exposure Limit (REL) of 50 ppm. TWA for carbon monoxide poses an immediate danger to life and health (IDLH) if the limit approaches 1,500 ppm.

ANSI

Z37.1 Allowable Concentration of Carbon Monoxide

U.S. Army Corps of Engineers

EM 385-1-1 (September 3, 1996)*Safety and Health Requirements Manual*

Section 9 Fire Prevention and Protection, Subpart 09.D Temporary Heating Devices, 09.D.12

American Conference of Governmental Industrial Hygienists (ACGIH)

The ACGIH recommends an exposure limit of 25 ppm TWA. For industrial lift trucks, they recommend 5000 cubic feet per minute (cfm) of exhaust ventilation per propane-fueled lift truck and 8000 cfm per gasoline-fueled lift truck.

American Society for Testing and Materials (ASTM)

ASTM C 288 Test Method for Disintegration of Refractories in an Atmosphere of Carbon Monoxide

ASTM D 3162 Carbon Monoxide in the Atmosphere (Continuous Measurement by Nondispersive Infrared Spectrometry)

ASTM D 3416 Total Hydrocarbons, Methane, Carbon Monoxide in the Atmosphere (Gas Chromatographic Method)

ASTM D 3669 Analysis for Carbon Monoxide Content of the Atmosphere

Consumer Product Safety Commission (CPSC)

The following CPSC publications deal with carbon monoxide:

The Senseless Killer, #4464

Carbon Monoxide Fact Sheet, #4466

Carbon Monoxide Poisoning From Camping Equipment, #5008

Carbon Monoxide Detectors Can Save Lives, #5008

Deaths From Burning Charcoal in Homes, Vehicles and Tents, #5012

CPSC and NKHA Stress Kerosene Heater Safety, #5052

International Organization for Standardization (ISO)

ISO 8186 Ambient Air—Determination of the Mass Concentration of Carbon Monoxide - Gas Chromatographic Method

ISO 8454 Cigarettes—Determination of Carbon Monoxide in the Vapour Phase of Cigarette Smoke—NDIR Method

ISO 8760 Work-Place Air—Determination of Mass Concentration of Carbon Monoxide—Method Using Detectors Tubes for Short-Term Sampling

Military Specifications (MIL)

MIL-B-23401 Burner, Carbon Monoxide and Hydrogen, Catalytic Type (Mark V)

MIL-D-3945 Detector Kit; Carbon Monoxide, Calorimetric

MIL-I-24772 Indicator, Oxygen, Combustible Gas, Carbon Monoxide and Hydrogen Sulfide, Portable (Metric)

MIL-I-3948 Indicator Tube, Carbon Monoxide

MIL-M-29198 Monitor, Carbon Monoxide, continuous and Alarm (For Breathing Air Systems)

MIL-STD-800 Procedure for Carbon Monoxide Detection

Society of Automotive Engineers (SAE)

SAE AS 412 Carbon Monoxide Detector Instruments

SAE J177 Measurement of Carbon Dioxide, Carbon Monoxide, and Oxides of Nitrogen in Diesel Exhaust

Underwriter's Laboratories (UL)

Underwriters Laboratories has a number of references relating to carbon monoxide detectors (marine use as well as single and multiple station). Refer to UL 1524 and UL 2034 series.

Other References

- David V. MacCollum, *Unvented Gas Heater Hazard Analysis*, report to the National Commission on Product Safety, March 3, 1970.
- Robert J. Crackel, "When you least expect it . . .," *Ohio Monitor*, December 1978.
- "Carbon Monoxide—Lethal But Controllable," *Ohio Monitor*, February 1983.
- Laura Avers, "Cleaning the Air," *Ohio Monitor*, November/December 1993.
- *Accident Facts, National Safety Council*, 1995 Edition.
- Dick Riley, "Creative Controls for Carbon Monoxide Emissions," *Occupational Safety & Health*, July 1995.
- Dr. Thomas Greiner, *Carbon Monoxide Poisoning: Health Effects*, ISU Extension Publication #AEN-166, Iowa State University, July 1996.
- Carla Merolla, "20 Items No Home Should Be Without," *NSC Safety + Health*, January 1997.
- Mick Hans, *Get the Facts on Carbon Monoxide,* NSC *Safety + Health*, March 1997.
- "Carbon Monoxide" *Hazard Information Newsletter*, Volume 2, Issue 10, January 1998.

Commentary

Carbon monoxide and other fumes are exceedingly dangerous. Toxic fumes can cause central-nervous-system damage and lung disease such as emphysema. Signs must be heeded when working with tools that have engines capable of producing carbon monoxide gas. Adequate ventilation must be provided and frequently checked in every maintenance location. Confined spaces are especially vulnerable to the hazards of toxic fumes, and workers should take extra precautions. Adequate breathing equipment should always be worn as a redundant safeguard, even when the possibility of a fume hazard seems relatively small.

Recommendations

Special conditions in the specs should identify all toxic gases that could foreseeably arise, and eliminate the hazards through the following specifications:

- List anticipated toxins
- Identify types of monitors to detect hazardous fumes
- Reference types of ventilation systems to control/eliminate toxic fumes
- Provide an independent air source for the worker
- Provide communication equipment to alert the worker or the safety personnel to a change in conditions

SECTION 5

Wind and Flooding

The Hazard Syndrome

In addition to the danger presented naturally by windstorms and high gusts, tall buildings have the capability to funnel ground winds into speeds that can endanger the public, creating a special danger to pedestrians. This is a design consideration. The potential effect of wind on unsecured materials during construction is an *armed* hazard. Wind hazards become *active* when previously secured materials become loose or are lost. The materials may be swept away, and a worker or a passerby can be struck, causing damage or injury. Alternatively, the compromised binding of materials that result from the force of the wind can present a danger to a worker who walks on top of or alongside the unsound materials.

Locations next to rivers or below the level of adjacent rivers or oceans also present *armed* hazards that need to be addressed at the time of design. The 2005 structural damage to New Orleans when the levees broke during hurricane Katrina illustrates the damage the *active* hazard of flooding can cause. The natural hazards of both wind and flooding need to be considered during the initial site investigation.

Scope

Natural environments with high winds or flooding are found all over the world. According to the Hydrologic Information Center, U.S., at least 9,564 people have been killed due to flooding in the years between 1903 and 2002 in the United States alone. While some architects and design engineers consider factors of the environment, some do not think to make design and methodology specifications to ensure optimum protection for the construction and maintainability of the building.

Peril from the Failure Mode

The perils of wind and flooding are commonly noted during hurricane activity, when buildings are blown over or inundated with water. These *natural* hazards can cause severe structural damage: wind can blow roofs off houses, shatter windows, and even collapse wooden structures. In many cases, the structural pieces removed from buildings by high winds continue to be blown, causing damage to other structures and colliding with any object in its path. Flooding can compromise structural integrity by waterlogging, thus damaging building materials and providing a breeding ground for molds and other spores and bacteria. In severe flooding, the pressure and force of incoming water is enough to wash some structures completely away.

Brief Summary of the Occurrence

The category-3-force (111–131 mph) winds and water forces of hurricane Rita slammed into the coastal region of Cameron Parish and demolished all but one of the area's libraries.

The main library in Cameron Parish was totally annihilated; only the bookmobile and previously evacuated electronic equipment remained. Library branches in three other towns were also destroyed.

Date of Occurrence

September 24, 2005

Source of Information

http://www.ala.org/ala/alonline/hurricanekatrinanews/katrinanews.htm

History of the Hazard

Natural disasters have always precipitated dramatic changes in structural components and lifestyle. The history of windstorms and flooding is well documented by the National Weather Service.

1803: The Congressional Act of 1803 provided assistance to a New Hampshire town following extensive fire damage. It is considered the first piece of U.S. disaster legislation, as well as the beginnings of FEMA. This act has been modified more than 100 times with *ad hoc* legislation in order to respond to other natural disasters, including hurricanes and floods.

Early 1900s: In Sacramento, CA streets were raised up in response to a series of floods between 1850 and 1910 in order to stop the repeated damage caused by water.

1917: A comprehensive flood-control project was passed in Sacramento when Congress authorized the Sacramento Flood Control System after a series of violent floods between 1902 and 1909.

1934: The Bureau of Public Roads was given authority to provide funding for highways and bridges damaged by natural disasters. In the same year, the U.S. Army Corps of Engineers was granted more authority for flood-control projects via The Flood Control Act.

1944: The Folsom Dam project was authorized to be undertaken by the U.S. Army Corps of Engineers. Folsom Dam was originally designed to provide in excess of a 500-year level of flood protection to Sacramento County.

1979: President Carter created FEMA, as it is known today, by merging many of the separate disaster-related responsibilities and agencies into the new Federal Emergency Management Agency.

November 29, 2005: The International Building Code was signed into legislation by Louisiana Governor Kathleen Blanco after hurricanes Katrina and Rita. The code will be applied to all new construction, as well as home or building repairs totaling over 50% of prehurricane valuation. This code was developed by the Virginia-based International Code Council and creates a uniform standard in an area that formerly had regional codes or, in some places, none at all. This code will increase building costs dramatically in some areas of Louisiana, and some experts say it will not be effective against flood damage. However, the code addresses roofs, wind anchors, bracing, siding and glazing, and affords increased wind protection.

Contributory Factors

The absence of specific flood- and wind-resistance building codes in some parts of the country leads to structures that do not adequately stand up to the weather. Some improper or inadequate building materials are used, especially by small or private contracting firms. Materials that are not water resistant and structures that are not designed to resist wind damage deteriorate rapidly under natural conditions.

Hazard Identification and Prevention Matrix

	Eliminate the Hazard		Guard the Hazard		Provide a Safety Factor		Provide Redundancy		Provide Reliability
	Hazard	Safety	Hazard	Safety	Hazard	Safety	Hazard	Safety	
Natural	Wind/ water								
Structural/ Mechanical		Levees/ dredge a deeper river channel		Structurally resistant materials, inherent bracing		Pump systems		Designated flood plain	90%
Electrical									
Chemical									
Radiant Energy									
Biological									
Automated Systems									

Applicable Standards

Due to the varied landscapes that span America, very few uniform structural-building codes are in place. However, the closest uniform building code is the aforementioned International Building Code developed by the Virginia Code Council. For a comprehensive listing and the latest updates on these building codes, please visit *http://www.iccsafe.org/*.

Commentary

Natural disasters, such as wind and flooding, are examples of where the concept of *designing out* a hazard takes on a new dimension. Obviously the pressure exerted by forces of nature can eventually raze any building if it is subjected to enough strain. However, architectural plans that anticipate natural hazards, create structures and facilities that last much longer than those built with plans that do not. In areas known to have heavy winds, shorter buildings with reinforced siding and fewer glass panels will fare better than buildings without these features. In areas prone to flooding, especially flash floods, inherently safe features should include basement pumps, waterproof materials, and sturdy foundations extending at least three feet below ground. These features should be incorporated into a design with minimal spaces that have the potential to trap water and breed mold.

Recommendations

Every building-site assessment should include a review of weather and flood records. Upstream draws and levees should never be considered fail-safe. Building in coastal areas below sea level, areas subject to hurricanes or tornados, along rivers, and on windy plains should require site-specific safeguards. Nature is not forgiving.

SECTION 6

Sanitation

The Hazard Syndrome

Sanitation is often overlooked at the time of design, planning, construction, and maintenance of the completed facility. Unsanitary conditions on construction sites and in completed buildings are not always given the attention they need. Though architects and planners address sanitation in the design plans, sanitation concerns do not often go far beyond a system to remove sewage. Little attention is given to the myriad facets of the creation and maintainance a clean construction site. Indeed, many sites have no further sanitation instructions than the ordering of portable toilets during the project's construction.

In many localities, water-collection ponds provide a breeding place for mosquitoes carrying the West Nile virus or a haven for mice and other rodents. Such pests give rise to the possibility of the hantavirus from deer mice, or the transmission of blood-borne pathogens from mosquito varieties. The avian flu that can be transmitted to humans is a result of contamination and has been the subject of recent concerns. The existence of the flu and other disease or damage-causing microbes is a *dormant* hazard. The discovery of infected pests at a facility is an *armed* hazard. If construction sanitation has been substandard, the presence of such pests may be traced to the time of construction. An *active* hazard arises when a person becomes infected from such a disease.

Scope

Sanitation hazards can take forms other than the improper disposal of waste. Rotting garbage and pools of standing water are breeding and feeding grounds for pests. Other sanitation problems include rusty nails, piles of scrap metal, or discarded materials that clutter and obstruct the job site. Such sharp objects are a source of tetanus infections and blood poisoning.

The broad category of sanitation conditions should be considered at the time of design, with a follow-up during construction planning, and an ongoing consideration during construction and maintenance of the completed facility for its entire life cycle.

Peril from the Failure Mode

Poor sanitation leads to a host of problems. Viruses and bacteria carried by pests that find a potential eating and breeding site among construction debris cause *biological* hazards. Biological hazards also include mold spores of both the toxic and nontoxic variety. The inadequate or improper disposal of toxic dust or other toxic substances creates *chemical* hazards that threaten health. The serious consequences of tripping over debris and having a work space dangerously cramped with unnecessary or scrap material do not compare with the life-threatening illnesses and infections that can arise from poor sanitation; yet falls from tripping over a pile of misplaced debris can result in a lost-time injury or worse.

Brief Summary of the Occurrence

The 1996 epidemic of West Nile virus that infected almost 400 people in southeast Romania was attributed to mosquitoes found living in the houses of the infected. Afflicted people living in apartments were infected by mosquitoes living in pools of water in the building's basements.

Date of Occurrence

1996

Source of Information

University of Chicago *Journal of Infectious Diseases*

History of the Hazard

1300s: The Bubonic Plague that infected the population of Europe under the name of the Black Death was able to spread due to egregious sanitation practices. This plague originated in Asia, traveled to Europe and Africa, and then back to Asia in its chain of devastation.

1700s: Cholera outbreaks were common in U.S. and European cities, mainly due to poor sanitation.

Contributory Factors

Many workers, supervisors, and construction managers see sanitation as a nuisance and fail to recognize the serious risks posed by its absence. There is little direct contact between medical research of this nature and the engineering profession that would result in safer sanitation design criteria. However, some sanitation hazards can be designed out using common sense. Pump and ventilation systems can eliminate standing pools in basements or between floors. Routine maintenance can usually catch new problems that arise when the building has been built.

Hazard Identification and Prevention Matrix

	Eliminate the Hazard		Guard the Hazard		Provide a Safety Factor		Provide Redundancy		Provide Reliability
	Hazard	Safety	Hazard	Safety	Hazard	Safety	Hazard	Safety	
Natural	Standing water								
Structural/ Mechanical	Trash piles			Drain water		Regularly dispose of all garbage in the proper containers			
Electrical									
Chemical	Accumulation of toxic substances			Use of pesticides (sparingly) when eggs, larvae, or spores are noticed		Designate a trained person to remove toxic substances			
Radiant Energy									
Biological	Bacteria, virus					Vaccination/ prescription		Insect sterilization programs	

Applicable Standards and Literature

At the time of design and planning, the U.S. Public Health Service is the best source of current sanitation data. Always consult local building codes for sewage requirements.

Commentary

Unsanitary conditions create a broad range of sources for infection, disease, and toxic reactions. Conventional standards for building codes are generally superficial requirements that address drinking water and toilet facilities. An industrial hygienist should be employed at the time of construction layout, and at periodic intervals throughout the construction process to make an analysis of site conditions. This assessment should be recorded in a log that monitors appropriate controls and changing conditions. Additional controls should be recommended as necessary for any additional sanitary concerns or observations that may arise.

Recommendations

Design engineers and architects should work with the aforementioned industrial hygienist during the design stages of the project. A conference with the construction manager and contracting engineer should be held at the onset of construction. Crew meetings should address specific sanitation problems, such as how to dispose of toxic material, and all disposal bins should be clearly marked.

The industrial hygienist should address specific environmental concerns, such as the potential presence of pools of standing water due to storms or humidity, at the time of project design.

Part III Mechanics of Inherently Safer Design

INTRODUCTION

Part II uses case studies to explore potential ways to prevent a hazard from becoming active, thus creating a safer, more cost-effective project. Part III brings together the actual working concepts presented in this book to "design in" safety for the total life of the facility. A soundly planned construction project evaluates the use of equipment and processes and chooses the safest alternatives, such as the use of safety appliances or alternate design features wherever possible. The construction industry is rapidly adopting new technology and should not be impeded by reliance upon traditional practices that are no longer safe or efficient.

MOVING AHEAD

Twenty years ago, the use of load-moment indicators (LMIs), which applied computer technology to ensure that a crane would lift loads only within its rated capacity at any boom configuration, was a significant step forward for industry safety. Even more dramatic changes are in store, including the use of Global Positioning Systems (GPS) technology for earth works, such as grading roadways and facility excavations. This technology is fast eliminating the traditional method of staking grade levels. More and more, concepts developed by aerospace technology are being applied to construction. Computer software has become the tool with which complex systems become programmable, thus minimizing the probability of human error. However, safety will always require proof that any "black box" technology functions properly with changing circumstances. On the leading edge of this technology are Todd Isaac of Raytheon Missile Systems and his collaborators Robert Konkle, Juan Fernandez, and Steve Dardini, in Tucson, Arizona. Their new system avoids the repetitive testing of the computer-chip circuits known as "firmware," (formerly known as hardware), and provides guidance on analysis based on the firmware's response to incorrect inputs. This new type of analysis provides an opportunity to insulate firmware from external stresses ranging from power failure to atmospheric electrostatic impulses. These advances are leading to a bright future for firmware designed for construction applications. Other future changes in construction may involve the biosciences to help control the waste in tunneling and mining. This new genetic engineering has the potential to develop ground-cover vegetation that will render toxic minerals inert and protect the public from hazardous chemical runoff.

Bringing together engineers from multiple disciplines in order to develop safety systems that integrate scientific knowledge into design concepts is the first step toward inherently safer construction. As these new concepts are developed, the mistakes of the past and present, covered broadly in Part II, become a starting point in developing inherently safer facility design, construction equipment, and industry processes. Examining design that has caused or exacerbated human error in the past encourages engineers to apply creativity and science for the control of the previously listed hazards.

PLAN FOR THE FUTURE

This third part addresses planning how new construction can adopt methods of incorporating this newly available technology. The key personnel in this endeavor are the specialists: licensed engineers who may be designers, managers of construction, or managers of the completed facilities. Engineers familiar with the many aspects of safety engineering are especially important players, as their input will be able to conquer human error with design that does not invite mistakes, tames natural hazards with innovative design, and monitors hazards to ensure control through the application of science and technology. A team composed of these knowledgeable and qualified people will be able to design and build a project free from hazards and harm, and ensure that both quality and efficiency remain throughout the life of the completed facility.

The concepts of safe design for construction embrace four main components:

1. The design of the structure to be built
2. The subsystems to be included in the structure (electrical, mechanical, plumbing, etc.)
3. The equipment involved in building the structure
4. The on-site facilities such as manlifts, elevators, concrete batch plants, and rock crushing/gravel processing or asphalt plants.

The design engineer has two routes in creating a plan that embraces all components of inherent safety. The first is to look for solutions, suggested in the 47 examples covered in Chapters 7 through 12. Each of these real-life examples has addressed options available to the design engineer using the five principles of inherently safer design. The methodology of examining the proposed design by defining a hazard in its dormant, armed, and active modes, matching it to the seven categories of hazards, and placing it onto a matrix of four possible design remedies, is the key to flushing hazards out of design before the work begins. With these five inherently safer design principles the engineer and the system safety specialist have the tools to ensure a successful solution to virtually any hazard that may arise.

The second choice is for the engineer, manager, and other construction specialist to comply with the ANSI/AIHA (American Industrial Hygiene Association) Z10 standard for the "Occupational Health and Safety Management System." This new standard provides management guidelines for assessment of risk, and an outline of a management system to evaluate construction processes. First published July 25, 2005, this important standard evolved over a period of six years, and is considered by the writers as a performance standard. While it is a benchmark for safety management, it includes few details as to how such performance will be achieved. The table of contents for this standard includes a long list of management functions. As it applies to the design function, the Z10 paragraph 5-1-B(a): Implementation and Operation states: "A minimum of 12 hours of training shall be given initially to engineers and safety practitioners in safety through design, to be followed annually with a minimum of six hours of refresher materials." This requirement brings the engineer back to the option of the five principles of inherently safer design. Training is a requirement for the new ANSI Z10 standard, in part because the standard itself does not

- state the nature of a given hazard;
- teach how to identify hazards;
- define how the failure mode occurs; or
- describe the actual peril of a hazard in terms of injury and property damage.

The real impact of this standard comes in the expansion of the engineer's scope of practice by effectively requiring that safety become a permanent requirement of design. An excellent review of the Z10 standard written by Fred A. Manuele, "Safety Management ANSI/AIHA Z10-2005," appeared in the February 2006 issue of ASSE's *Professional Safety.*

The International Organization for Standardization (ISO), the world's largest nongovernmental developer of standards, is also accentuating the issue of design safety. Based in Geneva, Switzerland, ISO has a membership of 48 countries. ANSI is the representative from the United States. This international organization has recently released standards that include:

ISO 12100-1 Safety of Machinery—Basic concepts, general principles for design. Part 1: Basic Terminology, Methodology

ISO 12100-2 Safety of Machinery—Technical Provisions

ISO 12121 Safety of Machinery—Principles of Risk Assessment

In 1997, the European Union started requiring the marking of "CE" on any equipment used commercially or industrially. CE certifies compliance to ISO 12100-1 via documentation by a responsible party. The CE stamp requires not only a level of expertise and knowledge in the field of machine safety, but also designates an authoritative party who will take responsibility for any subsequent safety issues that may arise.

This step forward for safe design was utilized in the United States in the early 1900s, when the licensing of professional engineers was intended to achieve a measure of safe design by requiring such qualified personnel to dictate the terms of a design or specification. Licensing was designed to give the engineer both authority and responsibility to use his or her knowledge. However, as years passed, politics allowed for engineer exemption from the licensing process. Many members of corporate management were excluded from the licensing requirement, and many policymakers assumed that management was competent to select qualified personnel for technical fields. Unfortunately, there have been numerous cases of the appointment of an unqualified designer or a layperson without the proper training for positions that require the technical knowledge an engineer provides.

The authority that licensing provides to an engineer has become more important than ever due to construction equipment that is designed and manufactured in the United States, but requires the CE designation in order to be marketed in Europe. The state licensing of engineers in the early 1900s occurred because engineering services were advertised and practiced by a great number of charlatans with no formal engineering training from accredited colleges. This lack of training led to many dangerous and inept designs, so the public decided upon licensing as a way to remove such people from professional practice. There is increasing international pressure to hold "Safety as a Paramount Design Function," as written in virtually every standard of engineering codes and ethics. In order to maintain our status as an international player, and manufacture products that other countries will buy, we must cultivate both the authority and the responsibility that the licensing of engineers provides.

PRACTICAL APPLICATION

Chapter 13 addresses specific aspects of safety that can be covered at the time of design. The importance of fully trained and knowledgeable engineers is evident in the ways that design alternatives and technological improvements could have been substituted to prevent the hazards in the examples. The examples in this chapter focus on one design flaw that could have been designed out, in order to illustrate specific ways a certain hazard can be controlled or prevented by design. An engineer familiar with the hazard-identification process is the most valuable tool on any construction project.

Chapter 14 addresses the construction-safety planning process by again using one specific design flaw to illustrate the intricacies and variables of the planning process, and give a concrete example of a hazard that may arise. Once a hazard is identified, careful planning is necessary in order to take the most effective steps to control it. Qualified engineers are again integral in the process of matching identified hazards to their appropriate safeguards, and designing each component of the project to work cohesively and safely with the overall project. The ideas in this section bring together the principles asserted in Part I and the examples in Part II to illustrate the multidimensional quality of comprehensive construction safety. A sound safety plan must extend beyond the responsibilities of the design engineer, architect, and construction manager to incorporate the safety precautions taken by equipment manufacturers, rental agencies, utility companies, and public-works offices. Often, smaller firms overlook important steps in this process, with tragic outcomes. Chapter 14 illustrates such mistakes made by the absence of planning within small organizations.

Chapter 15 provides an overview of the legal and economic conflicts that impede safer construction processes. The conclusion of this book provides an overview of how the legal and economic conflicts can be overcome with economic incentives to ensure safer construction.

13 Including Safety into Design Planning

INTRODUCTION

The first step in creating a safe facility is to design out hazards in the architectural phase. Far beyond designing the structure itself, architectural planning must bring together designers of the components within the structure, whether they are electrical, mechanical, fire protection, or special support features, to integrate these systems into a safe and cohesive project. Construction managers, operating and maintenance managers of completed facilities, as well as industrial and system-safety specialists should have a voice in architectural design meetings to evaluate the safest possible design options. Points to consider in this phase include built-in safety features that extend throughout the life of the facility, maintenance features, confined-space access, scaffolding options, structural soundness, and type and quality of materials used. No detail is too small to address, as the smallest details, such as the type of bolt connections, can rapidly turn into the largest problems. Remember, design safety begins with design, and design of a healthy facility begins with thorough planning by all involved in the design and construction process. Section 1 addresses the specific design flaw of the foreseeable misuse of sunshades and looks into how it could have been prevented.

Section 2 uses the example of improper design of an aerial lift to illustrate the subtle design flaws that can contribute to injuries on a construction site. Many pieces of equipment contain design flaws that have the potential to cause distress in ways that are seldom anticipated. This section emphasizes the process of designing or choosing the right equipment to construct the facility in the safest, most cost-effective fashion. The importance of the old adage that stresses the necessity of having the right tools for every job becomes evident when this section illustrates that equipment choice can literally make or break a construction project. Major considerations for equipment design planning include:

- Where the equipment will be used
- How it will be transported

- What categories and subgroups of hazards may be encountered by a given piece of equipment
- Who will operate the equipment (including the least skilled member of the operating team)
- Who will oversee and monitor its use (including those who may be unfamiliar with the equipment).

Integrating these considerations into the design process will provide a clear idea of how equipment can be used most efficiently and safely, decreasing both the cost and duration of its use.

SECTION 1

Architectural Design Planning

The Hazard Syndrome

Foreseeable maintenance practices must be considered when a building is designed. Thoughtlessly conceived design may necessitate dangerous work practices. Inherently safe design can usually eliminate the danger posed during maintenance tasks by specifying new components and installation methods. The building shown above was designed with sunshades that proved to be hazardous.

Scope

Most buildings are constructed to withstand the test of time, so many have a long maintenance life. From time to time they are remodeled to renew the facility with state-of-the-art fixtures and installations.

Peril from the Failure Mode

Serious injuries and fatalities may result where design encourages unsafe and dangerous maintenance or modernization practices. Structures that neglect to include sound scaffolding or anchor points in the design may cause injury to people who use unsound platforms as work areas when performing maintenance tasks. A sunshade may serve as a work platform that can fail and lacks fall protection. Though not designated as a work platform, the presence of the sunshade can be seen as a workable surface; therefore utilization as a work platform is a foreseeable misuse that should be avoided by design.

Brief Summary of the Occurrence

A window-replacement crew of four fell from a window sunshade. It collapsed while they were in the process of replacing a third-story window. A tension support member that had been tack welded to the sunshade grate instead of bolted gave way, resulting in the collapse of the sunshade. All other tension support members had been tack welded instead of bolted. The work crew did not have a fall protection system in place, so all of them fell 25–30 feet to the ground below. The crew foreman died. The other three members sustained a range of injuries. The sunshade had been used by building maintenance and contracted crews servicing the building (painting, window repair/replacement, window cleaning) since the building was constructed. This illustration shows the sunshade in detail.

Date of the Occurrence

October 16, 2001

Source of Information

State of Illinois, Cook County, # 02L6473

History of the Hazard

In this case, window washers had used the sunshades without railings as a work platform for some 30 years, and the architect made no provision for this use. The architect failed to consider foreseeable use and the well-established standards for fall prevention as previously discussed in Chapter 10, Section 1. Essentially, the design created an attractive nuisance of unsafe use.

Contributory Factors

When the windows were being replaced with modern triple-layer insulated window panels, the design did not consider:

- Future window washing on the building's exterior and the safety and accessibility issues connected with this method
- Provision for a window that could crank open to allow cleaning from inside the building
- Remodeling to specifications that allow maintenance work, such as window washing, to be accomplished from inside the building

Hazard Identification and Prevention Matrix

	Eliminate the Hazard		Guard the Hazard		Provide a Safety Factor		Provide Redundancy		Provide Reliability
	Hazard	Safety	Hazard	Safety	Hazard	Safety	Hazard	Safety	
Natural	Fall from elevation								
Structural/ Mechanical	Structural failure	Windows designed to crank open for cleaning on both sides from inside the building		Slope the sunshades so they would be unable to be used as work platform				Place warnings that the sunshades cannot support personnel	
Electrical									
Chemical									
Radiant Energy									
Biological									
Automated Systems									

Applicable Standards

OSHA, ANSI, and others have no mechanism to identify foreseeable unsafe use and options for inherently safer design.

Commentary

There is little textual material or training to address hazard identification and inherently safer design options at the time of architectural design. However, the architectural culture is rapidly changing to integrate principles of safety that start at construction and last throughout the life cycle of the building. Innovative architects are starting to look at building codes and safety features in conjunction with aesthetic and functional concerns. A solution that would have eliminated the hazard of insecure sunshades is the construction of a secure sunshade with guardrails surrounding it, calculated to hold the weight of a work crew. A design challenge, it is one of many creative solutions for which architects have the knowledge and skills to apply in the quest for safer buildings that are aesthetically pleasing to the eye.

Recommendations

Design-safety training should be developed as an engineering and architectural curriculum requirement necessary for any engineering or architecture degree. Many current engineers and architects have no notion of safe design alternatives simply because they have no prior experience with any such concepts.

To enhance previous education and introduce current engineers and architects to new trends in their field, requirements for continuing education and license renewal that involve training in hazard identification concepts and hazard control would be useful. Continuing education requirements are common practice for many fields, and help keep people skilled in their trade, increase their experience, and give them a competitive edge. Innovative design firms would benefit from workshops teaching the latest in design for safety ideas. Workshops for architectural designers to become knowledgeable in the fundamentals of the nature of hazard and inherently safer design concepts should be held on a regular basis.

SECTION 2

Equipment Design Planning

The Hazard Syndrome

The access into an insulated basket of a lineman's aerial lift is often unsafe, as it lacks handholds, foot steps, and guardrails to prevent falls from an elevated location. Traditionally, these aerial lifts are mounted on a flatbed truck with a bed roughly four feet above ground level. The articulated boom is folded, which raises the level another three feet that the line worker must climb. The top of the basket is usually three feet higher than the folded boom, raising the worker a good nine feet off the ground. Utility-line workers can climb into or out of these baskets up to four times a day.

Scope

The lack of safety considerations is a design flaw that puts thousands of utility workers at risk for a perilous fall from elevation. Any design that requires climbing onto a piece of equipment, and does not provide adequate handholds or guardrails, puts workers at a similar risk.

Peril from the Failure Mode

Any user of this type of aerial lift may be exposed to a fall of up to ten or eleven feet. Falls from this elevation can result in serious injury.

Brief Summary of the Occurrence

A utility lineman fell backwards while he was attempting to get out of the lowered basket. He struck his head on the pavement and became paralyzed, losing the use of his legs.

Date of the Occurrence

2004

Source of Information

A pending injury-claim filing (confidential).

History of the Hazard

The National Safety Council's "Study of Basket Accidents" was based on a four-year survey of accidents involving aerial-lift equipment. There were 38 documented occurrences between the years of 1967 and 1971 of utility-line workers being injured while entering or leaving the basket.

Contributory Factors

Most design flaws originate in a lack of anticipation of foreseeable use. Many designers of such equipment have never had to use the equipment they design. Though some machines are tested in a controlled environment, prolonged field use is the only indicator of the success and safety of a design. Some equipment hazards are not identified until after the design has been completed, leaving hazards wide open when exposed to human error.

Hazard Identification and Prevention Matrix

	Eliminate the Hazard		Guard the Hazard		Provide a Safety Factor		Provide Redundancy		Provide Reliability
	Hazard	Safety	Hazard	Safety	Hazard	Safety	Hazard	Safety	
Natural									
Structural/ Mechanical	No safe access	Steps, handholds, guardrails.		Nonslip stepping surfaces					100%
Electrical									
Chemical									
Radiant Energy									
Biological									
Automated Systems									

Applicable Standards

Neither OSHA nor the Society of Automotive Engineers provide specific guidelines on design or access to aerial-lift baskets.

Commentary

Simple functions such as access into and out of aerial lifts are often overlooked. When a piece of equipment serves a primary purpose, ergonomic considerations such as access and interior design are tempting to ignore. However, it is often the small, neglected details that lead to harm on a job site.

Recommendations

Equipment design should always cite all possible failure modes *before manufacture or purchase*, and list measures to control them. Use of a matrix to look at any potential harm-causing details can categorize the hazards and provide a visual list of alternate design solutions. Every list should be double-checked by a qualified engineer who has knowledge of design safety and hazard identification.

14 Including Safety into Construction Planning

INTRODUCTION

Construction safety planning is the focal point of achieving inherently safer construction. Conventional projects usually start when contractors advertise the design and specifications for bids. Review of the contracts often describes only the standard provisions known as "boiler plate" safety requirements, making the contractor solely responsible for safety with few clues to the proposed project's inherent hazards.

EARLY INVOLVEMENT

Construction safety planning should start with the cost estimates in the bidding process that include safe methods of construction in the contractor's proposal. The conventional bidding process immediately creates a disincentive to include safety costs, as competitive bidders may regard them as inessential to the construction process and ignore them. When the landowner and architect are presented a critical-path timeline showing each phase of the work and the anticipated hazards and necessary engineering controls, the resulting bid can then include such costs. This type of bid creates the groundwork for a trouble-free project that controls hazards by elimination, guarding, safety factors, and redundancy, rather than by second-guessing worker practices. Selection of the successful bidder by the landowner/architect should be based upon how anticipated hazards are controlled in the proposal.

Plan and Review

One of the most effective methods of developing construction management that will focus on the prevention of hazards is a prebid safety conference. Nearly fifty years ago the

Portland District of the U.S. Army Corps of Engineers initiated prebid safety conferences to discuss the safety requirements included in contracts during the construction of the Dalles Dam on the Columbia river. A major contract safety requirement stated that the bid proposal include a written safety plan. This requirement excited potential bidders. A number of prospective bidders detailed sound and well-planned proposals of safer construction methods. The opportunity to discuss the control of anticipated hazards before submission of bids via a written safety plan developed mutual understanding and coordination of control measures.

The next significant step in designing a safer project is the review of the safety plans. Increasingly, the selection of a successful bidder is no longer dependant upon the lowest bid. An emerging major factor of a successful bid is a well-defined safety plan that embraces the proper equipment and inherently safer methods. Such a bid usually proves to be more cost-effective, as fewer change orders are required to overcome the use of dangerous equipment and inadequate or dangerous methods of completing the work.

Any good safety plan includes a fast-track chain-of-command involvement for the resolution of unexpected hazardous conditions. These hazards could be identified by anyone associated with the project, either on or off the site. The plan should be reviewed to evaluate the listing of hazards and the quality of suggested controls: are hazards eliminated or guarded *before* the work begins on a particular phase of construction? Another key issue is the listing of the responsibilities of individual supervisors to remove hazards instead of relying on worker performance. These basic considerations can be used as criteria to determine the efficacy of a written safety plan and the qualifications of the bidding contractor.

Evaluate the Options

As bids are under evaluation, proposals by subcontractors need to be examined to identify the hazards that they may encounter. Electrical, mechanical, and other subsystems should be reviewed in order to identify any hazards, and a list of controls should be created for each plan. More importantly, allocation of duties and mapping of project resources must be performed. The construction equipment furnished by the general contractor, subcontractors or rental firms must have clearly defined limits of use, operator qualifications, and necessary safety features. Because subsystems such as elevators, escalators, personal and material hoists, heating and cooling systems, electric power, and other processes work in a common area, clearly defined times of the arrival and departure of crews and tasks lists of materials and processes can decrease the confusion on a work site and expedite the construction process. Task mapping creates a ready list of space, materials, and time. It also designates a chain of responsibility that can be referred to when needed. Suppliers, inspectors, and other off-site personnel will be able to get more information in less time, leaving a wide margin for the contractor to complete the work on schedule. The following chart designates the safety engineer with authority over all aspects of design and construction functions.

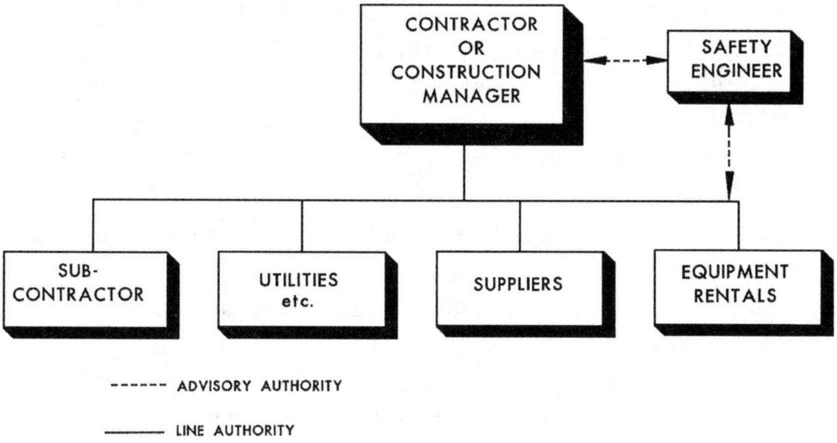

Plan Again

The next major step in construction safety planning is to hold a "prenotice to proceed" safety conference. This meeting is possibly the most important opportunity to reach agreements on how specific hazards will be controlled. This approach in defining the physical controls that will be installed before workers arrive at the work site is far more effective than assuming that safe worker practices will reliably overcome an unabated hazard. The notice to start construction should not be given until this conference has been held, and the minutes of this meeting should become part of the contract agreement. When this plan becomes official, the construction manager has stop-work authority when hazard-control measures are not in place.

Construction safety planning *before the formal notice to proceed* is fundamental to a safer construction project. Today many projects are conceived and built by a single engineering/building firm that specializes in turnkey projects. Some such projects include owner involvement because the facility houses the production process. Design-and-Build construction with owner involvement is the best way to achieve inherently safer construction, as its umbrella authority presents the opportunity to start "safety first" with the issue of design. For over fifty years, the U.S. Army Corps of Engineers has been both designing and supervising the construction of many hydroelectric navigation and flood-control dams. This dual authority has ensured both safe construction and safe life-cycle operation for these monumental projects.

Get Specific

The book *Construction Safety Planning* by the author of this text, published by John Wiley and Sons, provides a detailed review of project safety planning and insight on planning each phase of the project. The following list summarizes the points that construction safety planning should be sure to include.

1. Provide access that does not require fall protection to elevated surfaces.
2. Relocate powerlines and use proximity alarms and range-limiting devices to warn of their presence. Use insulation as a redundant safeguard to prevent the unwanted flow of current.
3. Overcome blind zones on large vehicles with near-object detection systems.
4. Isolate construction activities and equipment from public roadways.
5. Provide nighttime illumination of at least five candlepower for all construction-work areas.
6. All excavations and holes should be barricaded to prevent the hazard of pitfalls to both vehicle and pedestrian traffic.
7. Avoid overstressing and overloading temporary structures. Always stay within a reasonable safety factor when determining maximum weight for a structure. Post weight limits in conspicuous areas.
8. Avoid opportunity for entrapment in confined spaces, pinchpoints, and entanglements.
9. Keep operating instructions simple to eliminate human error. Design should not be error provocative.
10. Avoid reliance on warnings and training for a complex work process. Human behavior is unreliable when compared with engineering controls that could eliminate hazards.

Including safety in all design concepts places the engineer in his or her true position of developing facilities and machines safe for public use. It is the engineer who "subdues the earth" by providing inherently safer design. Working these designs into a framework of a sound construction plan and implementing this plan skillfully can create a productive, injury-free workplace.

Unfortunately, construction safety planning on small projects is rarely considered until the workers arrive at the work site. Many times, management personnel are not aware of a hazard until an injury occurs. This chapter uses the example of dangerous equipment to illustrate failure that could be eliminated with a construction safety plan.

SECTION 1

Construction Safety Planning

The Hazard Syndrome

Safety planning to ensure the use of recognized safeguards does not exist within many small construction activities. Optional safety requirements for the provision of expensive safety features (or the lack thereof) continue to make a number of smaller work sites excessively dangerous. Though now manufactured as an important standard feature, rollover protective structures (ROPS) are not always found on older models of road-rollers, compactors, or other heavy equipment.

Scope

A substantial number of smaller firms and construction projects rely upon older equipment that does not have recognized safety features such as ROPS. To this day, some rental equipment and lease purchases do not contain this lifesaving feature.

Peril from the Failure Mode

The absence of ROPS on road-rollers or compactors creates the life-threatening hazard of crushing due to upset. Unstable or graded ground can cause the equipment to tilt or fall. If a strong overhead canopy does not protect the operator, he or she will almost certainly be crushed to death. Litigation records include over ninety occurrences of serious injury or fatality from compactor or road-roller overturn.

Brief Summary of the Occurrence

Chapter 10, Section 6 detailed the hazard of traffic control. That section's "Occurrence" addressed a road-roller that overturned when it moved to the edge of a road to allow the passage of auto traffic. The operator of the roller was crushed to death. In the words of Paul Harvey, the "rest of the story" is the lack of ROPS on a road-roller acquired by lease purchase. At the time of transaction, the distributor had offered the small rural city that purchased the road-roller the option of ROPS. This protection came at an additional charge, so the city declined. The city's supervisor of public works did not want the additional expense of ROPS, as he had no knowledge of the proven need for this safeguard. A claim against the manufacturer was dismissed, even though they were aware of the need for ROPS, as ROPS are currently an OSHA requirement.

Date of the Occurrence

June 8, 1992

Source of Information

Case filed in Binghamton, New York, appealed and dismissed by the NY State Supreme Court

History of the Hazard

1958: ROPS was first required on tractors by the U.S. Army Corps of Engineers in the North Pacific Division.

1961: The North Pacific Division requirement was extended to include ROPS on certain equipment as a contract-pay item.

1990s: Most manufacturers provide ROPS as standard equipment on compactors and road-rollers to avoid liability.

Contributory Factors

The failure to adopt reasonable, proven safety appliances as standard equipment stems from several key issues. First, safety planners do not often perceive it necessary to incorporate state-of-the-art safety appliances or alternate design because such precautions are *not an accepted standard of care.* Principle Two addresses the importance of establishing and adhering to a standard of care. Once this groundwork is laid out, the steps necessary to meet the standard can be gauged. An appropriate budget can be allocated and pre-planning can commence.

It is crucial that the design firm and contracting engineer establish their own standard, because OSHA standards often fail to ensure the highest standard of care due to the practice of developing industry-wide consensus-oriented requirements. Another complication is that there is no immediate economic incentive for either the manufacturer or the purchaser to provide safety appliances or alternative safer design, as many of these features are often viewed as an unnecessary expense.

Hazard Identification and Prevention Matrix

	Eliminate the Hazard		**Guard the Hazard**		**Provide a Safety Factor**		**Provide Redundancy**		**Provide Reliability**
	Hazard	Safety	Hazard	Safety	Hazard	Safety	Hazard	Safety	
Natural	Upset/roll over								
Structural/ Mechanical		Design a plan that incorporates safety features into the cost of the contract		Acquire machines with adequate ROPS		Maintain new equipment, retrofit old equipment			100%
Electrical									
Chemical									
Radiant Energy									
Biological									
Automated Systems									

Applicable Standards

OSHA does not require ROPS on compactors or road-rollers, but the U.S. Army Corps of Engineers considers ROPS to be a contract-pay item.

The SAE has recommended design requirements that include ROPS on compactors.

Commentary

The exclusive remedy of workers' compensation insurance insulates the employer from liability and transfers responsibility for unsafe equipment to the supplier. Under current workers' compensation clauses, the manufacturer or supplier of equipment is responsible for providing necessary safeguards, even when a purchaser does not want them. Common law requires that the product be inherently safe for its intended use. The proven record of saving lives makes ROPS a reasonable safety appliance.

In a competitive market, "planning" for safety first is an oxymoron when reasonable safety appliances are omitted. The absence of immediate payback for safety features creates a strong disincentive to ensure safety planning that adopts reasonable proven safeguards such as ROPS. Principle Two, established in Part I of this text, instructs the engineer to *establish a standard of care* and states that "*Any hazard that has the potential for serious injury or death is always unreasonable and always unacceptable if reasonable design features and/or the use of safety appliances are available to prevent the hazard.*" If design engineers adhere to this standard in the construction-planning phase, many lives would be saved each year. In the case of ROPS, both engineering tests and years of situational application have proven that this design feature is a reasonable and reliable safeguard against crushing in the event of rollover. Currently, OSHA and ANSI compliance standards rarely live up to the standard of care that would make safety paramount on a job; therefore the engineer and construction manager must initiate a higher standard of care and ensure that it remains a priority throughout the job.

Small businesses and local governments generally operate on limited funds and are not often inclined to plan for safety to the highest standard of care by allocating funds for safety features. The U.S. Army Corps of Engineers found it possible to adopt safety planning as pay items on works projects by reimbursing the general contractor and subcontractors for adopting the use of safety appliances and alternate design. Today the leading general contractors who annually oversee multibillion-dollar projects are becoming increasingly receptive to funding for thorough safety planning and safety innovations. However, for the 80% of businesses and municipalities who engage in construction and have ten employees or fewer, the highest standard of care and safety is not a reality.

Recommendations

To become a grassroots function accessible to all firms, system safety at the highest standard of care needs an economic incentive such as a tax credits to provide immediate rewards for planning for, purchasing, and installing safety features.

15 The Economics of Inherently Safer Design

The previous fourteen chapters developed a methodology for inherently safer construction practices, and used examples to apply that methodology to the achievement of inherently safer construction by design. From an engineering viewpoint, the identification and prevention of hazards is an achievable reality. However, the implementation of safety research and development, marketing, and the purchase of safer design features require considerable money. Yet these aspects are only half of the implementation effort. The legislative environment creates a pervasive, arbitrary climate of legal constraints devoid of immediate economic incentives for safety. A major barrier to interest in and adoption of safer design features is the exclusive remedy of workers' compensation insurance carried by employers, which shelters the construction employer from liability for unsafe workplaces in every state in the nation. This protection extends to the contractor who can afford only outdated, dangerous equipment, and employs untrained laborers to engage in dangerous erection methods, as the contractor may be an underfinanced low bidder.

AN UNFAIR ADVANTAGE?

Construction demographics list 25% of construction workers as self-employed, with 80% of the remainder as employed by contractors with fewer than ten employees. This group usually lacks capital to invest in safety improvements. The construction industry employs only 5% of our nation's workforce, but sustains 20% of workforce fatalities, with no record of casualties in the self-employed sector. These deaths indicate an overexposure to undue dangers; a frightening fact, made worse by the knowledge that these dangers could be prevented or controlled with planning and technology.

Some 4000 years ago, any construction failures that resulted in the loss of life were punished swiftly and with vengeance. The Code of Hammurabi decreed that if a builder erected an unsound house and it collapsed and killed the owner's son, the builder's son would be put to death. This early law created a punitive approach to safety that is still felt today. The burden of compliance with safety laws and standards still falls to the builder. How can safety be a pass-through cost on a six-week contract, when it may take a decade to pay for safer new equipment? A popular option is to rent what is available—equipment that may be a dangerous substitute for the machine that should be used. From a safety perspective, the last choice is to make do with a worn-out relic; yet this state of affairs is often a reality of survival in a low-bid world. At the beginning of the twentieth century, workers' compensation served employees working on labor-intensive work sites, providing care for those who were maimed, or compensating the widows and children of those killed on a dangerous job. Our twenty-first century work site is machine-intensive. Structures to be built are complex and on fast-track completion schedules to reduce costs. The very face of labor has changed, and the culture of safety must correspond with the technological advances that define the modern work place. Currently, in the event of injury, claims are filed against third-party defendants who create the machines. These defendants in turn allege that many safety features are ineffective, a process that often delays industry acceptance of new safety features for up to three decades. This slow change to adopt new safety appliances or alternate designs does not represent the "safety first" idea that is integral to a successful work project in this day and age.

SHIFTING THE CULTURE

The rewards for the change to a safety-first culture could be monumental. Over half of the crippling injuries and deaths on a work site are attributed to defective design of either the construction equipment or the facility being built. If employers made safety a priority, workers' compensation claims would radically drop, with a corresponding drop in workers' compensation premiums; yet these rewards would not affect employers alone. Review of the social security payments for disability show some alarming costs. An estimated $1 billion per month in entitlements from the social security fund are spent on care for disabled construction workers, their families, or widows and children of the deceased. Fewer claims of this nature would save federal tax dollars as well. In addition, lost productivity, damaged equipment, and destroyed property are all common by-products of workplace injuries. Elimination or reduction of such injuries would result in savings in these areas as well.

The first paragraph of the U.S. Constitution states: "We, the people, in order to form a more perfect union . . . *promote the general welfare, to ourselves and our posterity.*" This decree provides authority for our congress to look upstream at the time of design and planning to reduce construction carnage. The Federal law of OSHA, Public Law 91-506 Section (2)(b) echoes the phraseology of our Constitution as follows: "and to provide for the *general welfare* to assure so far as possible every working man and woman in the nation safe and healthful working conditions."

Many sources of alternative fuels, sources of power generation, medical programs, and other national developments are funded by federal grants and tax credits. The health and well-being of America's workforce should be funded as well. In view of the savings in lives and the aim of a more prosperous economy, it is not unreasonable to grant a two dollar tax credit for every dollar spent on the voluntary development, distribution, and purchase of safety equipment and alternative safer designs for construction. Initial tax-credit funding for safety improvements during a five-year window of opportunity would allow new safety technology to take root and become accepted as necessary construction safety features.

DOING THE MATH

A quick review of a number-crunching exercise shows how estimated liability can triple the cost of a product, yet the same product with a safe design improvement raises the manufactured cost by only 10%. This type of analysis can provide reasonable insight when the following questions are asked:

1. How many units will be manufactured each year?
2. For how many years will this model be manufactured?
3. What is the anticipated life of each unit?
4. What is the number of hours that each unit will be used each year?
5. What is the cost to manufacture each unit?

Let's assign numbers to each of the above:

Estimated annual sales/unit production:	5000 units
Estimated years of manufacture:	10 years
Estimated hours of use each year per unit:	1000 hours
Estimated useful life of each unit:	20 years
Manufactured cost per unit:	$100

Now let's do a little multiplication to see how many hours of foreseeable use this product may have through its entire life cycle:

500 units per year × 10 years production = 50,000 units sold

50,000 units × 1,000 estimated hours of use each year per unit = 50,000,000 hours of use for the entire product line in one year.

50,000,000 hours of use × 20 years estimated useful life for each unit = 1,000,000,000 hours of use for the entire anticipated production line.

A lot can happen in an exposure of *one billion hours* to cause a hazard to become armed. Assume the product's hazard could become armed one hour in every 1,000,000 hours of use, which means during the life span of the products there would be 1000 hours in which some units would be armed and *could cause harm.* Now, assume that the armed hazard could be triggered into action, *causing harm* only one hour out of each 100 hours of armed use. This means a total of 10 hours somewhere in the entire product line that some of the units *could actually be causing harm,* or one (1) hour in one hundred million (100,000,000) hours of use. Such a remote possibility of serious injury or death from a product is not unusual. Because of the seemingly remote chances of such an incident, it is not uncommon for management to take the "head in the sand" attitude, so well portrayed in the "Dilbert" comic strip, and do nothing, instead of eliminating the hazard through improved design.

Now let's speculate that in this 10 hour window of time throughout the entire life cycle of the product line, two deaths and several crippling injuries occur. In today's world, the company will very likely be sued. The cost of a legal defense, any jury verdicts, and settlements could run to as much as $15,000,000 or more. If we were able to distribute this loss among the 50,000 units, we would have to add $300 to each unit manufactured, raising the $100 unit cost to $400. The price has quadrupled.

Reaping the Rewards

Just suppose that management decides to hire a safety-engineering consultant to provide technical assistance at the planning and design phase, and also provide follow-up during testing and throughout the life cycle of the product as needed. A consulting fee of $150 per hour for a reasonable 200 hour estimate for technical assistance would run $30,000, plus $20,000 for travel expenses, for a total of $50,000. For 50,000 units, the additional cost would be $1.00 per unit. Let's further suppose that the safety engineer sees that a simple design change will eliminate the hazard at the cost of $10.00 per unit. The cost of the product is now $111.00. Remember, these services and improvements have cost only $11.00 per unit and can be passed on to the purchaser, who also benefits by having a product that is reliable, more efficient, and safe for its intended use. This is a small price to pay when compared with the $300 cost per unit for death and injury if the hazard is never eliminated.

This example uses a product, but an unsafe method or process in construction or mining can yield the same results. On more expensive products, such as construction equipment, the cost of safety features is usually very modest—no more than 1%–2% of the manufactured cost.

The use of both engineering and economics principles can change the entire concept of construction safety by making it a profit center. The figures speak for themselves! In most instances, a simple hazard-cost analysis will show that *safe design reaps big profits!*

Safety Pays

The real economics of inherently safer construction have changed with technological advances. A world that relies upon machines and technology to do work can apply the same engineering skills that have created this progress to control hazards. Life was difficult as civilization evolved from a need to survive by one's wits to avoid nature's hazards. Today we are confronted by untold hazards that are beyond our ability to perceive. Every person, every worker, must continue to exercise due diligence and care; however, many machines and systems outpace human response time. Therefore, at the time of design, engineering advancements must hold safety paramount over productivity and utility. A machine or system designed without safety features, which relies upon user behavior modification to avoid hazards, only compromises the essence of professional engineering conduct. A machine or system that maims or kills is not profitable in any economic or human evaluation. Engineering skills can apply science and save lives. Make design safety a paramount culture in the construction industry and reap the benefits of an industry revolutionized.

Index